Market Design

Market Design

Auctions and Matching

Guillaume Haeringer

The MIT Press
Cambridge, Massachusetts
London, England

© 2017 Massachusetts Institute of Technology

All rights reserved. No part of this book may be reproduced in any form or by any electronic or mechanical means (including photocopying, recording, or information storage and retrieval) without permission in writing from the publisher.

This book was set in Palatino by Westchester Publishing Services.

Printed and bound in the United States of America.

Library of Congress Cataloging-in-Publication Data

Names: Haeringer, Guillaume, author.
Title: Market design : auctions and matching / Guillaume Haeringer.
Description: Cambridge, MA : MIT Press, [2018] | Includes bibliographical references and index.
Identifiers: LCCN 2017031384 | ISBN 9780262037549 (hardcover : alk. paper)
Subjects: LCSH: Auctions. | Markets. | Microeconomics.
Classification: LCC HF5476 .H24 2018 | DDC 381/.1–dc23 LC record available
    at https://lccn.loc.gov/2017031384

10 9 8 7 6 5 4 3 2

To Circe, Hermes, Nora, and Anna

Contents

Preface   xv

1 **Introduction**   1
   1.1   Market and Market Design   1
   1.2   Do We Necessarily Have Prices?   2
   1.3   More on Markets   3
         1.3.1   What a Market Needs to Work   4
         1.3.2   Commodities   5
   1.4   Market Design: First Example   6
         1.4.1   Feeding America   6
         1.4.2   The First Design   7
         1.4.3   The Problems...   9
         1.4.4   ...and a Solution   10
         1.4.5   Results   11

2 **Simple Auctions**   13
   2.1   Introduction   13
         2.1.1   Auctions: A Definition   14
         2.1.2   Auctions Are Everywhere   14
   2.2   Valuations   16
   2.3   Payoffs and Objectives   17
   2.4   Ascending Auctions   18
         2.4.1   The English Auctions   18
         2.4.2   Bids, Strategies, and Payoffs in the English Auctions   21
         2.4.3   Ticking Price   22
         2.4.4   Truthful Bidding   23
   2.5   Second-Price Auction   25
         2.5.1   The Essence of the English Auction   25
         2.5.2   The Vickrey Auction   26
         2.5.3   English versus Second-Price Auctions   28

2.6 First-Price Auction   29
    2.6.1 Definition   29
    2.6.2 Optimal Bids in the First-Price Auction   30
    2.6.3 Dutch Auction   34
2.7 Revenue Equivalence   36
2.8 Reserve Price   40
    2.8.1 Optimal Reserve Price   42
    2.8.2 Reserve Price versus Adding More Bidders   45

3 An Analysis of eBay   49
3.1 Introduction   49
3.2 eBay in Detail   50
    3.2.1 Proxy Bidding   50
    3.2.2 Bids and Bid Increments   51
    3.2.3 Updating Rules   51
3.3 eBay as a First-Price Auction?   54
3.4 Bid Sniping   54
    3.4.1 Amazon versus eBay   55
    3.4.2 Data Analysis   56
3.5 Reserve Price   59

4 The Vickrey-Clarke-Groves Auction   61
4.1 Introduction   61
4.2 The Model   61
4.3 The VCG Auction   63
    4.3.1 Computing the Optimal Assignment   63
    4.3.2 Calculating Prices   66
4.4 Incentives under the VCG Auction   67
4.5 Relation with the Vickrey Auction   68
4.6 The Complexity of the VCG Auction   69

5 Keyword Auctions   71
5.1 Introduction   71
5.2 Running Billions and Billions of Auctions   73
5.3 The Origins   74
    5.3.1 Payoff Flows   75
5.4 The Google Model: Generalized Second-Price Auction   76
    5.4.1 Quality Scores   76
    5.4.2 Truthtelling under the GSP Auction   78
    5.4.3 Equilibrium under the GSP Auction   78

		5.4.4	Assumptions about the Long-Run Equilibria 79

         5.4.4   Assumptions about the Long-Run Equilibria   79
         5.4.5   Refinement: Envy-Free Equilibrium   79
         5.4.6   The Generalized English Auction   88
   5.5   The Facebook Model: VCG for Internet Ads   91
         5.5.1   Comparing VCG and GSP Auctions   91
         5.5.2   The Rationale for VCG   94

6   Spectrum Auctions   95
   6.1   How Can Spectrum Be Allocated?   95
         6.1.1   Lotteries   95
         6.1.2   Beauty Contests   96
         6.1.3   Why Run Auctions?   97
   6.2   Issues   98
         6.2.1   General Issues   98
         6.2.2   Collusion, Demand Reduction, and Entry   99
         6.2.3   Maximum Revenue   101
         6.2.4   The Exposure Problem   102
         6.2.5   Winner's Curse   103
   6.3   The Simultaneous Ascending Auction   104
   6.4   Case Studies   106
         6.4.1   The U.S. 1994 PCS Broadband Auction   106
         6.4.2   Mistakes   109

7   Financial Markets   113
   7.1   Introduction   113
   7.2   Treasury Auctions   113
         7.2.1   Outline   113
         7.2.2   How Treasury Auctions Work   115
         7.2.3   Analysis   118
   7.3   Double Auctions   119
   7.4   Initial Public Offering   123
         7.4.1   Allocation and Pricing through Contracts   124
         7.4.2   Auctions for IPOs   124

8   Trading   127
   8.1   Stock Markets   127
   8.2   Opening and Closing   132
   8.3   High-Frequency Trading   135
         8.3.1   Market Structure   135
         8.3.2   Market Regulation   137

		8.3.3	Surfing on the Latency 139
		8.3.4	What Is the Matter with High Frequencies? 141
		8.3.5	A Flawed Market Design? 143
	8.4	Alternative Market Designs 145	
		8.4.1	Slowing Down Markets? 145
		8.4.2	Frequent Batch Auctions 147

9 The Basic Matching Model   149

9.1 The Basic Matching Model   149
    9.1.1 Preferences   150
    9.1.2 Matching   152
    9.1.3 Stability   153
9.2 Algorithms and Mechanisms   156
    9.2.1 Algorithms?   156
    9.2.2 Matching Mechanism   159
9.3 Finding Stable Matchings   159
    9.3.1 The Deferred Acceptance Algorithm   160
    9.3.2 Deferred Acceptance and Stable Matchings   162
9.4 Preferences Over Stable Matchings   164
    9.4.1 Musician-Optimal and Singer-Optimal Matchings   164
    9.4.2 Proofs   165
9.5 Incentives with the Deferred Acceptance Algorithm   167

10 The Medical Match   173

10.1 History   173
10.2 The Many-to-One Matching Model   175
    10.2.1 Preferences in the Many-to-One Matching Model   176
    10.2.2 Matchings and Stability in a Many-to-One Matching Model   179
    10.2.3 Finding Stable Matchings   182
    10.2.4 One-to-One v. Many-to-One Matchings: Similarities and Differences   184
10.3 Why Stability Matters   186
    10.3.1 A Natural Experiment   186
    10.3.2 Unraveling in the Lab   187
10.4 The Rural Hospital Theorem   190
10.5 The Case of Couples and the Engineering Method   192
    10.5.1 A Very Complex Problem   192
    10.5.2 When Theory Fails   193
    10.5.3 Fixing the NRMP   194

11   Assignment Problems   197
    11.1   The Basic Model   197
        11.1.1   Public versus Private Endowments   198
        11.1.2   Evaluating Assignments   198
    11.2   Finding Efficient Assignments   199
        11.2.1   Serial Dictators   199
        11.2.2   Trading Cycles   200
        11.2.3   Implementing Allocation Rules   205
        11.2.4   Individual Rationality and the Core   210
    11.3   Mixed Public-Private Endowments   212
        11.3.1   Inefficient Mechanisms   213
        11.3.2   Two Efficient Solutions   217

12   Probabilistic Assignments   223
    12.1   Random Assignments   223
        12.1.1   Preliminaries   223
        12.1.2   The Birkhoff–von Neumann Theorem   225
        12.1.3   Evaluating Random Assignments   227
    12.2   Random Serial Dictatorship   229
    12.3   The Probabilistic Serial Mechanism   231

13   School Choice   239
    13.1   The Many-to-One Assignment Model   239
        13.1.1   Preferences versus Priorities   239
        13.1.2   The Model   241
        13.1.3   Assignments   241
        13.1.4   Stability and Efficiency   242
    13.2   Competing Algorithms   245
        13.2.1   The Role of Each Side of the Market   245
        13.2.2   The Deferred Acceptance Algorithm   246
        13.2.3   The Immediate Acceptance Algorithm   248
        13.2.4   Top Trading Cycles   250
    13.3   The Problem with the Immediate Acceptance Algorithm   256
    13.4   Applications   257
        13.4.1   The Boston School Match   257
        13.4.2   The New York City School Match   259

14   School Choice: Further Developments   263
    14.1   Weak Priorities   263
        14.1.1   The Problem   263
        14.1.2   Efficiency Loss   264

14.1.3 The Student-Optimal Assignment with
Weak Priorities   265
14.1.4 Restoring Efficiency   266
14.1.5 How to Break Ties If You Must   272
14.2 Constrained Choice   276
14.2.1 Issues   277
14.2.2 From Very Manipulable to Less Manipulable   279

15 Course Allocation   283
15.1 Preliminaries   283
15.2 Bidding for Courses   284
15.2.1 The Bidding and Allocation Process   285
15.2.2 Issues: Nonmarket Prices and Inefficiency   286
15.2.3 Deferred Acceptance with Bids   288
15.3 The Harvard Business School Method   290
15.3.1 The Harvard Draft Mechanism   290
15.3.2 Strategic Behavior   291
15.3.3 Welfare   293
15.4 The Wharton Method   295
15.4.1 Approximate Competitive Equilibrium from
Equal Incomes   296
15.4.2 The Wharton Experiment   298

16 Kidney Exchange   303
16.1 Background   303
16.2 Trading Kidneys   305
16.2.1 Trades versus Waiting List   306
16.2.2 The Kidney Exchange Algorithm   306
16.2.3 Chain Selection Rules   309
16.2.4 Efficiency and Incentives   311
16.3 On the Number of Exchanges   313

Appendix A: Game Theory   317
A.1 Strategic Form Games   317
A.1.1 Definition   317
A.1.2 Pure and Mixed Strategies   318
A.2 Extensive Form Games   319
A.2.1 Definition   320
A.2.2 Strategies   321
A.2.3 Imperfect Information   322

    A.3   Solving Games   324
        A.3.1   Dominated and Dominant Strategies   324
        A.3.2   Elimination of Dominated Strategies   326
        A.3.3   Nash Equilibrium   328
    A.4   Bayesian Games: Games with Incomplete Information   330
        A.4.1   Introductory Example   330
        A.4.2   Definition   331

Appendix B: Mechanism Design   335
B.1   Preliminaries   335
B.2   The Model   337
    B.2.1   Mechanism   337
    B.2.2   Implementing Social Choice Functions   339
    B.2.3   Direct versus Indirect Mechanism   340
B.3   Dominant Strategy Implementation   341
    B.3.1   The Revelation Principle   341
    B.3.2   The Gibbard-Satterthwaite Theorem   342
    B.3.3   The Vickrey-Clarke-Groves Mechanism   343
B.4   Bayesian Mechanism Design   344
    B.4.1   Bayesian Incentive Compatibility   344
    B.4.2   Trading: The Myerson-Satterthwaite Theorem   345

Appendix C: Order Statistics   349
C.1   Expected Highest Valuation   349
    C.1.1   Obtaining the Cumulative Density Function   350
    C.1.2   Obtaining the Probability Density Function   350
    C.1.3   Calculate the Expectation   350
C.2   Expected Second-Highest Valuation   351
C.3   Conditional Expectation of the Highest Valuation   352
C.4   Changing the Upper and Lower Bounds   353

Notes   355
References   365
Index   369

Preface

This book is an introduction to market design. It aims at providing students a broad overview of questions related to the design and analysis of market mechanisms.

Why take a course on market design or, put differently, why take a course on market mechanisms? Students in economics or business learn a great deal about markets: Where do supply and demand come from? How do equilibrium prices change when the market structure or some parameters change? But few students are exposed to simple, basic questions like: How are prices calculated? Does the way a price is determined matter? Does the procedure for applying to a school or a college have an impact on *which* school or college I will attend? How do we proceed when monetary transactions are not permitted, such as for organ donations? If one believes that students should be exposed to such questions, which are fundamental to economics, this textbook is a first step in that direction.

This book was written with several objectives in mind:

• Make the material as accessible as possible. Auctions or matching models can easily become involved, requiring a level of mathematics that not all students have. Special effort has thus been made to minimize the formal descriptions of the models and, whenever some formalism is introduced, carefully explain the mathematics to make it understandable for the majority of students. Minimizing the mathematics required emphasizing the intuitions and engaging in sometimes lengthy explanations and detailed resolutions of examples. Also, many results are presented without a proof.

• Over the past three decades or so, there have been a growing number of situations where economists have been able to help policy makers or regulators design market institutions, most notably auctions and matching markets. There have been successes and failures, and much has been learned. Successes are helpful because they can show, among other things, that theoretical works are not for theorists only, that the models economists study can be helpful. Studying failures is also helpful, because they show students that economists are aware of the limits of their models and are willing to learn from real-life, concrete problems.

Too often, students complain that microeconomics is too abstract and disconnected from reality. Market design is a formidable opportunity for instructors to show how theory can help to solve existing problems, which in turn can generate new theoretical questions.

The book alternates, as much as possible, theory and real-life applications or case studies, putting aside, unfortunately, many theoretical results and empirical studies. Such a selection has been more severe for theoretical results. Priority has been given to those results that have an immediate, concrete application and/or are considered fundamental results but not too difficult to illustrate and explain. There are thus many interesting and well-known results that are not presented in this book.

Although the book puts a strong emphasis on the dialogue between theory and applications, some parts are purely theoretical.

Theory can be very helpful in that it gives some structure to our thinking. It allows for an in-depth analysis of the essence of a problem. Because auctions and matching theory can be seen as applications of mechanism design, a textbook without some presentation of mechanism design and game theory, even if succinct, would be incomplete. I tried to make the book accessible for students with very little knowledge of game theory and mechanism design, but readers will also find two chapters that offer a general review of elements of game theory and mechanism design that are related to the themes explored in the book (Appendices A and B). Although these two chapters depart a little bit from the philosophy of this book, they follow the tradition at work in it by presenting the basic material, giving as many explanations and illustrations as possible.

Several people provided invaluable help with this project. First and foremost, I thank Emily Taber from MIT Press for her constant support (and patience). Thanks to her, I have been fortunate to be assigned an extraordinary set of reviewers. The suggestions and comments they made on earlier versions of the manuscript contributed significantly to the coherence of this book. I am also thankful to David Cristales and Bradley Wells for their proofreading (and checking consistency across the various sections and chapters). Of course, I am solely responsible for any remaining errors. Like many textbooks, this work started as lecture notes for my market design course at Baruch College. Creating a new course has been possible thanks to the help I received from Armen Hovakimian, Sebastiano Manzan, Judith Tse, and Ashok Vora. I also benefited from the feedback and comments I received from my students at Baruch College and, thanks to

Yeo-Koo Che, at Columbia University, where I also taught a market design course. I also want to express my gratitude to the many scholars who contributed, indirectly, to this book. Many aspects of this book have been shaped by converstations I had with (in alphabetical order) Atila Abdulkadiroğlu, Chris Avery, Yeon-Koo Che, Onur Kesten, Paul Milgrom, Al Roth, Tayfun Sönmez, and Utku Ünver (this list is necessarily and unfortunately incomplete). Last but not least, I am greatly indebted to my family for the constant support I received.

New York, April 4, 2017

1 Introduction

1.1 Market and Market Design

The way goods or services are exchanged or distributed, one of the key questions that economics aims to address, depends on many factors. Some of these factors are not under our control, but some are, for instance the rules that govern how individuals or firms interact. Those rules are usually not accidental; they have been *designed* by, among others, policy makers or regulators. Economists usually refer to those rules as *the market*. A market is thus, in a very broad sense, an institution where goods and services are exchanged or traded. Put differently, the role of a market it to determine *who* gets *what*. **Market design** is the area of economics that is particularly interested in *how* who gets what is decided. It studies the extent to which different protocols (*how*) can yield different outcomes. As its name suggests, market design focuses on the *design* of market institutions, and to do that we may need to study markets from various angles. But to design markets we also need to study how they work. Understanding how markets work is necessary if we want to fix them when they do not perform well.

Traditionally markets are thought of as institutions that determine the allocation of goods and services through *prices*. The main role of a market is to set prices of the goods and services to be exchanged, which will in turn determine who gets what. To see this, consider the standard supply-and-demand framework. The demand for a good simply indicates, for any price of the good, what quantity the consumers would like to buy. Similarly, the supply indicates, for any price of the good, the quantity that the sellers would like to sell. An equilibrium in this context is defined as the price at which the demand and the supply meet: at an equilibrium price, the quantity that the buyers would like to buy is exactly the same as the quantity that the sellers would like to sell. In other words, the notion of equilibrium is part of the trading protocol (*how*) in the sense that it implicitly describes the allocation (*who gets what*): goods and services have a price, so being allocated a good means paying its price. In this framework, most of the analysis done in economics consists of studying, for instance, how the equilibrium changes when some parameters change (e.g., a sales tax), or how the equilibrium depends on the number of buyers and/or sellers.

We have just argued that in the "standard" case the price is what determines who gets what. So is the price the definitive answer for how who gets what is decided? Not completely, for we still have to explain *how* we reach an equilibrium, how prices are set.

Consider, for instance, the owner of, say, a Monet painting, who wants to sell it. The supply is easy to characterize in this case: the seller has perhaps a minimum price, below which she does not want to sell the painting. Above that price, the supply is constant: there is only one such painting. The demand also is not too difficult to characterize. We can imagine that there are many buyers, each with a maximum price. For instance, a million people are willing to pay at most $10,000, but only a thousand people are willing to pay at most $1,000,000, and so on. The equilibrium in that case would be the price at which there is only one person left willing to pay that price.

This looks easy, but observe that this approach is not very well suited to giving us a sharp prediction. Suppose, for instance, that the second-highest price a buyer is willing to spend is $5,000,000, and suppose that the highest price a buyer is willing to pay is $20,000,000. It is not difficult to see that *any* price between $5,000,000.01 and $20,000,000 is an equilibrium price. So which price will prevail?

1.2 Do We Necessarily Have Prices?

In some situations, however, prices may not be sufficient to determine who gets what, and the problem of college admission is a good example of such a situation. At the outset, college admission looks like a problem that would fit market analysis. On the one side, we have colleges that offer a service, education (and a degree), and on the other side we have prospective students who seek that service. That service carries a price: the tuition.

The tuition is certainly an important part of students' decisions, but it is not the only part, and neither is it for the colleges. If it were, we would see colleges raising or decreasing their tuition fees until the number of applicants any college receives matches its enrollment capacity, a far cry from reality, because for colleges the price is not sufficient to determine enrollment. It is more a question of *choosing* and *being chosen* than setting tuition fees. Alice, a student, will go to college A because, among all colleges that wanted her, college A is her most preferred. Similarly, college B accepted Bob's application because the college wants to enroll him.

The observation we just made does not imply that economics has nothing to say about college admission. Rather, we may need to consider different approaches than the traditional market analysis where prices are enough to determine transactions.

But in some cases we may even go further, considering situations where transactions are made without monetary payments, for instance barter. In many cases, the absence of monetary transactions is not a design decision but rather a constraint faced by the market designer. Two examples that will be treated in this book will illustrate this point. Consider first the problem of assigning students to public schools (elementary, middle, and high schools). This problem can be seen as a standard allocation problem: we have on the one side schools, which offer a service (education), and on the other side students (or their parents), who have to be enrolled at a school. This problem looks similar to the problem of college admission, except that usually public schools are free, meaning students do not pay tuition. How to assign students to schools while taking into account parents' preferences and school districts' constraints is one of the questions addressed by market design. Another instance of a problem where transactions are made without transfer payments is the allocation of organs for transplant. In almost all countries, human organs cannot be sold or bought (i.e., traded with a transfer payment). Yet policy makers design rules that determine how patients can obtain organs. A constraint for a market designer in that context is that the assignment mechanism does not involve prices.

The economics profession is split with respect to the very definition of a market. Some economists believe that markets necessarily involve prices, and thus a market is essentially a protocol or mechanism that helps elicit prices. Under this approach, the assignment of students to public schools or the assignment of organs to patients cannot be considered instances of markets. Other economists instead define a market as simply an institution in which goods and services are traded (or assigned, exchanged, etc.). With this approach, we can have situations where trades are made along with transfer payments and other situations where transactions occur without payments.

1.3 More on Markets

As we have already intuited, different types of markets may not need to work the same way. Putting different situations under the same umbrella does not mean that their inner workings must be identical. Each market has a particular *design*, a set of rules (implicit or explicit) that agents on each side of the market follow and that will determine the transactions.

One crucial question then is, how do we come up with a particular design? Thinking that the design will "arise naturally" is not a good way to start. It is true that in most cases we can quickly see what type of design is the most appropriate, but the exact details often matter, and sometimes the market needs to be fixed.

1.3.1 What a Market Needs to Work

If a market "works," it means that some trades are realized. For this to occur, several conditions must or should be met.

First of all, we need "enough" actors from both sides. Each individual in the market will seek to make the best transaction (from his or her point of view). For a buyer, the presence of many sellers increases competition (and thus reduces prices) and allows the opportunity to forge a good opinion about which deals are good. Similarly, a seller will benefit if there are many buyers, because it increases the likelihood of finding the right buyers. In economics, when we talk about the number of actors in a market, we use the term **market thickness**. Having a thick market is not necessary, of course, but it helps a lot. A seller facing many buyers (and "meeting most of them") will be able to have pretty good knowledge of what the demand looks like. The same happens for a buyer. The converse of a thick market is a **thin market**. In a thin market, there are few buyers and sellers, which can have negative effects, such as increasing price volatility or reducing competition (and thus diminishing the possibility of finding the perfect deal for buyers or sellers).

A problem that can occur once a market is thick is that it becomes congested. In a **congested** market, what is important is not that there are too many buyers and sellers but rather that the sellers do not have time to review all of the preferred purchase prices, or bids, of the potential buyers. So, congestion in a market is like a traffic jam. In the housing market, websites like trulia.com or zillow.com or brokers are here to help reduce (a little bit) congestion. These websites act like meeting places for buyers and sellers. In a commodity market, congestion is easily reduced, thanks to the price. Having a public price gives immediate information to buyers and sellers about what they can (or cannot) hope to do. When the market is not a commodity market, goods are differentiated among sellers, and congestion can kick in very quickly. Instead of having to consider one unique good, I have to consider many different options at the same time.

Once market thickness and congestion are no longer an issue, a third problem can occur. When facing a possible transaction, an individual must decide what to do. What do I say or claim when I approach someone for a possible transaction? If a deal is proposed to me, should I take the offer or should I wait and hope to get a better deal? For instance, if I am a buyer, should I buy now or should I wait for a better deal (taking the risk that I may not find something better later)? For the seller, a similar problem arises. Do I sell my product to this buyer for the requested price or wait for a better offer? Sometimes it is easy to see what we should or should not do. For instance, if you are bargaining to buy some antique, you do not want to announce the maximum price you are willing to pay before the seller announces a price, for otherwise the price that will be proposed is likely to be high. But in other circumstances determining the right decision can be tricky.

Obviously, we do not want to participate in a market where we know in advance that we will get ripped off, unless we are obliged to do so. In a more general way, we may want to avoid markets that are too complicated. In other words, most people prefer to participate in markets that are safe and simple.

1.3.2 Commodities

Economics talks about markets for goods (or services), but it helps to distinguish the goods or services exchanged that are commodities from those that are not. A **commodity** is simply a good that can be sold or bought in batches that can all be considered basically the same. For instance, the baker who made the bread you bought probably does not know who grew the wheat that went into the flour he or she used to make the bread. However, the baker certainly (at least we hope!) knows the type and quality (what we call the *grade*) of the flour that was used. In the market for flour, two bags of flour of the same type and of the same grade are considered identical. That is, flour is a commodity. Today, this concept of commodity may seem natural for us, but this has not always been the case.

Today, flour is traded as a commodity, but that is fairly recent. Until the first half of the nineteenth century, this was not the case. The most common type of flour is made from wheat. Since each field of wheat is slightly unique (because of the soil, sun exposure, watering, farming techniques, and other factors), each bag of wheat (or flour) is likely to be slightly unique. Until the mid-nineteenth century, wheat was sold by sample. That is, a buyer would take a small sample and evaluate it before making an offer (if any) to the seller.

The Chicago Board of Trade changed all that in 1848 by establishing a unique trading place for grains. Wheat became a commodity when people started to create categories of wheat, separating lots as a function of the type (winter or spring, hard or soft, etc.) and the quality. That way, two lots of wheat coming from different farmers but ending up in the same category would be deemed identical. Buyers would no longer have to worry about which farmer was behind which bag of wheat. With this new trading place, the price of wheat became a sufficient statistic to clear the market.

One may imagine that today all crops are traded through exchanges like the Chicago Board of Trade. That was not the case, at least until very recently. Until 2008, crops in Ethiopia were traded the way American wheat had been before 1848. Buying coffee, sesame, or teff meant that you needed an agent, who would first take a sample, evaluate it, and then bargain with the seller. If you needed more coffee from other sellers, the same operation of testing and then bargaining had to be repeated again.

Grain markets in Ethiopia had other problems. First of all, distrust was widespread. Accordingly, people would only negotiate with a handful of persons, people they knew and could trust. In other words, from the perspective of a farmer or a broker, the market was thin. Making the market thicker was not safe: late payment (if there was any payment at all) was a common practice. Also, local producers were poorly informed about the real price of the crops, so they were easily manipulated by brokers.

The lack of confidence and market thickness are the perfect ingredients for having nonoptimal transactions and a fragile market. Small events are enough to disrupt the market by halting all transactions. If I see that the rainy season is two weeks late and I do not fully trust the farmers I am transacting with, how can I be sure that I will be paid in due time? Lack of a well-functioning market is believed to have been the main cause of the 1984 famine in Ethiopia, where over 1 million people died from starvation. Most of those people died in the north of the country, where a drought severely reduced crop yields. The saddest part of the story is that in the southwest part of Ethiopia there was at that time an excess of food.

To avoid such events occurring again (and to help the country's economy grow), the Ethiopian Commodity Exchange (ECX) was created in 2008, on the same principle as the Chicago Board of Trade. ECX managed to impose payments to be made at $T+1$, one day after the transaction is agreed on.

Among other things, ECX "simply" set up a reliable system under which grains could be traded as commodities, thus making the market thick. By requiring sellers to first store their grain in one of ECX's warehouses and requiring buyers to deposit money in a special account, ECX managed to make market participation safe for both sellers and buyers.

1.4 Market Design: First Example

We review here the case of a market that was not well designed but that was eventually fixed: the distribution of food at Feeding America. This section summarizes the diagnosis of the situation at Feeding America as well as the new market design that was implemented by Canice Prendergast and his colleagues from the America's Second Harvest Allocation Task Force.[1]

1.4.1 Feeding America

Feeding America is the third-largest not-for-profit organization in the United States (after the Red Cross and United Way) and consists of a network of food banks.[2] Its main mission is to distribute food to the poor via its branches

Introduction

(churches, community centers, soup kitchens, and others). Feeding America regularly receives food donations (mostly from producers) that it seeks to distribute to different food banks. The "market" we will be interested in is the one involving Feeding America's headquarters and the regional food banks affiliated with Feeding America. According to our definition, it is a market because:

- There is a demand: the regional entities of Feeding America, spread across the country. There are a bit more than 200 such entities. These entities look for food that they will distribute to the poor (via local branches).
- There is a supply: Feeding America's headquarters, which receives food donations (e.g., several trucks of frozen chicken) that they need to allocate to the regional entities.

We have here a market design problem: how to allocate the food received by Feeding America to the regional food banks. Whatever design we can come up with, there is an important constraint that must be taken into account: there cannot be monetary transactions. The food has been donated; the purpose is not to resell it. Also, selling food to the food banks would create an additional problem: the food banks would have to raise enough money to buy the food.

1.4.2 The First Design

To understand how Feeding America initially designed its allocation procedure, it is important to have in mind one of the main concerns it has: any food that has been donated has to be redistributed, and waste has to be minimized. Besides the fact that wasting food is not desirable from a public health standpoint, it can also have adverse economic effects. If Feeding America fails to find some recipient for a donation it received, the donor may be less inclined to donate again in the future. So if Feeding America is proposing to give some food bank a truckload of food, the food bank should have an incentive to take it.

Another concern is that the distribution of food across the network of food banks has to be fair, in the sense that it cannot always be the same entities receiving all (or most of) the food. But, of course, all food banks are not equal. Some entities are located in highly populated areas, while others operate in rural areas, some operate across a large radius, and so on. In other words, some food banks "should" receive more food than others.

These two constraints (food banks should not refuse proposed food donations, and the asymmetric needs across food banks) led Feeding America to develop a simple design to allocate food: construct a queue of the food banks, and when Feeding America receives some food, it proposes to donate it to the first food bank in the queue. If this food bank refuses it, it is offered to the second bank in the queue, and so on. This is the general principle of the mechanism Feeding

America came up with, but we need to give more details to understand how it works.

The first problem is the construction of the queue. This is easily solved. First, Feeding America calculates, for each food bank, two variables:

- the number of pounds of food that it *should* receive, calculated using comparisons across food banks and taking into account the population size of the service area;
- the number of pounds of food that it has received so far.

Using these two parameters, Feeding America can easily calculate how much food each food bank needs, and list the food banks in order of greatest need.

The second problem is how to create incentives for food banks not to waste donated food. To this end, whenever a food bank is offered food, it would be asked if it wants the food. There is one important reason why Feeding America only proposes food to the food bank and does not impose it: it does not know everything about each food bank. Indeed, food banks usually have other sources of donations besides Feeding America, mostly from regional sources. Consequently, Feeding America may not know exactly what each food bank needs. Moreover, consumption habits vary across regions, which adds an additional layer of complexity when one tries to guess the needs of a particular food bank. Also, transportation costs accrue to the food bank. The benefit of a food donation for a food bank depends then on many variables: whether it needs the food, whether it can store it, whether it can get it shipped, and other factors. One way of improving food allocation could be to have each food bank report all the relevant variables to Feeding America, but with more than 200 food banks it would considerably increase the complexity of Feeding America's task.

So, each time Feeding America receives some food donation, the food bank at the top of the queue is asked whether it would like the food proposed.

- If the answer is yes, the food bank will be liable for the transportation costs, and the number of pounds of food proposed is added to the "received pounds" entry corresponding to the food bank.
- If the answer is no, Feeding America would add up the number of pounds of food proposed (although rejected) to the "received pounds" entry corresponding to the food bank.

The incentive to accept food comes with the consequence of saying no. If a food bank refuses some food, then it automatically decreases the amount of food it needs. As a consequence, the food bank's rank in the queue drops, meaning that the food bank would have to wait several days or weeks before being proposed another load of food.

Introduction 9

Counting the food as given when a food bank is actually refusing it is not as unfair as one may think when considering food that can be stored. Unless the food bank's warehouse is full and cannot store any more food, the cost of having more food to store may not be very high. The only decision variable is then the transportation cost. And if the warehouse is full, then the food bank does not need food and can afford to wait. From a food bank's perspective, the best strategy in most cases is to accept the food, because whether it accepts it or not, its position in the queue will be the same. With such a mechanism, Feeding America is thus almost sure that no food donation will remain without a recipient.

There are two caveats with this procedure, however. First, the incentives put in place make sense when the food can be stored for a long time. If the donation is produce or fresh products with a close expiration date, forcing a food bank to accept it may result in some waste if the bank already has a sufficient quantity of this food in its warehouse. To solve this problem, Feeding America would not increase the tally of "received pounds" when a food bank refuses produce. If a food bank refuses a truckload of, say, clementines, it should be because they already have lots of them. Forcing the food bank to accept the offer would just make it more likely that the clementines end up rotting. Spoiling food is not the mission of Feeding America, so a food bank cannot be forced to take a product it knows will be spoiled.

The second issue relates to the distance the food had to travel. Since the main decision variable for a food bank is transportation cost, proposing some food that has a high shipping cost would amount to an unfair offer. So if the food received was too far away from a food bank, that food bank would be skipped in the queue.

1.4.3 The Problems...

So far, so good. Feeding America receives food donations, and set up a system to distribute them to the regional food banks. The system is apparently designed so that food waste is minimized, and the distribution seems to be fair.

But in fact the allocation procedure was far from efficient. Prendergast reports that food banks receive on average about 20% of their food from Feeding America, and the latter has very little knowledge of what the other 80% consists of. The incentives put in place were too strong, forcing food banks to accept food they did not really need. Another source of inefficiency was that the system does not fully take into account variations in regional diets. But perhaps one of the most important issues is that Feeding America was treating all types of food equally through this allocation procedure.[3] This is reason for concern because some types of food are more valuable than others (e.g., pasta vs. pickles). Some food may also be undesirable from a transportation perspective. A truck of frozen meat contains much more food than a truck (of similar volume) of potato chips.

So, overfilled warehouses and misallocation of food turned out to be common among food banks. In more general terms, Feeding America failed to incorporate the differences between food banks in terms of their needs. Because Feeding America did not know the specific needs of each food bank, misallocation of food was common. For instance, a bank could have accepted a truck of cereal in spite of not needing it whereas there was another food bank desperately seeking cereal. The problem behind this was a poor system of information revelation: The procedure put in place did not permit food banks to reveal their needs exactly.

In economics, information revelation is closely linked to the notion of *choice*. But for choice to be expressed fully, we need a trade-off, which is usually introduced through the existence of a budget constraint. If I have no budget constraint, I will say that I want both an apple and an orange, and thus I am not revealing which fruit I prefer. But if I can take only one fruit, my choice will reveal what I prefer. The problem is not only that food banks cannot communicate to Feeding America what they need but also to that they cannot communicate the "intensity" of their needs. Absent a budget constraint, an agent will request all types of items, and thus the choice becomes uninformative.

But a budget is problematic for Feeding America, because it implicitly introduces money into the equation. The solution proposed by Prendergast and his colleagues from the University of Chicago is to have "fake money" that would be distributed to the food banks.

1.4.4 ... and a Solution

Setting up an auction for the food is a very good way to have food banks express their preferences (through their bids). It (theoretically) completely eliminates the problem of not knowing precisely the needs of the various food banks. But it is a bold move. To some extent, an auction is tantamount to implementing a competitive market mechanism, where the richest are the more likely winners. This is at odds with the mission of Feeding America and the food banks. Markets are not always perfect, in that they tend to favor the richest, strongest, or the most powerful, and the proceeds of a competitive market are usually not distributed evenly. There are winners and losers. A food bank distributes to the poor, to those who are left out by the markets. Using a market mechanism at Feeding America looked initially like setting the fox to guard the henhouse. As one food bank director said to one of the Chicago economists who set up the auction: "I am a socialist. That's why I run a food bank. I don't believe in markets. I'm not saying I won't listen, but I am against this."

The auction system put in place at Feeding America works as follows. First, each food bank is allocated some amount of fake money, called "shares." The distribution of shares is made using the same principle as in the previous system.

Introduction

That is, the neediest food banks would receive the most. Then, twice a day, Feeding America runs an auction for the food it aims to distribute. The auction is pretty simple it's a *first-price sealed-bid auction* (see section 2.6 for a more detailed description). In this auction, each bidder submits a bid, but the bid is not public (hence the "sealed-bid" term). At the end of the auction, the winner is the bidder with the highest bid, and the winner pays his bid (hence the term "first-price"). As we will see in chapter 2, a first-price auction may not be the best auction format, but it has the advantage that it is relatively easy to understand: I bid, and if I win, I pay my bid. Not making bids public was a necessity to avoid "sniping," waiting until the last minute to put in a bid, so that competitors do not have the time to counterbid. Some food banks are big, with lots of people working there, and thus could afford to have someone permanently monitoring the auction. But small food banks may not have this capacity. So avoiding bid sniping has the effect of leveling the playing field in the bidding mechanism.

At the end of the day, the proceeds of the auctions (i.e., the shares) are redistributed among the food banks, using the same formula as for the initial allocation: the more a food bank needs food, the more shares it receives.

Food banks are also allowed to bid jointly. This turns out to be very helpful for small food banks, which may not need a whole truckload of some food. Several food banks can thus bid jointly and split the food between them if they win (and share the transportation costs). Allowing for joint bidding undeniably makes the system more flexible. Without this, the food proposed by Feeding America would mostly consist of *indivisible goods*, using the economics terminology. Indivisible goods create problems because I may refrain from acquiring some good simply because I cannot get the quantity I want, and because of my constraints I am better off without the good than with too much of it. This is particularly relevant for food banks that may have limited storage capabilities.

1.4.5 Results

The auction put in place at Feeding America had great benefits. First and foremost, Feeding America is now able to make a more precise distinction between the types of food that are highly demanded (e.g., meat, fish, or peanut butter) and those that are not (e.g., produce, sugary drinks, potato chips). Having such information is important because it allows Feeding America to better target its donors. Since the auction system was put in place, the supply of food has drastically increased, going from about 250 million pounds per year to more than 350 million pounds.

The implementation of the auction mechanism together with the use of fake money also brought interesting questions for economists. One issue was related to the volatility of prices. Even though the shares used by the food banks are not real money, the new currency may still be subject to the same types of problems real-world currencies experience, such as inflation. Having stable prices is desirable for the auction at Feeding America, because it helps food bank managers to have a precise idea of what amount they should bid for each type of food. This problem is particularly acute in this case because there is no exchange rate between shares and dollars. Bidding in an auction using a currency we are used to can already be tricky. It is certainly more complex when we are using a currency we are not used to.

2 Simple Auctions

This chapter is devoted to the most basic auctions. The auctions we will consider here are restricted to the case when there is only one object to be sold. For most people, some of the auctions that we will study in this chapter will already be familiar, and perhaps the most widely known is the one we see in movies or in the news, such as the auctions for art: there is an auctioneer who runs the auction and takes the bids of the participants, and at some point a winner is declared. We will see that even this very simple auction needs to be defined in a precise way. Slight, apparently innocuous variations can influence the behavior or the bidders. This chapter is thus an opportunity to provide a precise definition of different auctions and explain in detail what we mean by a bid.

2.1 Introduction

In the basic microeconomic model, we learn that the price of a good is obtained when the supply and the demand meet. In practice, buyers and sellers often make little adjustments in their demands and their supplies until the equilibrium price is reached. What is important to note is that for these adjustments to take place it must be that the encounter between buyers and sellers is repeated over time rather than just a one-time event. For instance, each day or each week the manager of a grocery store can adjust the price of the goods she sells, and buyers can try different stores or change the quantity of the goods they purchase, depending on their prices. Such adjustments are particularly relevant and not very costly when the situation is repeated over time. But in some situations such adjustments cannot be made, and we need to find other ways to reach an equilibrium. Auctions are a common way to sell or buy goods when:

- sellers and/or buyers have little knowledge of what would be the "right" price;
- there is scarcity: fixed supply (e.g., an art painting, a piece of land);

- the quality or quantity of the good to be sold or bought changes very frequently (e.g., electricity, fish);
- transaction frequency is low.

In a market (i.e., when we face a demand and a supply), one crucial question is how prices are formed (if monetary transactions are allowed), and how the final allocation is determined (who gets what). Any rule or way to determine prices and the allocation is called a **market mechanism**. It gives a final allocation for any set of agents with all their characteristics (their preferences, budget, production capacities, and so on) and the environment (the type of goods or items to be exchanged).

Auctions are thus nothing but one particular class of market mechanism. In particular, they also serve as a **price discovery** mechanism, a "device" to discover the price(s) at which we may sell or buy a good.

2.1.1 Auctions: A Definition

The basic definition of an auction is when we have one side of the market making bids (generally the buyers) that are used to determine a final price (or several prices) and an allocation. This definition is vague because there are many different types of auctions. In a more general way, an auction can be defined by addressing the following points:

- **Bidding format rules** (the form of the bids)

  For instance, a bid can consist of a price, a price and a quantity, a quantity only, or a list of items (when several objects are sold at the same time).

- **Bidding process rules**

  When does the auction stop? What information is given to the bidders? How many times can bidders bid? Are there any special conditions to allow a bidder to counterbid?

- **Price and allocation rules**

  How are the winners of the auctions chosen, and what are the final prices?

2.1.2 Auctions Are Everywhere

There is evidence that auctions have been used throughout history. One of the earliest examples is the sale of women through auctions in Babylon in 500 BC. On March 28, 193 AD, the Pretorian guards killed Emperor Pertinax and then ran an auction to choose the next emperor. The winner was Didius Julianus (who unfortunately for him ruled only two months). Perhaps the most famous auction houses today are Sotheby's and Christie's, founded in 1744 and 1766, respectively. Until the nineteenth century, slaves were often sold through auctions.

Simple Auctions

Today we find auctions everywhere in our daily lives. Sometimes the auctions involve individuals (when we buy something on eBay, for instance), but many times auction participants are organizations.

- The best-known auctions are those on eBay, art sales (e.g., Sotheby's), house foreclosures, or those for secondhand cars.
- Auctions have started to be used for event tickets (concerts or games).

One of the main problems in the sale of event tickets is the presence of the so-called *ticket scalpers*, people who, as soon as possible, buy event tickets (at the official, low price) with the purpose of selling them back on a secondary market (via websites like stubhub.com or ticketmaster.com) at a high price. Some ticket scalpers manage to make several million dollars a year by doing this. The problem can be so severe that for some events almost all the tickets are initially sold to ticket scalpers.

One key observation is that the price at which tickets are officially sold does *not* correspond to the *market price*, which is the price people are really willing to pay for the ticket. Rather, the market price is the price the ticket scalpers will set when reselling the tickets. One idea that is being used to diminish the impact of ticket scalpers is to sell tickets through an auction. That way the price at which the ticket is sold corresponds to or is close to the market price, which means the resale will not be profitable (and thus we avoid the presence of ticket scalpers).

- The Treasury sells its bonds, bills, or notes through an auction.

The problem for a government wanting to raise money on financial markets is to determine the interest rate that will be offered to investors. An auction is a perfect tool to determine the interest rate of Treasury bonds.[1] We will analyze Treasury auctions in detail in chapter 7.

- Most of the ads (if not all) we see when browsing the Internet are auction outcomes where the bidders are the advertisers.

We will analyze the market for Internet ads in detail in chapter 5.

- There are auctions that run every day, such as for electricity or fish markets.
- The construction and maintenance of many public facilities (e.g., bridges or tunnels) are decided through auctions. In such auctions, called "procurement auctions," the bidders are the sellers, and the buyer is the public authority (e.g., city, government). The sellers compete by proposing a price for the facility they will build and/or operate, and the seller will generally pick the seller that proposes the lowest price.
- Daily quotations at stock exchanges and IPOs are also auction-based mechanisms.

- Some business schools let students choose their courses through an auction. In many business schools, students are offered a wide variety of courses and must choose which courses they want to attend. One of the main problems for students is that many courses have a limited capacity: there are not enough seats for all the students who want to attend a course. The way some business schools assign students to courses is to run an auction.[2] At the beginning of the year, students are allocated some amount of tokens, or fake money, and they use it to bid for courses. The students who enroll in a course are then those who bid the highest for that course.

- In many countries, cell phone bandwidth is allocated to carriers through an auction.

Auctions are used in all of the cases listed (and many more), but the auction format varies from one case to another. What the differences are and why we use one auction format instead of another one are among the main questions addressed by auction theory.

When designing an auction many details need to be addressed:

- **For the Seller**

  Can bidders bid several times? If there are multiple items to be sold, is it better to sell them all at the same time or one at a time (and in which sequence)? Does the seller have an interest in who wins the auction? How does the seller establish a reputation (if she plans to sell again in the future)? Do we put in a minimum opening bid and/or a reserve price (and what amount)? How long should the auction run?

- **For the Buyer**

  Is resale allowed? (I may indeed value the item differently if resale is allowed, which could affect my bidding strategy.) Should my strategy depend on whether another auction will be run in the future? Do I care about who will win the auction?

2.2 Valuations

In an auction, a bidder's **value** (or **valuation**) for an object is the maximum amount that the bidder is willing to pay for the object (or objects) being sold. Bidders' values are of course unknown to the seller. Otherwise the seller could just propose the object to the bidder with the highest value having a price equal to that bidder's value (or slightly below).

Auction theory makes the distinction between different cases, depending on how much bidders know about their values and how bidders' values are related. The simplest case is that of **private values**. This refers to the situation where each bidder knows her valuation but does not know the values of the other bidders.

An implicit assumption in this case is that knowing the values of other bidders does not affect how much a bidder values the object.[3] The private value case is particularly relevant when we consider pure consumption goods.

The opposite case is when bidders do not have precise knowledge of their values. In this case, bidders typically have only an imprecise signal of how much they value the object. In such situations, bidders first obtain an estimate or a *signal* about the value of the object, and the uncertainty about the value comes from the fact that the signal does not give perfect information. A typical example is when bidders' interest in the object depends on the resale value of the object. The uncertainty about the resale value implies then that each bidder only has an estimate of how much she is willing to pay for the object. Note that in such a case the values that bidders attach to the good, although unknown to them, are related. That is, values are **interdependent**.

An extreme case of interdependent values is when the value of the good is the same for each bidder values are **common**. In the **pure common-value** model, all bidders have the same valuation but each bidder has a different signal about what the actual value is. A classic example of such a situation is when a government is selling drilling rights for an oil field. The value of the drilling right is the same for each bidder, as it is essentially the market value of the oil that will be extracted.[4] Oil companies (the bidders) usually perform their own geological surveys to obtain an estimate of the size and quality of the oil field, and thus each has a different estimate (signal) about the actual value of the oil field.

When studying auctions, it is usually assumed that bidders' values (for the private value case) or bidders' signals about their values (for the interdependent case) are drawn according to some probability distribution. Those draws can either be independent or correlated. In this chapter, we will assume that bidders' valuations are private and drawn independently from each other.

2.3 Payoffs and Objectives

In most cases, buyers only care about whether they win or lose, and if they win, how much they have to pay.[5] If a buyer wins an auction, his or her payoff is his or her valuation minus the price paid. If a buyer loses the auction (she does not get anything), the payoff is 0. The payoffs for a buyer in an auction are then easily defined:

$$\text{payoff} = \begin{cases} \text{valuation} - \text{price paid} & \text{if win} \\ 0 & \text{if lose} \end{cases} \tag{2.1}$$

For a seller, the payoff is simply the revenue of the auction, the price paid by the winner. It is generally assumed that the seller's objective is to maximize her payoff; that is, to sell at the highest price. In an auction, the best way to achieve this consists of selling to the bidder that has the highest valuation, because that bidder is the one who can afford to pay the highest price (whether the price is equal to her valuation is a different issue). For the economist (or the auction designer), having an auction such that the winner is the bidder with the highest valuation is efficient because it maximizes the *surplus*.

In economics, the surplus of a consumer is the difference between the price that consumer is willing to pay to obtain a good and the price she actually pays (if a consumer does not buy the good, the surplus is zero). So, in our setup, the surplus of a buyer is simply the payoff we defined in equation (2.1). For a seller, the surplus is the difference between the price at which the seller is selling and the minimum price at which the seller would agree to sell. For most cases we will see in this book, the minimum price will be assumed to be 0, but sometimes that price is strictly positive. For the sake of simplicity, we assume here that the seller is willing to sell at any positive price, so her surplus is simply equal to the price at which she sells the good.

The total surplus is then the sum of the buyer's and the seller's surpluses. Let i be the bidder who wins an auction, with v_i her valuation and p the price bidder i has to pay to the seller. We then have

total surplus = surplus of bidder i + surplus of seller

$$= v_i - p + p = v_i.$$

We can see that the total surplus does not depend on the price paid by bidder i. It only depends on the valuation v_i. This implies that for an auction to be efficient, for an auction to maximize the total surplus, the winner must be the bidder with the highest valuation.

2.4 Ascending Auctions

The first type of auction is perhaps the most famous one. We start with a (relatively) low price, potential buyers bid on the item, raising the price, and the winner is the one with the highest bid. An auction of this type is called an **ascending auction**. As we will see, this is a very rough description. We need to be more precise regarding the exact details of the auction, and, of course, details will matter.

2.4.1 The English Auctions

When considering simple contexts where there is only one object to be sold, ascending auctions are usually called English auctions. There are many aspects that one must take into account when designing an English auction: Can bidders bid anonymously? Are there maximum or minimum limits on bids? Auction

Simple Auctions

theory usually considers one particular version of the English auction, which we will describe in section 2.4.1.2. It is nevertheless interesting to present different versions of the English auction, as it will allow us to see how slight variations in the design of an auction can bring different insights and influence bidders' strategies.

Perhaps one of the most important aspects in an English auction is who announces the prices. Is it the bidders who propose prices or is it the auctioneer who proposes a price and the bidders simply say yes or no? The answer to this question defines two broad variants of English auctions.

2.4.1.1 When bidders announce prices An English auction where bidders announce prices is called an **English outcry auction**. The auctions we usually see in movies are outcry auctions. In general, bidding is not anonymous: each bidder can see who the other bidders are. In some cases, the auctioneer can set a rule for minimum increments between bids or a bid format. For instance, the auctioneer can require that any bid must be at least \$100 more than the previous bid (or the starting price) or that any bid must be a multiple of, say, \$1000. Those rules are generally imposed to make the auction more lively and not last too much time. (Imagine an auction of an art painting where bids increase by 1 penny!) Another classic feature of outcry auctions is that **jump bids** are allowed. By this we mean that there is no upper limit on bids. We can thus have an auction where the auctioneer requires that bids increase by at least \$50 and we observe the following bid sequence: \$400, \$450, \$500, \$550, \$800, \$850. Here the fifth bid is a jump bid: instead of increasing by the minimum amount required by the auctioneer (from \$550 to \$600), a bidder made a jump of \$250.

2.4.1.2 When bidders do not announce prices The opposite case from the English outcry auction is an auction where the prices are announced by the auctioneer and not the bidders. The general structure of such an auction is the following. For any proposed price, bidders have to say whether they accept or refuse the proposed price. As soon as a bidder refuses the proposed price, she is no longer participating in the auction. The price increases until there is only one bidder left.

A well-known auction that fits this description is the so-called **Japanese button auction**. In this auction, each bidder has access to a remote control that has only one purpose: signaling, as the price goes up, whether the bidder is still bidding or has stopped bidding. More precisely, bidders who are still willing to buy the object at the displayed price have to press a button on their remote control. A bidder is considered not participating in the auction the moment she stops pushing the button. If a bidder stops pressing the button, she cannot press it again later. A bidder is said to be **active** if she presses the button and **inactive** otherwise. The auction ends as soon as there is only one active bidder left, meaning when all but one bidder has stopped pushing the button on their remote control.[6]

Auctions that follows the preceding description are called in the literature English auctions.

Remark 2.1 The simplicity of the Japanese button auction has made it the workhorse of auction theory. In the literature on auctions, when authors write English auction they generally are referring to the Japanese button auction. We will follow this tradition here.

We can easily imagine variations of this auction. For instance, it can be that for each new price it suffices that one bidder accepts the proposed price in order for the auctioneer to propose a new, higher price. The auction stops as soon as no bidder accepts the new price, and the winner is the last bidder who accepted a price (and she pays the price she accepted).

Note that there are several aspects that are left unspecified. One aspect that can be important is whether bidders' decisions are anonymous. Having the possibility to bid anonymously can be important in some contexts, especially when bidders' valuations are interdependent. We can consider, for example, an art auction where an expert bidder has a good knowledge of the "real" value of the object to be sold and is known to be cautious. If that bidder bids, say, $100, then the other bidders can deduce that the object is worth at least $100. So a bidder whose valuation is only, say, $60 may update her valuation if she sees that the expert bidder is bidding $100. Obviously, in such a case, our expert bidder would prefer to bid anonymously. We will see in chapter 3 when studying eBay's auctions how bidders' behaviors change when bidders have interdependent valuations. We will see that, when bidders' values are private, the issue of whether bidders can bid anonymously has no impact.

In general, we consider situations where there is only one bidder left at the end of the auction, meaning there is only one bidder with the highest bid. It may happen, however, that all the remaining bidders become inactive at the same price. In this case, all those bidders can be considered winners, and the unique winner will be chosen randomly.

Example 2.1 Suppose that we have three bidders, Alice, Bob, and Carol. Their valuations are $v_{Alice} = 8$, $v_{Bob} = 10$, and $v_{Carol} = 12$.

Suppose that Alice and Bob become inactive when the price is 6 and 9, respectively, and Carol is planning to drop out of the auction when the price is 10. So, when the price is 9, Bob drops out and there is only one bidder left, Carol. So the auction stops and Carol pays 9 for the good. Her payoff is then $v_{Carol} - 9 = 12 - 9 = 3$.

Simple Auctions

Suppose now that Bob and Carol both plan to drop out when the price is 9.1, and let us assume that 9 is the price proposed by the auctioneer just before the price 9.1. So for the price at 9 Bob and Carol are both active, but for the next price, 9.1, Bob and Carol are no longer active. Each one will be selected as the winner with probability $\frac{1}{2}$. So Carol's payoff is

$$\frac{1}{2} \text{ Carol payoff if selected} + \frac{1}{2} \text{ Carol payoff if not selected.}$$

If Carol is selected, she buys the good (and pays 9), and if she is not selected, she does not buy the good (it will be Bob). So her expected payoff is

$$\text{Carol's expected payoff} = \frac{1}{2} \times (v_{\text{Carol}} - 9) + \frac{1}{2} \times 0 = \frac{1}{2} \times (12 - 9) = \frac{3}{2}.$$

For Bob, his expected payoff is

$$\text{Bob's expected payoff} = \frac{1}{2} \times (v_{\text{Bob}} - 9) + \frac{1}{2} \times 0 = \frac{1}{2} \times (10 - 9) = \frac{1}{2}.$$

2.4.2 Bids, Strategies, and Payoffs in the English Auctions

A crucial difference between the Japanese button and the English outcry auctions is with respect to the bids. As we said, in the English auction, bidders do not decide or announce a price; it is the auctioneer's role. Instead, each time the price changes, the bidders simply have to say whether they are still active or not. If not, they are considered as no longer participating in the bidding process. So, what is a bid in this auction?

A bid for a bidder in an English auction is a *price target*. It is the last price for which the bidder was still active. So the winner is the bidder whose price target is the highest, the bidder with the highest bid. This is because all the other bidders have dropped out at lower prices (meaning their bids/price targets were lower).

Example 2.2 Three bidders, Alice, Bob, and Carol, with bids of $80, $90, and $100, respectively (we do not specify the valuations because they are not needed to describe the English auction). This means that:

- Alice drops out as soon as the price is strictly greater than $80.
- Bob drops out as soon as the price is strictly greater than $90.
- Carol drops out as soon as the price is strictly greater than $100.

For our purposes, suppose the price increment is 1 penny. At the price $90, Alice has already dropped out she is no longer participating. However, Bob and Carol are still active. The next price is $90 + $0.01 = $90.01. At this price, Bob drops out, and we now have only one bidder left, Carol. So Carol is the winner, and the price she pays is the last displayed price, $90.01.

Observe that in the English auction the auctioneer will know at the end of the auction the bids of all bidders except the winner. Indeed, all the losing bidders in the auction stop bidding when the price reaches their bid. So observing the price at which they become inactive tells us what their bids are. In contrast, we cannot observe the bid of the winner, because the displayed price does not reach the winner's bid. In example 2.2, the auctioneer can learn that Alice's bid is $80 and that Bob's bid is $90, but the auctioneer cannot learn Carol's bid. Had her bid been, say, $110 or $1000, the auction would still have stopped at $90.01. The only thing that the auctioneer can learn about Carol is that her bid is at least $90.01.

2.4.3 Ticking Price

In an auction, the **ticking price** is the amount by which the price has to go up (if it is an ascending auction) or down (if it is a descending auction—see section 2.6.3). The ticking price is also sometimes called the **price increment rule**.

For some auctions, bidders are just asked to submit a bid once. In this case, the price does not decrease or increase, and thus there is no ticking price. This is the case, for instance, with the Vickrey auction that we will see in section 2.5.2. However, there are auctions where the price changes over time, such as the English auction or the English outcry auction. The ticking price is then simply the rule that says by how much the price changes at each moment. In auction theory, it is usually assumed that the price increases continuously.

The ticking price rule may depend on the price level. For instance, a rule could be

- for any price strictly below $10.00, the ticking price is $0.50; or
- for prices strictly above $9.99, the ticking price is $1.00.

If the auction starts at, say, $5.00, then the possible prices we can see during the auction are $5.00, $5.50, $6.00, ..., $9.50, $10.00, $11.00, ... This means that we cannot see, say, the price $6.25.

In the English outcry auction, there are no ticking prices, because the bids consist of price proposals by the bidders, and they can make the bids they want. In the English (Japanese button) auction, there is a ticking price: bidders do not propose prices during the auction; they just say whether they are still bidding.

To see how the ticking price works, suppose that there is an item and three bidders, Alice, Bob, and Carol, whose bids are $7.00, $10.00, and $12.00, respectively. Let us say that the auction starts at the price of $6.00 and that the ticking price is $0.50 whenever the price is less than $10 and then $1 if the price is higher than $10. Table 2.1 shows how the auction unfolds. The first column displays the sequence

Simple Auctions

Table 2.1
Bidding with ticking prices in the Japanese button auction

Displayed price	Alice	Bob	Carol
$6.00	•	•	•
$6.50	•	•	•
$7.00	•	•	•
$7.50		•	•
$8.00		•	•
$8.50		•	•
$9.00		•	•
$9.50		•	•
$10.00		•	•
$11.00			•
$12.00			•

of prices, and the other columns display the behavior of the bidders. The symbol • says that the corresponding bidder is active for the corresponding price. For instance, when the price is $9.00, Alice is not pressing the button, while Bob and Carol are still pressing the button.

We can see that Alice is still bidding when the price is $7.00, because it is not above her bid. However, when the price jumps to $7.50, Alice becomes inactive, she stops bidding, while Bob and Carol are still active. When the price is $10.00, Bob is still bidding (for the same reason that Alice was still bidding when the price was $7.00). When the price jumps to $11.00, Bob becomes inactive. The auction then stops at the price $11.00 because for that price there is only one active bidder left, Carol. Table 2.1 displays what would happen for the price $12.00, but we will never see such a price because the auction stopped before that price, at $11.00. So, the result of the auction is that Carol wins the auction for the item and gets to pay $11.00, the last displayed price.

When studying auctions from a theoretical perspective, ticking prices play a small role. They are in fact here just to "break ties" between the winner(s) and the other bidders. Whether the ticking price is $0.50, $0.01, or $1.00 (or any other amount), the result would remain unchanged: Carol wins the auction, and the price is "just above" the bid of the last bidder who stopped before the winner.

2.4.4 Truthful Bidding

One of the first questions we may ask is whether it is optimal for a bidder to bid her valuation, whether it is optimal to be truthful. In the English auction, truthful bidding is defined very simply: it consists of pressing the button until the displayed price equals one's valuation. We have the following result.

Result 2.1 Truthful bidding is a dominant strategy in the English auction.

Recall that a dominant strategy is a strategy that outperforms any other strategy, for any possible strategic choice by the opponents (see appendix A).

Proof For the sake of simplicity, suppose that my valuation is $100. Call p the displayed price of the auction and p^* the winning price (i.e., p^* is the last price displayed during the auction). We distinguish between several cases.

Case 1 *I keep pressing the button when $p > \$100$.*

This case means that at some moment the displayed price was above 100. Since in an English auction the price can only increase, we have $p^* > \$100$.

If I am the winner, then I pay the price p^* (the exact number does not matter as long as we know it is more than $100). So, my payoff is $\$100 - p^*$, which is negative. If I lose the auction (i.e., I stopped pressing the button before the auction finished), then my payoff is 0.

So, continuing to press the button when the price goes above $100 can bring me at most a payoff of zero, and in some cases a negative payoff. Observe that if I stop pressing when $p = \$100$, then I lose the auction. The winner is the bidder who is the last one still pressing the button (i.e., the winner is the bidder who *never* stops pressing the button). In this case, my payoff is always 0, so stopping at $p = \$100$ is better than stopping later.

Case 2 *I stop pressing the button for some $p < \$100$.*

Suppose, for instance, that I stop at $p = \$80$. Since I stop pressing, I lose the auction, and thus my payoff is 0, no matter at what price the auction stops.

Consider now what would happen if I bid truthfully, meaning I keep pressing the button until $p = \$100$. If p goes above $100, I already know from case 1 that I stop pressing the button, so I lose the auction, and my payoff is 0. But, if the price is $\$80 < p < \$100$, and I am the only bidder left, then I win. Suppose that $p^* = \$90$. Then my payoff is $\$100 - \$90 = \$10$, higher than if I stop at $p = \$80$. So, planning to stop pressing the button when the price reaches my valuation is better than planning to stop at a lower price. ∎

The preceding result (and its proof) is valid for the English (Japanese) auction. But to see how details matter, take the apparently similar English outcry auction. We will see that truthful bidding is not necessarily an optimal strategy.

Example 2.3 Truthful Bidding in the English Outcry Auction   Again, suppose that my valuation is $100. In the auction room, Mr. Grinch does not like prices that are between $50 and $80. If the displayed price is between these two prices, he bids $1000. Otherwise, he never bids.

Simple Auctions

The question here is not whether Mr. Grinch's strategy is good for him but rather whether truthful bidding is a *dominant* strategy for me; that is, whether it is my best strategy *no matter how* my opponents decide to bid.

In the English outcry auction, truthful bidding consists of making gradual increments in my bids (i.e., always bid a slightly higher amount so that I am still in the race until my bid is equal to my valuation).

In this case, the price would eventually reach $50, and because Mr. Grinch is around, it will jump to $1000. To prevent Mr. Grinch from influencing the auction, at $49 I should make a jump bid and bid $81. This is a good strategy for me, but it is not truthful bidding.

2.5 Second-Price Auction

2.5.1 The Essence of the English Auction

In the end, what is an English auction? When we compare the English and the English outcry auctions, they look very similar. Yet we already know that tiny details in how they operate make those auctions somewhat different.

Nevertheless, we would like to think that they have something in common. The idea here would be to define an auction that captures the idea that the price is ascending and gives the same result as the other auctions under "normal conditions" (i.e., without the weird cases like Mr. Grinch in the English outcry auction). To this end, consider the following simple example.

Example 2.4 Suppose there are three buyers: Alice, Bob, and Carol. Their respective valuations are $v_{\text{Alice}} = \$50$, $v_{\text{Bob}} = \$70$, and $v_{\text{Carol}} = \$100$.

Now consider the English (Japanese button) auction, and let us try to simulate what will happen. Recall that in this auction it is a dominant strategy to bid truthfully. This means that Alice will bid $50, Bob will bid $70, and Carol will bid $100. As we have explained, those bids are price targets: they are the maximum prices at which these bidders will be pressing the buttons on their remote controls. For instance, as soon as the price is above $50, Alice will stop pressing the button and become an inactive bidder.

Suppose that the ticking price is $1: the price always goes up by $1 (it does not matter how often). Suppose that the auction starts at the price $30. At this price, all bidders press the button. They are active. Then the price increases. When the displayed price is $50, we still have all three bidders continuing to press the button. However, when the price reaches $51, Alice drops out; she is no longer bidding.

When the price is $70, we still have two buyers bidding: Bob and Carol. But when the price is $71, only Carol is left, for Bob has dropped out.

So in this auction the winner is Carol, and she pays $71. Observe that the price she pays is *not* equal to her bid: she bids $100 (her price target) but ends up paying only $71.

Example 2.4 suggests an important property of the English auction: the winner is the bidder with the highest bid, but the price she pays is the price corresponding to the second-highest bid (+ the ticking price).

So here is the insight: in an ascending price auction, under "normal conditions," we can expect the winner to be the bidder with the highest bid, but the price that the winner pays is equal to the second-highest bid, not the winner's bid.

2.5.2 The Vickrey Auction

The Vickrey auction is an auction that is aimed at capturing what we have just described for an ascending price auction: the winner is the bidder with the highest bid, and the price she pays is equal to the second-highest bid.

The Vickrey auction is defined in a much more simple way than the English auction. It just needs bidders to bid once, and there is no need to "spend some time" following the auction. Formally, the Vickrey auction is described as follows:

1. Each bidder secretly submits a bid to the auctioneer. It is a *sealed-bid auction*.

2. The winner is the bidder with the highest bid; the other bidders lose the auction.

3. The price paid by the winner is equal to the *second-highest bid*. The other bidders do not get anything and pay nothing to the auctioneer.

This auction is called a **sealed-bid second-price** auction. The auction is sealed-bid because when a bidder submits his or her bid, she does not know the bids of the other bidders. This is not the case with the ascending price auctions, where, during the course of the auction, bidders can observe whether there are other bidders bidding at least as high as the current displayed price. Notice that since bids are submitted "secretly," they need not be submitted at the same time. All that matters is that bids are not disclosed until the auctioneer has received all of them. The reason that the Vickrey auction is a second-price auction is obvious from its description.

Remark 2.2 The auction originally defined by William Vickrey is more general than the Vickrey auction we saw in chapter 2. In his initial treatment, Vickrey considered auctions of multiple *identical* items.[7] See chapter 4 for the generalization of the second-price auction to multiple items (but not necessarily identical).

What if there is more than one bidder submitting the highest bid? In this case, as for the English auction, we have several winners. One of them will be chosen at random. Formally, the bidders' payoffs are defined as follows. For each bidder

$i = 1, \ldots, n$, let v_i and b_i be bidder i's valuation and bid, respectively. Let $b_{\max} = \max\{b_1, \ldots, b_n\}$ be the highest bid. The second-highest bid is denoted $b^{(2)}$. So a bidder's payoff is

0 if $b_i < b_{\max}$,

$$\frac{1}{|\{j : b_j = b_{\max}\}|}(v_i - b^{(2)}) \quad \text{if } b_i = b_{\max}.$$

The number of highest bidders is $|\{j : b_j = b_{\max}\}|$, so $\frac{1}{|\{j:b_j=b_{\max}\}|}$ is the probability that a max bidder is awarded the good.

It turns out that in the Vickrey auction it is very easy to characterize what is the best strategy a bidder can use: submit one's true valuation.

Result 2.2 Truthful bidding is a dominant strategy in the Vickrey auction.

Proof Suppose that my value is $100 and I bid $110. Let $b be the highest bid submitted by a rival.

Case 1 $b \leq \$100 < \$110$

Then overbidding does not matter: whether I bid $100 or $120 I win, but I pay only $b. That is, I am indifferent between bidding $100 and $110.

Case 2 $\$100 < \$b < \$110$

In this case, I win, but I pay $b > 100$. By overbidding, I have a negative payoff. Had I bid only $100, I would have lost the auction and thus received a payoff equal to 0. So in this case I am better off bidding my valuation rather than overbidding.

Case 3 $\$110 < \$b$

So I lose the auction and my payoff is 0. Had I bid $100, I would also have lost (and also received a 0 payoff). So in this case I am indifferent between bidding $100 and $110.

To conclude, there are some cases in which I am indifferent between overbidding and bidding truthfully, and a case where I would prefer to bid truthfully. So, bidding truthfully dominates overbidding.

The proof is incomplete, as we have not shown that bidding less than my valuation is also a dominant strategy (but the argument is similar). ∎

The preceding result has important implications. Since bidders have the same optimal strategies in the English and the Vickrey auctions, the outcomes of the auctions are the same (i.e., who the winner is and what price is paid). It follows then that the seller obtains the same revenue in the English and the Vickrey auctions.

2.5.3 English versus Second-Price Auctions

Strictly speaking, the Vickrey and English auctions are not completely equivalent, because in their game-theoretic representations, bidders' strategy sets are not equivalent (see appendix A for a review of game theory concepts). To see this, observe that both auctions can be modeled as extensive form games where the first player to play is Nature, who chooses a valuation for each bidder. After Nature's move, things differ between the two auctions.

In the Vickrey auction, once a bidder observes her valuation, she has only one possible action: choose the bid amount. The important point here is not what the bidder does (picks an amount) but rather that a bidder takes only *one* action. This is where things differ from the English auction, where a bidder can take several actions during the course of the auction. Each time a new price is proposed, the bidder has to pick an action. From a game-theoretic perspective, there are many more histories in an English auction. To see this, let us assume that the starting price is 0 and the ticking price is $1. In this case, there is the history where the auctioneer proposed only $1. There is another history where the auctioneer proposed first $1 and then $2. Such histories only describe what the auctioneer is doing.

If bidders can observe what the other bidders are doing, then the histories will become even richer in the English auction. To see this, suppose there are three bidders, Alice, Bob, and Carol, and that they can each observe, for any price that is proposed by the auctioneer, whether the other bidders are still active or not. For simplicity, assume that the starting price is 0 and that the ticking price is $1. One possible course of action is that Alice drops out when the price is $2 (and she is the first bidder to drop out), and another course of action is that Alice drops out when the price is $5. In this case, a strategy for Bob can consist of two choices:

• Drop out for the price $4 if Alice drops out when the price is $2 (if Carol has not dropped out before).

• Drop out for the price $7 if Alice drops out when the price is $5 (if Carol has not dropped out before).

Our intention here is not to discuss whether such a strategy for Bob is optimal (we already know the answer from result 2.1) but rather to show that, from a strategic point of view, a strategy in an English auction can be a more complex object than a strategy in the Vickrey auction.

For the case of private values, this strategic difference between these two auctions is inconsequential. It does matter, however, when values are interdependent. In the second-price auction, there is no way for a bidder to learn other bidders' bids during the auction. This contrasts with the English auction, where, as the

Simple Auctions

proposed price increases, a bidder can learn about her opponents' valuations (or signal about her value) as she sees them dropping out of the auction.

2.6 First-Price Auction

2.6.1 Definition

The first-price auction is a static auction that looks similar to the Vickrey auction, with the only difference being the price paid by the winner. This auction is described as follows:

1. Each bidder secretly submits a bid (a price) to the auctioneer. It is a *sealed-bid auction*.
2. The winner is the bidder with the highest bid. The other bidders lose the auction.
3. The price paid by the winner is equal to the *highest bid*. The other bidders do not get anything and pay nothing to the auctioneer.

As for the English or the Vickrey auctions, if the highest bid has been submitted by several bidders, then we have several winners. In this case, the bidder that buys the item is chosen randomly among those winners. Formally, assume there are n bidders indexed from 1 to n, and let us denote by v_i the valuation of bidder i and by b_i her bid. Then $b_{max} = \max\{b_1, \ldots, b_n\}$ is the highest bid, and the set $\{j : b_j = b_{max}\}$ is the set of all bidders j that bid the maximum bid b_{max}. The number of the highest bidders is $|\{j : b_j = b_{max}\}|$, and each of them is selected with equal probability. Therefore, the payoff of bidder i is

$$0 \quad \text{if } b_i < b_{max},$$

$$\frac{1}{|\{j : b_j = b_{max}\}|}(v_i - b_i) \quad \text{if } b_i = b_{max}.$$

Here again, the payoff is in *expected terms* because if a bidder is not the unique highest bidder, she obtains the item with a probability less than one.

Example 2.5 Suppose that we have three bidders: Alice, Bob, and Carol. Their valuations are $v_{Alice} = 8$, $v_{Bob} = 10$, and $v_{Carol} = 12$.

Suppose first that we have the following bids:

$b_{Alice} = 6$, $b_{Bob} = 7$, and $b_{Carol} = 10$.

Then Carol is the unique winner. She gets the good with probability 1, and thus the payoffs are 0 for Alice and Bob (they lost the auction) and $12 - 10 = 2$ for Carol.

Now consider the following bids:

$b_{Alice} = 6$, $b_{Bob} = 8$, and $b_{Carol} = 8$.

Now we have two winners, Bob and Carol. So they each get the good with a probability of $\frac{1}{2}$. Alice's payoff is 0 (she lost the auction). Bob's expected payoff is $\frac{1}{2}(10-8) = 1$, and Carol's expected payoff is $\frac{1}{2}(12-8) = 2$.

As we have already discussed, the English auction is equivalent to a second-price auction (the winner will stop at the second-highest bid), and thus it is *not* equivalent to the first-price auction.

The description of the first-price auction immediately raises the question of how much a bidder should bid. Since the winner pays her bid, there is little incentive to bid truthfully (bidding one's valuation). To see this, recall that the payoff of the winner is *valuation − price paid*. If the winner bid her true valuation, then the payoff is 0. This is not very interesting for the winner. Obviously, a bidder does not want to bid higher than her valuation, as she would have a negative payoff if she won. The interesting option is to bid *less* than one's valuation. But how much less?

2.6.2 Optimal Bids in the First-Price Auction

To determine how much a bidder should bid in a first-price auction, it is useful to be a bit more formal. To begin with, observe that in any auction bidders have private information: their valuations. That information is private because a bidder does not know the valuations of the other bidders.[8] We then have an incomplete information game (see appendix A) where

- the players are the bidders;
- the strategy set of each bidder is the set of all possible prices (bids);
- a bidder's private information (her type) is her valuation; and
- the payoff function is such that the winner's payoff is her *valuation minus her bid* and is 0 for the other bidders.

The Vickrey auction is also an incomplete information game, but solving it is relatively easy because we have seen that each bidder has a dominant strategy (bidding one's true valuation). In the first-price auction, choosing the optimal bid is not trivial. We need to analyze the Bayesian equilibria of this game.

For the sake of simplicity, let us assume that we have n bidders and that all bidders' valuations are drawn randomly and independently between 0 and 100.

A *strategy* of a bidder in this context is a function that gives for each possible valuation a bid. Let us denote by s_i the strategy of bidder i. If v_i is the valuation of bidder i, she will bid $s_i(v_i)$. For instance, s_i could be such that $s_i(20) = 15$ and $s_i(10) = 8$, which means that bidder i plans to bid 15 if her valuation is 20 and 8

Simple Auctions

if her valuation is 10. We will make the following assumptions (again, in order to simplify our analysis):

- **Assumption 1** Each bidder uses the same strategy: for any pair of bidders, i and j, $s_i = s_j$. In other words, two bidders with the same valuation will bid the same amount. So now we simply write s to denote the strategy used by any bidder (we drop the subscript that refers to the bidder).
- **Assumption 2** The bidding function s is *strictly* increasing. So if bidder i's valuation is higher than bidder j's valuation, then bidder i bids higher than bidder j. Also, this implies that if two bidders have different valuations, then they necessarily have different bids.
- **Assumption 3** For each bidder $i = 1, \ldots, n$, $s(v_i) \leq v_i$. We already discussed that in section 2.6.1. Note that this implies that a bidder with a 0 valuation will bid 0: $s(0) = 0$.

Since there is no dominant strategy in the first-price auction, we need to look for an equilibrium: a bidding strategy for each bidder such that no bidder can be better off by bidding according to a different strategy. When bidders' valuations are private and distributed uniformly, we can derive a precise characterization of bidders' equilibrium strategies.

Result 2.3 When there are n bidders and their valuations are uniformly distributed, and bidders' bidding strategies satisfy assumptions 1, 2, and 3, then a bidder with valuation v bids in equilibrium

$$s(v) = \frac{n-1}{n} v.$$

The proof of result 2.3 is not too difficult. To begin with, notice that assumption 2 has an important implication: since the bidder with the highest valuation is the one with the highest bid, *the bidder with the highest valuation wins the auction*.

Consider now a bidder, say bidder i, whose valuation is v_i. What is the probability that this bidder wins the auction? Following our previous discussion, it is simply the probability that bidder i's valuation is the highest. So, what is the probability that v_i is the highest valuation?

The valuations are drawn at random between 0 and 100, so this means that the probability distribution of the valuations is the *uniform distribution*. This means that any two numbers have the same probability of being drawn. What is the probability that I take a value that is between 0 and, say, 70? With the uniform distribution, the answer is very simple: it is 0.7 (i.e., 70%). More generally, the probability that I draw a number between 0 and x (with $x \leq 100$) is simply $\frac{x}{100}$.

So let us consider our bidder i who has the valuation v_i, and suppose that i's bid is b. Let us assume that all the other bidders' strategies satisfy assumptions 1, 2, and 3, and let $s(\cdot)$ denote the bid function of those bidders. Bidder i wins the auction whenever her bid is the highest; that is, if $b > \max_{j \neq i} s(v_j)$. Since all the other bidders follow the same strategy, the highest bid among those other bidders is made by the bidder with the highest valuation. Since bidders' valuations are obtained from the same distribution, the highest valuation among all bidders except bidder i is the highest valuation among $n-1$ valuations. Let x be this highest valuation. So the bid made by the bidder with the highest valuation is $s(x)$.

Now assume that the other bidder's strategy, $s(\cdot)$, consists of bidding a *fraction* α of one's bid (The letter α is called "alpha," the equivalent of a in the Greek alphabet). So a bidder with valuation v bids αv, and thus the highest bidder with valuation x, bids αx. We now show that in this case bidder i maximizes her expected payoff by also using this strategy.

We know that bidder i wins if her bid is higher than that of the bidder with valuation equal to x; that is, $b > s(x)$. Since $s(x) = \alpha x$, we have

$$b > \alpha x \quad \Leftrightarrow \quad \frac{b}{\alpha} > x.$$

Therefore, we obtain

$$Prob(i \text{ wins}) = Prob(b > \alpha x)$$

$$= Prob\left(x < \frac{b}{\alpha}\right)$$

$$= Prob\left(\text{valuations of all bidders except bidder } i < \frac{b}{\alpha}\right)$$

$$= Prob\left(v_1 < \frac{b}{\alpha}\right) \times Prob\left(v_2 < \frac{b}{\alpha}\right) \times \cdots$$

$$\times Prob\left(v_{i-1} < \frac{b}{\alpha}\right) \times Prob\left(v_{i+1} < \frac{b}{\alpha}\right) \qquad \text{(we skip bidder } i\text{)}$$

$$\times \cdots Prob\left(v_n < \frac{b}{\alpha}\right)$$

$$= \underbrace{\frac{b}{100\alpha} \times \frac{b}{100\alpha} \times \cdots \times \frac{b}{100\alpha}}_{n-1 \text{ times}} = \left(\frac{b}{100\alpha}\right)^{n-1},$$

Simple Auctions

where the fourth equality comes from the fact that bidders' valuations are drawn independently from each other. Now that we have the probability that bidder i wins when bidding b, we can give the formula of i's expected payoff when bidding b (with her true valuation being v_i) as

$$\left(\frac{b}{100\alpha}\right)^{n-1}(v_i - b). \tag{2.2}$$

We now just have to find the value b that maximizes this expression. Note that if $b=0$ or if $b=v_i$, then the expected payoff is 0. If $0<b<v_i$, then the probability of winning is strictly positive and the net payoff, $v_i - b$, is also positive. The maximum value is obtained when taking the derivative with respect to b and finding the value b such that the derivative is equal to 0.[9] If we derive equation (2.2) with respect to b, we have

$$(n-1) \times \left(\frac{b}{100\alpha}\right)^{n-2} \times \frac{1}{100\alpha} \times (v_i - b) - \left(\frac{b}{100\alpha}\right)^{n-1}$$

$$= \frac{1}{100\alpha}\left(\frac{b}{100\alpha}\right)^{n-2}\left((n-1)(v_i - b) - b\right). \tag{2.3}$$

We want this expression to be equal to 0, so this gives

$$(n-1) \times (v_i - b) = b \quad \Leftrightarrow \quad b = \frac{n-1}{n}v_i. \tag{2.4}$$

In other words, for bidder i with valuation v_i, the bid that maximizes her expected payoff is $\frac{n-1}{n}v_i$.[10] Observe that bidder i's optimal bid does not depend on the value of α! It follows that if we set $\alpha = \frac{n}{n-1}$, then it is optimal for bidder i to follow the same strategy as the other bidders. So the bid function we were looking for is

$$s(v_i) = \frac{n-1}{n}v_i, \tag{2.5}$$

which completes the proof of result 2.3.

Equation (2.5) has an easy interpretation. It turns out to be the *expected second-highest valuation conditional on v_i being the highest valuation*. We detail this observation in section C.3 of appendix C).

More generally, for a bidder i with valuation v_i, we can summarize i's optimal bid in a first-price sealed-bid auction as

$$s(v_i) = E(\max_{j \neq i} v_j \mid v_j \leq v_i \text{ for all } j \neq i), \tag{2.6}$$

where the "$E(\cdot)$" is the *expectation*. This equation reads as follows. The first part to look at is the end,

$| v_j \leq v_i$ for all $j \neq i)$,

which means that bidder i considers that she has the highest valuation (the "conditional" part of the expectation): the valuation of any other bidder j, v_j, is smaller than the valuation of bidder i, v_i. Then we look at the highest of those values,

$$\max_{j \neq i} v_j.$$

Since we are considering the case when v_i is the highest, the highest valuation among all other valuations (i.e., taking all indices j such that $j \neq i$) will be the second-highest valuation. Then we take the expectation thereof. So, for a bidder with valuation v, her optimal bid in the first-price sealed-bid auction consists of bidding what she expects the second-highest valuation to be if she supposes that v is the highest valuation among all bidders.

Example 2.6 There are ten bidders, and let us suppose that each bidder's valuation is between 0 and 100. We consider Alice, a bidder whose valuation is 60. Alice knows her valuation, but she does not know the valuations of the other bidders. She knows, however, that each bidder's valuation is a random number between 0 and 100.

Alice's bid only makes sense for her if she wins. So she has to think that her valuation is the highest, that the valuations of all bidders are between 0 and 60. Alice's optimal bid is given by equation (2.5):

$$s(v_{\text{Alice}}) = \frac{n-1}{n} \times v_{\text{Alice}} = \frac{9}{10} \times 60 = 54.$$

So Alice will bid 54, which is what she expects the second-highest valuation to be if she takes for granted that her valuation, 60, is the highest.

Observe that the optimal bidding strategy we have derived in equation (2.5) does not depend on the interval $[0, 100]$. For instance, if all the valuations are randomly taken from an interval $[0, 1]$ or $[0, 10]$, we would obtain the same formula.

2.6.3 Dutch Auction

Until now, all the nonsimultaneous auctions we have seen were ascending price auctions: the price goes up until a winner is declared. There is an auction that does just the opposite: the starting price is (very) high and the price goes down until there is a winner. Such auctions are called **descending price auctions**, and one of the best-known versions is the **Dutch auction**, where:

- There is a "clock price" that displays a price that is decreasing over time.
- The auction stops as soon as someone says "Mine!"

In practice, the Dutch auction can go (very) fast, as it uses bidders' nervousness (so that they bid higher than their value). The Dutch auction (or its variants) is often used in the daily market for fresh commodities such as fish or flowers. The Dutch auction can be described in a very simple way, but determining one's optimal strategy is not trivial: for which price will I say "mine"?

To this end, consider first the English auction. At any time, there is a history, which consists of the different prices proposed by the auctioneer and the list of bidders who have dropped out (and at which prices). That is, the description of a price is not enough to provide a complete description of a history in an English auction. For example, we could have an English auction with a starting price of $10, and after some time the price is $50. But at that price we could have two different histories:

(a) Mr. Jones stopped bidding when the price was $25.00, and Mrs. Smith stopped bidding when the price was $45.00.

(b) Mr. Jones stopped bidding when the price was $35.00, and Mrs. Jones did not stop bidding.

Knowing that the current price is $50 does not tell me which history, (a) or (b), is the correct one (and it could be that the real history is neither (a) nor (b)). So the set of potential histories in the English auction can be very large.

In the Dutch auction, we do not have this problem: the price uniquely defines a history. If the price dropped from, say, $100 (the starting price) and is now $50, the only thing I know is that nobody has moved yet. In the Dutch auction, moving means shouting "mine," and the auction stops immediately. Had someone moved before the price reached $50, the auction would have stopped, and thus we would not have reached the price $50 (it would have stopped at the price for which someone said "mine").

It follows then that in a Dutch auction a strategy is simply choosing a price at which I will shout "mine." That is, I just have to choose a bid, and not a bid conditional on the history as in the English auction.[11]

The question now is how to know the optimal bid in a Dutch auction. Since I just need to decide at what price I will shout "mine," I can decide it just *before* the auction starts. This implies that we can perfectly study bidders' behaviors in a Dutch auction by assuming that they indeed choose their bids (i.e., their strategies) before the auction starts. But then, since the winner is the bidder with the highest bid, and the price is the winner's bid, the Dutch auction is equivalent to the first-price sealed-bid auction. So, we then have the following result.

Result 2.4 The Dutch and first-price sealed-bid auctions are strategically equivalent.

Result 2.4 then implies that in a Dutch auction the optimal bid for a bidder is simply the expected second-highest valuation conditional on having the highest valuation, given in equation (2.6).

2.7 Revenue Equivalence

So far, we have seen four main types of auctions: the English auction (Japanese button), the second-price sealed-bid auction (Vickrey), the Dutch auction, and the first-price sealed-bid auction.

We also know that the English and the Vickrey auctions are equivalent: the revenue for the seller and the bidders' optimal strategies and payoffs are the same across both auctions. Similarly, the Dutch and the first-price sealed-bid auctions are equivalent. However, the Dutch and the English auctions are not equivalent: the bidders use different strategies. In the English auction, the bidders bid truthfully, while in the Dutch auction they have strong incentives to bid below their valuations.

From the point of view of the seller, which auction format is the best? It turns out that in expected terms the English and the Dutch auctions are the same. This result is known as the **revenue equivalence theorem**.

A rapid intuition for this result is the following. Consider the second-price auction. We know that in this auction bidders bid their valuations. For some given valuations of the bidders, the seller's revenue is equal to the second-highest valuation. So the expected revenue for the seller is simply the *expected value of the second-highest valuation*. When n valuations are distributed uniformly between 0 and 100, the expected value of the second-highest valuation is given by the following formula (see equation [C.7] in appendix C):

$$100 \times \frac{n-1}{n+1}. \qquad (2.7)$$

So equation (2.7) is the *expected revenue* of the seller in the second-price auction.

Now consider the first-price auction. In that auction, the winner will be the bidder with the highest valuation, so the seller's revenue is the expectation of the highest bidder's bid. We know from result 2.3 that each bidder will bid the expectation of the second-highest valuation (conditional on having the

Simple Auctions

highest valuation). So the expectation of the highest bid is the expectation of the second-highest valuation.[12]

In the case of n valuations distributed uniformly between 0 and 100, a bidder with valuation v will bid $\frac{(n-1)}{n}v$. Since the winner pays her bid, the expected revenue of the seller is

$$\text{Expectation (highest bid)} = \text{Expectation}\left(\frac{(n-1)}{n}\text{ highest valuation}\right)$$

$$= \frac{(n-1)}{n}\text{ Expectation (highest valuation)}.$$

So we need the expectation of the highest valuation, which is given by the following formula (see equation (C.2) in appendix C):

$$100 \times \frac{n}{n+1}.$$

Therefore, the expected revenue of the seller in the first-price auction is

$$\frac{(n-1)}{n} \times 100 \times \frac{n}{n+1} = 100 \times \frac{n-1}{n+1}. \tag{2.8}$$

Equations (2.7) and (2.8) are identical! So we have the following result.

Result 2.5 When bidders' valuations are private and uniformly distributed, the expected revenue of the seller is the same in the first-price and second-price auctions.

Result 2.5 is no coincidence. It is in fact a direct corollary of the following more general result.

Result 2.6 (Revenue Equivalence Theorem) Assume that there are n bidders, and their valuations are private and identically and independently distributed (from a distribution over an interval $[\underline{v}, \bar{v}]$ that has a strictly increasing cumulative probability distribution).

Then any auction such that

(i) the winner is always the bidder with the highest valuation and

(ii) the bidder with the lowest valuation $\underline{v}$ has a zero expected payoff

yields in equilibrium the **same expected revenue** for the seller, and the expected price made by any buyer is the same.

In the numerical applications we saw earlier, for the bidders, valuations we had $\underline{0}$ and $\bar{v} = 100$, and bidders' valuations were distributed uniformly. Result 2.6

allows for a more general assumption. The assumption that the cumulative probability distribution is *strictly increasing* means that there bidders, valuations can be *any* number between $\underline{v}$ and $\bar{v}$.

Proof Consider *any* auction. For that auction, each bidder has a (private) valuation, and we consider what happens in equilibrium. Recall that an equilibrium is simply a specification of a strategy for each bidder where the strategy of a bidder depends on her valuation. Note that the strategy of any bidder i cannot depend on the valuation of any other bidder (because each bidder only knows her valuation and not the valuations of the other bidders).

For any bidder i, let $U(v_i)$ be the *expected* net payoff that bidder i will obtain in equilibrium when her valuation is v_i. So we can write that bidder's expected payoff as

$$U(v_i) = v_i Prob(v_i) - E(\text{payment if she wins when valuation} = v_i), \qquad (2.9)$$

where $Prob(v_i)$ is the probability that i wins the auction when her valuation is v_i. From condition (i) in result 2.6, this is the probability that v_i is the highest valuation. Note that this does not depend on the auction we are using. The second part of equation (2.9) is the price i pays if she wins, which does depend on the auction we are using. Since we do not know exactly what auction we are considering, we cannot be more specific.

Suppose now that bidder i with valuation v_i does not use the bid that is optimal for a valuation v_i but rather the bid that is optimal for another valuation, say $\tilde{v}$.[13] Of course, in equilibrium, no bidder should do that. This is from the very definition of an equilibrium, which says that in equilibrium no player/bidder has an incentive to change her strategy. Denote by $W(\tilde{v}, v_i)$ the expected payoff of bidder i if

- she bids as if her valuation were $\tilde{v}$;
- her true valuation is v_i.

Recall that, in equilibrium, when bidder i's true valuation is v_i, her expected payoff is $U_i(v_i)$. So we have

$$U(v_i) \geq W(\tilde{v}, v_i),$$

where

$$W(\tilde{v}, v_i) = v_i Prob(\tilde{v}) - E(\text{payment if she wins when valuation} = \tilde{v}).$$

Notice that in the first term the probability of winning (when behaving as if the true valuation were $\tilde{v}$), $Prob(\tilde{v})$, is multiplied by v_i and not $\tilde{v}$. This is because we are changing i's strategy (and thus the probability that she wins), not her payoff

Simple Auctions

function. Whether or not she uses the right or optimal strategy, she still considers the object's true value to be v_i, and thus her expected gain is $v_i Prob(\tilde{v})$. Only the probability of winning and the price bidder i has to pay are affected by her bidding behavior.

Clearly, we can rewrite the previous equation as

$$W(\tilde{v}, v_i) = v_i Prob(\tilde{v}) - E(\text{payment if she wins when valuation} = \tilde{v})$$
$$+ \tilde{v} Prob(\tilde{v}) - \tilde{v} Prob(\tilde{v}).$$

Rearranging terms, we obtain

$$W(\tilde{v}, v_i) = \underbrace{\tilde{v} Prob(\tilde{v}) - E(\text{payment if she wins when valuation} = \tilde{v})}_{= U(\tilde{v})}$$
$$+ v_i Prob(\tilde{v}) - \tilde{v} Prob(\tilde{v})$$
$$= U(\tilde{v}) + (v_i - \tilde{v}) Prob(\tilde{v}).$$

Therefore, $U(v_i) \geq W(\tilde{v}, v_i)$ is equivalent to

$$U(v_i) \geq U(\tilde{v}) + (v_i - \tilde{v}) Prob(\tilde{v}).$$

Since $\tilde{v}$ can be any valuation, we can replace it by $v_i + dv$, where dv is a very small amount. So we still have

$$U(v_i) \geq U(v_i + dv, v_i) + (-dv) Prob(v_i + dv). \tag{2.10}$$

Suppose now that a bidder has a valuation that is just equal to $v_i + dv$. In equilibrium, that bidder should not bid mimicking a player whose valuation is v_i, so we obtain (reversing the roles)

$$U(v + dv) \geq U(v_i) + (dv) Prob(v). \tag{2.11}$$

Reorganizing equations (2.10) and (2.11), we obtain

$$Prob(v_i + dv) \geq \frac{U(v_i + dv) - U(v_i)}{dv} \geq Prob(v_i). \tag{2.12}$$

We now take the limit $dv \to 0$. In this case, the fraction in the middle of equation (2.12) is just the derivative of $U(\cdot)$ evaluated at v_i,

$$\frac{dU_i}{dv} = Prob(v). \tag{2.13}$$

Equation (2.13) simply says that the marginal utility of a bidder when her valuation increases is equal to the probability of winning with that valuation, which is simply the probability that this bidder has the highest valuation.

So we know the formula of the derivative of the function U: it is equal to $P(v)$. Hence, we just have to *integrate* that function,[14]

$$U(v_i) = U(\underline{v}) + \int_{x=\underline{v}}^{v_i} Prob(x)dx, \tag{2.14}$$

where $U(\underline{v})$ is the payoff of the bidder with the lowest valuation (who will lose for sure).

Equation (2.14) implies the following. Take any auction. We just need to know the expected payoff of the bidder with valuation $\underline{v}$ to deduce the expected payoff of any other bidder.

We are almost done. Take any two auctions satisfying the assumptions of result 2.6. The probability distribution of bidders' valuations does not depend on the auction, so for any bidder her *expected payoff function* is the same for the two auctions (where the function is with respect to the bidder's valuation). But a bidder's expected payoff function is also given by equation (2.9), which means that for the two auctions the *expected payment* is the same. It is the same for all bidders, so the expected revenue of the seller is the same for the two auctions. ∎

2.8 Reserve Price

In any auction problem, it is assumed that bidders have a valuation of the object to be sold. So far, we have implicitly assumed that the seller is willing to sell at any positive price. This is clearly not the case for many situations, and it is easy to think of a case where a seller prefers not to sell if the price is too low.

One way to proceed is that, once the auction ends, the seller decides whether to proceed with the sale by comparing the final price and her valuation. But doing so can be risky because it undermines the reputation of the seller (or that of the auction house running the auction). Nobody would like to win an auction and learn later that the seller had canceled the auction. It is not difficult to modify an auction so that the final price is at least above the seller's valuation. The way to proceed is by introducing a **reserve price**. Such a price is simply a minimum price at which bids will be taken into account. For instance, if the reserve price is, say, $10, then any bid must be at least $10.

The introduction of a reserve price is then an excellent way for a seller to take into account her valuation in an auction. Since bids must be above the reserve price, the seller is guaranteed not to sell the object at a price below the reserve price.

But a reserve price has another twist: it is somehow equivalent to allowing the seller to place a bid on the item she is selling. This feature is particularly interesting when we consider the Vickrey auction. In a second-price auction, the price paid

Simple Auctions

by the winner when there is a reserve price is not necessarily equal to the second-highest bid. Instead, the price paid is the *maximum price between the reserve price and the second-highest bid*. Formally, given the bids of the bidders, let $b_1^{(n)}$ and $b_2^{(n)}$ denote the highest and the second-highest bids and r the reserve price set by the seller.[15] So we have

$$\text{seller's revenue} = \begin{cases} 0 & \text{if } b_1^{(n)} < r \\ r & \text{if } b_2^{(n)} \le r \le b_n^{(n)} \\ b_2^{(n)} & \text{if } r < b_2^{(n)} \end{cases} \quad (2.15)$$

Thus, we have three cases under which to determine the seller's revenue. If the highest bid is lower than the reserve price, the good is not sold. If only the highest bid is higher than the reserve price ($b_2^{(n)} \le r \le b_n^{(n)}$), then the winner pays the reserve price. Finally, if the second-highest bid is higher than the reserve price, then we have the "standard" second-price auction, where the winner pays the second-highest bid.

Example 2.7 For a second-price auction, Alice, Bob, and Carol are the only bidders. Alice's, Bob's, and Carol's valuations are $10, $15, and $20, respectively.

Without a reserve price, we know that it is a dominant strategy to bid one's valuation. So Alice bids $10, Bob bids $15, and Carol bids $20. The winner is Carol, and she pays $15.

If there is a reserve price, it is not difficult to see that it is still a dominant strategy to bid one's valuation as long as it is higher than or equal to the reserve price (the proof is essentially the same as that of result 2.2). If the valuation is below the reserve price, then the bidder is indifferent between bidding (her valuation) or not participating to in the auction (which is the same as bidding $0).

If the seller puts a reserve price of, say, $8, then Alice, Bob, and Carol bid their valuations, and the price is the same as when there is no reserve price, $15 (and Carol is still the winner).

Suppose now that the seller decides that the reserve price is, say, $17. In this case, the second-highest bid is below the reserve price, so the winner pays the reserve price, $17.

An immediate rationale for introducing a reserve price in a second-price auction is that the seller can protect herself against low revenue. A famous example is the case of New Zealand in 1990, when the government ran several second-price auctions to sell radio spectrum. In one auction, a bidder placed a bid for NZ$10,000 but ended up paying only NZ$6, which was the second-highest bid. In another auction, the winner paid only NZ$5000 (the second-highest bid), while its bid was NZ$7 million.

A second, more complex rationale is that it widens the range of auction mechanisms that the auctioneer can use. By playing with the reserve price, the seller can, in principle, extract more revenue from the buyers. Here, we have to understand that the seller can be strategic by setting a reserve price that is *different* from her valuation. In a second-price auction, the total surplus is the bid of the winner minus the valuation of the seller (i.e., the minimum reserve price the seller would impose). If the reserve price is low, then there is a part of the surplus that always goes to the winning buyers: her valuation minus the second-highest valuation. In example 2.7, that surplus is $20 − $15 = $5 (the difference between Carol's and Bob's valuations).

By appropriately choosing the reserve price, the seller can extract that additional surplus, thereby increasing her revenue. In the extreme case, the seller manages to set a reserve price equal to the highest valuation. In this case, the auction amounts to a *take-it-or-leave-it* offer from the seller to the bidder with the highest valuation. In example 2.7, that would be setting a reserve price of $20. In this case, Carol would accept the offer (i.e., bid $20) and obtain a surplus equal to 0, and the seller's revenue would be the highest possible: no buyer would accept to paying a price higher than $20.

Introducing a reserve price is then an excellent tool for the seller. Of course, setting the right reserve price is a delicate exercise because the seller does not know a priori the buyers' valuations. In example 2.7, if the seller sets a reserve price higher than $20, she would not sell the object, ending up with a payoff of 0.

2.8.1 Optimal Reserve Price

In a Vickrey auction, the **optimal reserve price** is the reserve price that maximizes the seller's expected revenue.

As an illustration, let us first study the optimal reserve price that a seller must set when bidders' valuations are uniformly distributed on [0, 100]. Before starting, note that in the presence of a reserve price it is still a dominant strategy for the bidders to bid their valuations in the second-price auction (it is not difficult to check this; the trick simply consists of using the reserve price as if it were a bid by some bidder).

We first consider the case when there is only one bidder, whose valuation is denoted v. In this case, according to equation (2.15), the seller's revenue is either 0 (the bidder's bid is less than the reserve price) or equal to r (the bidder's bid is less than the reserve price). The probability that the sale does not occur is the probability that the bidder's valuation is less than r. So the expected revenue for the seller is

$$\text{expected revenue} = Prob(v \leq r) \times 0 + Prob(v > r) \times r = r \times (1 - Prob(v \leq r)).$$

Simple Auctions

With the uniform distribution on $[0, 100]$, we have $Prob(v \leq r) = \frac{r}{100}$, so the expected revenue of the seller is

$$\text{expected revenue} = r \times \left(1 - \frac{r}{100}\right).$$

Maximizing this expected revenue is tantamount to maximizing $r \times (100 - r)$, which gives $r = 50$. That is, when there is only one buyer, the best reserve price the auctioneer can set is $r = 50$, which gives an expected revenue of 25. In fact, it can be shown that when bidders' valuations are taken from the same probability distribution but independently from each other, then the optimal reserve price does not depend on the number of bidders. So, in our case, for any bidder, the optimal reserve price is 50.

Setting aside the problem of choosing the right reserve price in a Vickrey auction, a fundamental question for auction theory is to characterize the auction that maximizes a seller's expected revenue. In other words, what is the **optimal auction**? That question was answered in two fundamental contributions by Roger Myerson on the one side and John Riley and William Samuelson on the other side.[16]

Result 2.7 (Optimal Auction)   Let F be the cumulative distribution (and f the corresponding density function). If bidders' valuations are drawn independently from the distribution F, then a seller's optimal auction consists of choosing and running a Vickrey auction with the reserve price r that satisfies $\psi(r) = 0$, where

$$\psi(r) = r - \frac{1 - F(r)}{f(r)}. \qquad (2.16)$$

In auction theory, the function ψ in result 2.7 is called the **virtual valuation**.[17] Myerson showed that in equilibrium the expected payment of a bidder is equal to her expected virtual valuation.

To understand what a virtual valuation is, let us consider the case when there is only one buyer. Suppose that the seller makes a "take-it-or-leave-it" offer to the buyer at a price p. This means that the buyer has only two possible decisions: accept and pay p or refuse (and thus no transaction is made). Clearly, the buyer will refuse as long as her valuation v is less than p. The probability that this happens is $Prob(v \leq p) = F(p)$. So the buyer accepts with a probability equal to $(1 - F(p))$.

For our purposes, it is useful to view the function $1 - F(p)$ as the "quantity" that our buyer would buy. In other words, buying with a probability of, say, 30% is "similar" to buying only 30% of the good. To be fair, we are only saying that the probability that our buyer accepts the proposal is *similar* to a demand function in "standard" microeconomics.

If $q(p) = 1 - F(p)$ is the demand function, then the inverse demand function is $p(q) = F^{-1}(1-q)$. In other words, if the seller wants to sell the quantity q, then the maximum price that can be asked is $p(q)$, or equivalently $F^{-1}(1-q)$. Therefore, the revenue function for the seller is

$$R(q) = p(q) \times q = q \times F^{-1}(1-q). \tag{2.17}$$

Now, our seller wants to maximize her profit. In a monopoly, we know that the profit is maximized when the marginal revenue is equal to the marginal cost. Here the seller (the auctioneer) has no production cost, because he already has the good. Thus, the marginal cost is zero. So the optimal solution for our seller is to find the price such that the marginal revenue is equal to 0 or, equivalently, the price such that the virtual valuation is zero: p such that $\psi(p) = 0$. We then take the derivative of equation (2.17) with respect to q and get

$$\text{marginal revenue} = R'(q) = F^{-1}(1-q) - \frac{q}{F'(F^{-1}(1-q))}. \tag{2.18}$$

Using the fact that $q(p) = 1 - F(p)$ and $p = F^{-1}(1-q)$, equation (2.18) is equivalent to

$$MR(p) = p - \frac{1 - F(p)}{f(p)} = \psi(p). \tag{2.19}$$

The message of equation (2.19) is simple: the virtual valuation of a buyer, $\psi(p)$, captures the marginal revenue brought by that buyer.

Now, if bidders' valuations are uniformly distributed on $[0, 100]$, then the cumulative distribution is $F(v) = \frac{v}{100}$, and thus $f(v) = \frac{1}{100}$. In this case, the virtual valuation is

$$\psi(v) = v - \frac{1 - \frac{v}{100}}{\frac{1}{100}} = 2v - 100. \tag{2.20}$$

Recall that here the profit is maximized when the price is such that the marginal revenue is equal to 0; that is, when v is such that $\psi(v) = 0$. The solution is $v = 50$.

Remark 2.3 Here we presented only Riley and Samuelson's result, which is in fact a simpler version of Myerson's result. Riley and Samuelson made the assumption that the good is always awarded to the highest bidder and that all bidders' valuations are drawn from the same distribution. In contrast, Myerson considered a broader class of allocation mechanisms (e.g., the winner may not be the highest bidder) and also allowed bidders' valuations to be drawn from different distributions (we then talk about *asymmetric auctions*). Also, Myerson's approach is more

Simple Auctions

general in the sense that for him *any* selling mechanism (whether an auction or not) is feasible. He approached the question of the optimal design problem as a *mechanism design* problem (see appendix B).

2.8.2 Reserve Price versus Adding More Bidders

The first message of result 2.7 is that the Vickrey auction is the best auction we can hope for. Since we know that many auction mechanisms are actually equivalent from the seller's perspective (from the revenue equivalence theorem, result 2.6), the result is simply telling us that there is little need to consider other types of auctions. This is interesting for bidders, because it is a much easier auction to participate in.

The second message is that somehow everything boils down to choosing the optimal reserve price. But a surprising result due to Jeremy Bulow and Paul Klemperer suggests that in the end it may not be that important.[18]

Result 2.8 (Bulow and Klemperer)   When valuations are private and independent, a second-price auction with $n+1$ bidders gives a higher expected revenue than an *optimal mechanism* with n bidders.

So, according to Bulow and Klemperer's result, we should not worry about computing the optimal reserve price. It suffices to bring in one additional bidder and run an English or Vickrey auction.

As an illustration, let us compare the expected revenue of the seller with an optimal reserve price with one bidder and the expected revenue of the Vickrey auction with two bidders (but no reserve price). We saw in section 2.8.1 that with one bidder whose valuation is drawn randomly between 0 and 100, the optimal reserve price is 50, which yields an expected revenue of 25. The expected revenue of the Vickrey auction is given by equation (2.7). So with two bidders that gives an expected revenue equal to $100 \times \frac{2-1}{2+1} = \frac{100}{3}$, which is higher than the expected revenue with an optimal reserve price and one bidder only.

To explain the intuition behind result 2.8, we present an argument due to René Kirkegaard.[19] In the auction literature, an auction is **constrained optimal** if, among all the auctions where the object is sold with *certainty*, it maximizes the expected revenue.

A key observation is that Myerson's optimal auction that we saw in section 2.8.1 is not constrained optimal, for the simple reason that the object is not sold with certainty. Indeed, if the reserve price is higher than the highest valuation, then no bidder buys the good. That is, there is a nonzero probability that the good is not sold.

We want to compare the Vickrey auction with n bidders and an optimal reserve price with the Vickrey auction with an additional bidder (but without a reserve price). For simplicity, call the first and the second auctions the "Myerson

n-auction" and the "Vickrey $(n+1)$-auction," respectively. The trick to do that is to consider a third auction, which works as follows:

Step 1 Run the Myerson auction with the initial n bidders. If the good is not sold, then go to step 2.

Step 2 Sell the good to the new bidder at a price equal to 0.

Let us call this auction the "Kirkegaard $(n+1)$-auction."

Clearly, with the Kirkegaard auction, the good is sold with certainty.[20] Also, we obviously have

expected revenue Kirkegaard $(n+1)$-auction

$\quad$ = expected revenue Myerson n-auction.

Myerson showed that in our context (i.e., buyers have private, independent valuations), an auction is constrained optimal if, and only if, the bidder with the highest marginal revenue wins. We saw earlier that the marginal revenue is given by the virtual valuation, and Myerson's result holds when the virtual valuation function ψ is strictly increasing. So this means that in a constrained optimal auction the winner is the bidder with the highest valuation.

The Kirkegaard auction is clearly not constrained optimal. Suppose that the highest valuation among the first n bidders is lower than the reserve price but higher than the valuation of bidder $n+1$. A constrained optimal auction would require selling the object to the bidder with the highest valuation, who is in the group of the first n bidders. So the expected revenue of a constrained optimal auction with n bidders is necessarily higher than that of the Kirkegaard auction. However, it is not difficult to see that the Vickrey $(n+1)$-auction is constrained optimal: the winner is the bidder with the highest valuation, and the good is clearly sold with probability 1. So we have

expected revenue Vickrey $(n+1)$-auction

$\quad >$ expected revenue Kirkegaard $(n+1)$-auction

$\quad =$ expected revenue Myerson n-auction.

Problems

1. Why is there usually a ticking price for English auctions but not for Vickrey auctions?

2. Suppose there are three bidders and their valuations are drawn randomly and independently between 1 and 5. Find the symmetric equilibrium of the first-price auction.

Simple Auctions

3. We consider an English auction. There are three bidders for the item to be sold. The valuation of the first bidder is $v_1 = 4$, the valuation of the second bidder is $v_2 = 10$, and the valuation of the third bidder is $v_3 = 18$. The ticking price is 0.1, and the starting price is 1. Who is the winner of the auction, and what is the price paid by the winner?

4. Describe an ascending auction where the auctioneer announces the price (i.e., not an English outcry auction) and where the speed at which bidders react matters (obviously it cannot be the Japanese button auction).

5. The English and the Vickrey auctions are strategically equivalent, and the revenue for the seller is the same. Yet, in terms of the *information* received by the seller/auctioneer, they are different. What is this difference?

6. A third-price auction is an auction where the winner is the bidder with the highest bid and the price paid is the third-highest bid. Is it a dominant strategy in this auction to bid one's valuation? If yes, provide a sketch of the proof. If not, provide an example.

7. A **uniform-price auction** is an auction where all the goods sold through the auction (or units of the good) are sold at the same price. Suppose there are k units of a good to be sold, and each bidder wants at most 1 unit of the good. We run a $(k+1)$th price, uniform-price auction. That is, the k highest bidders win, and they all pay the same price, the $(k+1)$th bid. We assume that there are at least $k+1$ bidders, $k \geq 2$.

In this auction, is it a dominant strategy to bid one's valuation? If yes, provide a sketch of the proof. If not, provide an example.

3 An Analysis of eBay

3.1 Introduction

eBay is certainly the most well-known auction place on the Internet. At its inception, eBay had to address the problem of how to run an auction on the Internet. More precisely, the exact rules of the auctions had to be defined, and an auction format had to be chosen. The basic questions one may ask in this case are:

- What price does the highest bidder pay?
- What information is posted about the current bidding?

The most obvious and natural choice in this case would be to run a first-price auction such as an English outcry auction. This means that the winner pays her bid, and the price that is posted on eBay during an auction is simply the highest bid.

But this would work smoothly only if all bidders are logged on at the same time and follow the auction in real time. Clearly, following an auction that lasts several days is easier on the Internet than in an auction house. We cannot spend all of our time at the auction house (and it may be far away for many people), while we have access to the Internet almost everywhere and at any time. Nevertheless, having 24/7 access to the Internet does not mean that we are necessarily constantly on the Internet, let alone watching an auction. So we would need to make the auction "safe" for buyers in the sense that they do not need to be constantly following the auction they are participating in.

Also, a first-price auction may not be very interesting for buyers. It is indeed likely that the bidding game will end with the winner bidding her valuation, which means that the auction winners obtain a zero net payoff. The prospect of making a zero payoff is not very attractive for the buyers. Most (if not all) buyers would prefer a system where the price they pay is lower than their valuations.

So a natural choice is to run a second-price auction. But, in this case, what price should be displayed? If we display the highest bid, it may not give other bidders incentives to bid, which may yield low revenues for the seller. To see this, consider the example where there are two bids so far, $1 and $100. If my value is $50, I will not bid (I know I will lose). So the auction would stop with the highest bidder winning by paying only $1. Displaying the first price in a second-price auction is a disaster for the sellers, for it diminishes buyers' incentives to bid.

So the solution would be to run a second-price auction in a dynamic setting:

- The auction can last several days or weeks (set by the seller).
- Potential buyers can bid whenever they want.
- The displayed price is the second-highest bid (+ an increment, discussed later).
- The winner is the bidder with the highest bid, and he pays the price that is displayed when the auction ends.

The eBay auction format then mixes some aspects of the English auction (it is dynamic) and Vickrey auction (it is a second-price auction). Some people call this auction the **California auction**.

3.2 eBay in Detail

3.2.1 Proxy Bidding

When bidders enter bids in an eBay auction, they are actually entering not just a bid but rather a maximum bid. eBay understands this as the maximum price the bidder is willing to pay for the item. eBay also assumes that a buyer prefers to pay the lowest amount possible. The general idea is that eBay will bid on behalf of the bidder, always trying to win with the lowest possible price, up to the maximum amount specified by the buyer. Having a third party bidding for a bidder is called **proxy bidding**.

The information displayed by eBay during an auction consists of:

- information related to the auction: the time left before the end of the auction, shipping information, a description of the item, and other details;
- **current bid**: the price the winner will pay (if there are no further bids); if no bid has been made yet, then eBay displays instead the **starting bid** (the minimum price set by the seller);
- an input box where the bidder can type her bid, saying

"Enter X or more."

The amount X is the **minimum bid**.

How eBay's proxy bidding works boils down to how the *current bid* and the *minimum bid* change during an auction. To explain this, we first need to discuss the bid increment rule set by eBay.

Table 3.1
eBay bid increments

Current bid	Bid increment
$0.01–$0.99	$0.05
$1.00–$4.99	$0.25
$5.00–$24.99	$0.50
$25.00–$99.99	$1.00
$100.00–$249.99	$2.50
$250.00–$499.99	$5.00
$500.00–$999.99	$10.00
$1000.00–$2499.99	$25.00
$2500.00–$4999.99	$50.00
$5000.00 and up	$100.00

3.2.2 Bids and Bid Increments

In any auction, the auctioneer has to announce what prices are admissible. For instance, in an art auction, the auctioneer can decide that any bid or price must be a multiple of $1000. This means that we can have a bid of, say, $5000, but we cannot have a bid (or a price) equal to, say, $5500, because this is not a multiple of $1000. In the case of eBay (for the United States), the unit is the cent. So a price of, say, $4.97 is feasible on eBay but not a price of, say, $2.9999.

A second feature that eBay has to decide is the "ticking price," called **bid increments** by eBay. They are given in table 3.1. For instance, if the current bid is $12, then the corresponding bid increment is $0.50. This means that the next current bid will be at least equal to $12 + $0.50 = $12.50 (we will see in section 3.2.3 that this may not always be the case).

3.2.3 Updating Rules
3.2.3.1 Minimum bids
The formula for the minimum bid is relatively simple:

$$\text{minimum bid} = \begin{cases} \text{starting bid} & \text{if nobody has placed a bid yet} \\ \text{current bid} + \text{increment} & \text{if there is at least one bid} \end{cases}$$

Suppose that for some item the seller decides that the auction starts at $10. This means that any bid must be at least $10. That is, the starting bid is $10, and the minimum bid must be at least $10.

Now consider another auction, for which there are already some bids. So eBay now displays the *current bid* (instead of the *starting bid*). Suppose that the current bid is $20. The bid increment for $20 is $0.50 (see table 3.1), so the *minimum bid* is $20 + $0.50 = $20.50.

3.2.3.2 Current bid The general rule to update the current bid consists of taking the second-highest bid so far and then adding the corresponding bid increment. There are two issues here. First, we need to state precisely what we mean by the "second-highest bid," and second, we need to say what happens when the second-highest bid + the bid increment is larger than the highest bid.

At any time, eBay ranks the bid by following two criteria. First, bids are ranked according to their value. So a bid of, say, $50 is ranked above a bid of, say, $34. But what if two bidders bid the same amount? In the English or the Vickrey auctions, we pick a winner at random among the highest bidders. Instead of doing this, eBay will rank the bids according to the date and time at which eBay received those bids, starting with the oldest bid.

To see how this works, suppose that we have the following bids (and that there are no other bids).

Bidder	Bid	Time at which the bid was submitted
Alice	$50	10:00 a.m.
Bob	$40	10:10 a.m.
Carol	$50	10:20 a.m.

Then, in this case, the bid ranking will be as follows.

Date	Highest bid (bidder)	Second-highest bid (bidder)
10:00 a.m.	$50 (Alice)	—
10:11 a.m.	$50 (Alice)	$40 (Bob)
10:21 a.m.	$50 (Alice)	$50 (Carol)

We are now ready to describe the formula used to update the current bid,

$$\text{current bid} = \begin{cases} \text{starting bid} & \text{if nobody has placed a bid yet} \\ \text{second-highest bid} \\ \quad + \text{increment} & \text{if second-highest bid} + \text{increment} \leq \text{bid} \\ \text{bid} & \text{if second-highest bid} + \text{increment} > \text{bid} \end{cases}$$

Case 1   Nobody Has Placed a Bid Yet

Consider the first, easiest case, when nobody has placed a bid yet. Suppose that the staring bid is $10 (so the minimum bid is also $10). The first bid is from Alice, and she bids, say, $12. Then right after Alice has submitted her bid, the auction will display as the current bid $10 (and the minimum bid will thus become $10 + $0.50 = $10.50). Note that if Alice bids instead $10, $20, or $1000, we would still have the current bid equal to $10 (and the minimum bid would still be equal to $10.50).

Case 2   Second-Highest Bid + Increment ≤ Bid

The second case is the most common. Let us see how it works. Suppose that the current bid is $20 (so the minimum bid is $20 + $0.50 = $20.50) and the highest bid is $30, from Bob. Note that the other bidders cannot know that Bob's bid is $30. They can only know that the highest bid is at least $20. Now Carol posts a bid of $21. She thus becomes the second-highest bidder.[1] Bob's bid of $30 is understood as an authorization he gave eBay to bid on his behalf, but eBay has to bid the lowest amount necessary for him to win, up to $30. For Bob to win, he must bid more than $21. The bid increment for that amount is $0.50, so the lowest bid with which Bob can win is $21.50. That is less than $30, so eBay places a bid for Bob of $21.50, and this becomes the new current bid.

Observe that if Bob is the only bidder, then it means that the starting price was $20 (this would be case 1). If Bob is not the only bidder, then it means that the second-highest bid is $20 − $0.50 = $19.50.

Case 3   Second-Highest Bid + Increment > Bid

As we said, the general rule to compute the current bid is to take the second-highest bid and add the corresponding increment. It may turn out, however, that this sum is higher than the highest bid. Since we cannot ask a bidder to pay more than her bid, eBay in this case simply uses the highest bid as the current bid.

Suppose that the current bid is $40 and that the highest bid is $50, by John. At $40, the bid increment is $1, so the minimum bid must be $40 + $1 = $41. Now comes Erin, who places a bid of $49.87. She is now the second-highest bidder. If we add the corresponding bid increment, we obtain $49.87 + $1 = $50.87, which is higher than John's bid. So eBay will display a current bid equal to $50. If instead Erin posts a bid of $50.01, the same reasoning would hold. She would be the highest bidder (and John the second-highest bidder). Since $50 + $1 > $50.01, the current bid is set to be equal to $50.01.

Case 4   A New Bid Equal to the Highest Bid

When a bidder places a bid that is just equal to the current highest bid, then the highest and the second-highest bids are identical (but the highest and second-highest bidders are not the same!). The rule to compute the current bid is similar to the one in case 3.

Assume that the current bid is $110 and the highest bid is $150, submitted by Frank. So the minimum bid in this case is $110 + $2.50 = $112.50. Suppose now that Gina places a bid of $150 (i.e., after Frank), so she becomes the second-highest bidder. Since the highest and the second-highest bids are identical, we cannot add the bid increment to the second-highest bid (for otherwise Frank would have to pay more than his bid). So the current bid jumps to $150 and the minimum bid becomes $150 + $2.50 = $152.50.

3.3 eBay as a First-Price Auction?

Many people have the misconception about eBay that on eBay they pay their bid (if they win). The reason for that misconception is that most people only bid the minimum amount that eBay is asking.

Suppose that for some item on sale the current highest bidder is Alice and her bid is $10.00. eBay's price increment rule says that the lowest bid that is higher than $10.00 is $10.50 because the increment must be at least $0.50. This means that if another bidder comes after Alice has submitted her bid, the minimum bid that bidder must submit is $10.50. Suppose that Bob is bidding and, like most people, he places the minimum bid, $10.50. So now Bob is the current highest bidder. What is the price that eBay will display on its web page? To compute that bid, eBay does the following:

- Take the second-highest bid, $10.00 (Alice's bid).
- Add the price increment, $0.50.

So the price that eBay is displaying (the "current bid" on the web page) is

second-highest bid + price increment = $10.00 + $0.50 = $10.50.

If the auction stops just after Bob's bid, the price that he will pay will be $10.50. So Bob will think that he pays his bid; that is, he may thus think that eBay is a first-price auction, while he actually pays Alice's bid + price increment. If Bob had bid, say, $20.00 instead of $10.50, then he would still have to pay $10.50.

In fact, eBay is bidding on behalf of the two bidders, but always using the minimum increment that is necessary to win. Since Alice bid $10.00 and Bob's bid is higher, eBay will choose a bid for the second bidder that is higher than $10.00, but among all possible bids eBay will pick the smallest that is necessary for Bob to win.

3.4 Bid Sniping

Bid sniping is a technique that consists of bidding just before the auction ends so that the other bidders do not have time to react and counterbid. The very notion of bid sniping is thus relevant only when there is a hard deadline, a date (and time) at which the auction closes no matter what the bids are and when they were submitted.

3.4.1   Amazon versus eBay

From 1999 to 2006, Amazon had an auction website (like eBay), but the auction rules were slightly different. Both eBay and Amazon used an English ascending auction with proxy bidding, but the rules for ending the auction were different. On eBay, the auction ends when we reach the deadline set by the seller at the outset. On Amazon, the seller also had to set a deadline, but it was extended by 10 minutes whenever there was a bid in the last 10 minutes. So on Amazon there was in principle no way to strategically manipulate the auction by bidding late. This does not mean that there should be no late bidding on Amazon, but, unlike on eBay, bidding late so that the other bidders do not have time to react was not feasible on Amazon.

There are various hypotheses we can formulate about why people engage in bid sniping and in which cases it is profitable to do so. But one thing is certain: bid sniping is done on purpose. This occurs, for instance, when bidders want to avoid bidding wars with other bidders, especially those that we may call "incremental bidders," those bidders who revise their bids often but always with the minimum amount needed to head the auction (see section 3.3). By bidding late, we diminish the length of the bidding war, which yields lower final prices.

Another reason one may want to bid late is to protect one's information. In the case of interdependent or common values, a bid may reveal to other bidders some information relative to how valuable the object is. For instance, if I bid a high amount very early on some item, I am revealing that this object is very valuable to me. But if bidders have common values, it means that it is also valuable to the other bidders, who may then want to revise their bids. There is thus a strong incentive to bid late for bidders who are identified as being "experts," bidders who are known to have a better knowledge of the "true" value of an item.

But bidding late may also occur for nonstrategic reasons as well. Some people may procrastinate and simply decide to postpone their bidding to a later date, or some bidders may also be unaware of the inner workings of the auction. We may also face what social scientists call the **endowment effect**, which consists of giving an item a higher value when we have it than when we do not have it. During the course of an auction, a bidder may be temporarily winning and may thus start thinking that the object is hers. If later the bidder is no longer winning, she may be subject to this endowment effect and give a higher valuation to the item than before (and thus bid higher).

We can then formulate some hypotheses about the differences between eBay and Amazon when considering late bidding. On Amazon, we should see less strategic late bidding than on eBay. However, when we consider nonstrategic late bidding, there should not be any significant difference between eBay and Amazon. The hypotheses are summarized in table 3.2.

Table 3.2
Predictions about late bidding with Amazon and eBay

Hypothesis	Predicted observation
Strategic hypothesis	
• Avoid war against incremental bidders	More late bidding on eBay than on Amazon
• Protect one's information	
Nonstrategic hypothesis	
• Procrastination	No difference between Amazon and eBay
• Search engines first present soon-to-expire auctions	
• Bidders unaware of proxy bidding	
• Valuation increases (endowment effect)	
• Bidders do not like bids hanging	

Table 3.3
Number of bidders

	Computers	Antiques
Amazon	595 bidders	340 bidders
eBay	740 bidders	604 bidders

3.4.2 Data Analysis

Alvin Roth and Axel Ockenfels (2002) analyzed data from 480 auctions from eBay and Amazon between October 1999 and January 2000.[2] They considered auctions about laptops (+ monitors) and antiques. Laptops and monitors are items for which we have mostly private values but also for which it is quite easy to know what would be the "right" price for a given model.[3] Antiques, on the contrary, have a large common value component. For instance, it is easier to resell an antique than a laptop after a few years, so bidders are also likely to be interested in how much other people value it. Also, antiques are typically items for which there is a clear difference between an "expert" and the other potential buyers.

Roth and Ockenfels only considered auctions where there were at least two bidders. In total, their data comprised 480 auctions with 2279 bidders. For each possible combination of object (computers or antiques) and platform (eBay or Amazon), they analyzed 120 auctions. The number of bidders for each combination is given in table 3.3.

The data also contain bidders' ratings. Both Amazon and eBay have rating systems that allow bidders and sellers to establish a reputation. The rating systems

An Analysis of eBay

differ between the two platforms. On Amazon, buyers and sellers can rate each other on a 1 to 5 star scale, and on eBay individuals can only give negative (−1), positive (+1), or neutral feedback (0), and the overal rating of a bidder or a seller is the sum of the positive and negative ratings. Amazon displays for each bidder the total number of feedback responses she received, while eBay only displays the aggregate number (the number of positive evaluations minus the number of negative evaluations). So in principle there is no way to infer from a bidder's aggregate feedback on eBay the *number* of feedback responses that bidder received. Roth and Ockenfels note, however, that no bidder has a totally negative feedback, suggesting that for both platforms the feedback number can serve as a proxy for the bidder's experience.

The main piece of data Roth and Ockenfels were interested in was how long before the end of the auctions bidders must submit their bids. On eBay, this information is unambiguous, but not on Amazon, because by design all bids in an auction that has ended were submitted at least 10 minutes before the end. In other words, raw data from Amazon cannot contain any last-minute bidding. To circumvent this problem, Roth and Ockenfels calculated for each bid the *hypothetical* deadline, which is the time at which the auction would have ended had the 10 extra minute rule not been applied when the bid was submitted. The following example illustrates how those hypothetical deadlines are calculated.

Example 3.1   Suppose that an auction is initially scheduled to end at 11:00 a.m. The data for that auction show that only two bids were submitted. The first bid was by Alice, at 10:55 a.m. That bid triggered the 10 minute extension rule, so once Alice submitted her bid, the auction had a new deadline, set at 11:10 a.m. For the data analysis, Alice's bid is registered as a bid submitted five minutes before the deadline; in other words, the hypothetical deadline that corresponds to Alice's bid is 11:00 a.m.

The second bid was by Bob at 11:09 a.m. Again, that bid triggered the 10 minute extension rule, so once Bob submitted his bid, the auction had a new deadline, set at 11:20 a.m. For the data analysis, Bob's bid is registered as a bid submitted one minute before the deadline; that is, the hypothetical deadline that corresponds to Bob's bid is 11:10 a.m.

Figure 3.1 shows the difference between eBay and Amazon for antiques and laptops. The figure shows clearly that late bidding is more common on eBay than on Amazon. As table 3.4 shows, the percentage of auctions that had a bid submitted in the last five minutes is considerably higher for eBay than for Amazon. This difference between the two platforms is still maintained when looking at the last minute and the last ten seconds of an auction.

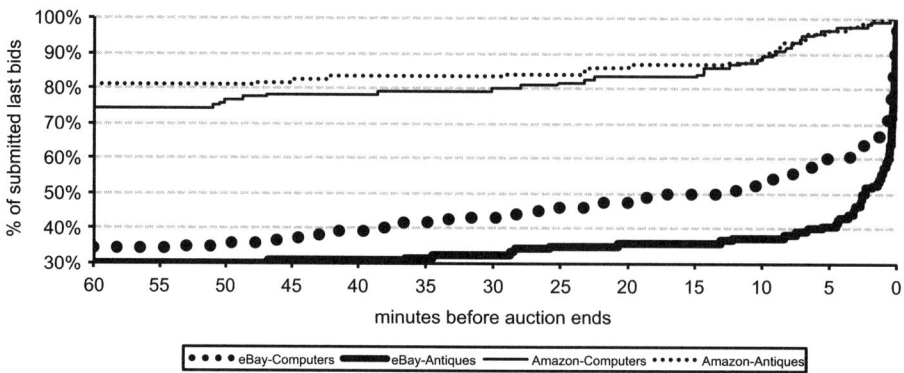

Figure 3.1.
eBay versus Amazon. Thick lines: eBay (grey = antiques, black = computers), thin lines: Amazon (grey = antiques, black = computers). Source: Roth and Ockenfels (2002).

Table 3.4
Percentage of auctions with a bid in the last five minutes

	Computers	Antiques
Amazon	3%	3%
eBay	40%	59%

Roth and Ockenfels also did several statistical tests that confirm that the observed differences in bidding behavior between antiques and computers and between eBay and Amazon are significant. More precisely, they observe that there is significantly

- more late bidding on eBay than on Amazon for computers;
- more late bidding on eBay than on Amazon for antiques;
- more late bidding on eBay-antiques than on eBay-computers.

However, the difference in late bidding behavior between antiques and computers on Amazon is not significant. Using the bidder feedback data, Roth and Ockenfels also observe that on eBay there is a significant correlation between experience and late bidding: the more experienced a bidder is, the more likely it is that she bids close to the end of the auction.

In a survey conducted with auction participants, bidders also confirmed Roth and Ockenfels's initial intuitions that late bidding decreases the risk of starting a bidding war (and thus keeps the price low) and is also viewed as a tool to limit information sharing with competitors.

3.5 Reserve Price

On eBay, the seller can set a reserve price, which is the minimum bid that a buyer has to put in to win the auction. If the highest bid at the end of the auction is below the minimum price, then the auction is canceled. The reserve price is different from the starting price. While the starting price is always public, the reserve price can be secret. It is up to the seller to decide whether the reserve price is public (i.e., everybody can see it) or secret.

There are many reasons why a seller would be better off by putting in a reserve price. But we can outline a few motives. When the reserve price is public, announcing a reserve price is a way for the seller to signal to the buyer that the item has some minimum value. Setting a lower starting price can then attract buyers. It is also a way for a seller to make trials to guess what would be the right price without actually selling the item.

When the reserve price is secret, it becomes equivalent to a "shill bid," having a "fake" buyer place a bid equal to the reserve price. That way, whenever there is a buyer placing a bid below the reserve price, it will be outbid by the reserve price. One reason for doing that is to trigger bidding wars and/or to signal buyers that there are "other buyers" interested in the item (and thus signal that the item is valuable).

Problems

1. Alice, Bob, Carol, and Denis participated in an auction on eBay. The starting price was $5. Table 3.5 gives the history of the bids. For instance, on December 3 at 8:35:14, Alice submitted a bid of $6. Give the history (with the date and time) of the current bid displayed on the auction's web page.

Table 3.5
Submitted bids

Bidder	Submitted bid	Date stamp of bid submission
Alice	$6	Dec-03 08:35:14
Bob	$16	Dec-05 11:15:04
Carol	$12	Dec-03 17:47:14
Denis	$7	Dec-04 10:05:06
Denis	$8	Dec-04 10:05:46
Denis	$9.65	Dec-04 10:06:02

Table 3.6
Bid history

Bidder	Current bid	Date stamp of bid
Denis	$9.00	Feb-07 08:07:24
Carol	$8.50	Feb-06 19:31:28
Erin	$8.00	Feb-06 23:02:56
Carol	$5.57	Feb-06 19:31:28
Bob	$5.07	Feb-05 11:43:30
Carol	$5.00	Feb-06 19:31:10
Bob	$3.75	Feb-05 11:43:30
Carol	$3.50	Feb-06 19:30:54
Bob	$3.25	Feb-05 11:43:30
Carol	$3.00	Feb-06 19:30:50
Bob	$2.75	Feb-05 11:43:30
Carol	$2.50	Feb-06 19:30:45
Bob	$2.25	Feb-05 11:43:30
Carol	$2.00	Feb-06 19:30:40
Bob	$1.75	Feb-05 11:43:30
Carol	$1.50	Feb-06 19:30:34
Bob	$1.25	Feb-05 11:43:30
Alice	$1.00	Feb-03 06:20:51
Alice	$0.99	Feb-03 06:20:51
Starting price	$0.99	Feb-02 21:43:03

2. Table 3.6 gives the history of current bids that have been displayed for some auction on eBay. The first column is the name of the bidder, the second column is the current bid that was displayed at some time on the auction's page, and the last column is the date stamp of the bid: the date at which the corresponding bidder submitted his or her bid (i.e., the "maximum amount" that eBay is asking on the auction's web page).

Retrieve each bidder the bid (or the bids) he or she submitted, and for each submitted bid the date and time at which the bid was made.

3. Most bidders focus on amounts when bidding on eBay: multiples of $10 or $5, or simply integers. Alice is a bidder who uses the following strategy. Whenever she comes up with an integer for a bid, she adds one penny. So, when she decides to bid, say, $10 or $25, she instead bids $10.01 or $25.01. Bob has a strategy similar to that of Alice, but instead of adding one penny he adds a random amount between 40 and 70 cents. So, when he decides to bid $10 or $25, he instead bids $10.56 or $25.47. Why do you think Alice and Bob adopt such strategies?

4 The Vickrey-Clarke-Groves Auction

4.1 Introduction

Until now, all the auctions we have seen have been auctions designed for selling only one item. There are, however, many cases when there are several items for sale *at the same time*. For instance, timber auctions often consist of the sale of various lots of wood, and buyers typically only buy several lots (but not all of them). Another case of an auction with multiple items is the auction for bandwidth (spectrum auction). The usual configuration is that of a public authority selling several licenses for mobile phone operators. Even if each operator only wants to buy one license, such auctions cannot be considered as auctions with a single item because operators typically want to bid on several licenses so as to maximize their chances to obtain *a* license. One problem when buyers bid on several items but only want one is how to decide which item a buyer is awarded if she won several items.

A classic solution for such auctions is known as the **Vickrey-Clarke-Groves (VCG) auction**, which is, as we will see, a generalization of the Vickrey auction. In fact, it is sometimes simply called the **generalized Vickrey auction**.

4.2 The Model

The VCG auction is, at the outset, like any other auction. We have a set of potential buyers (the bidders) who have valuations. Valuations in a VCG setup differ from the previous cases in sense that since bidders are participating in a sale of several items, and since some bidders may want to buy several items at the same time (such as a combination of items), bidders' valuations are no longer described by a single number. In a VCG setup, a bidder valuation is a function from the set of all possible combinations of items to the real numbers.

Formally, suppose that there is a set $N = \{1, 2, \ldots, n\}$ of bidders, and there is a set $X = \{x_1, x_2, \ldots, x_n\}$ of sellers. The valuation of a bidder i is then a function, denoted $v_i(\cdot)$, that specifies how much bidder i values any possible combination of items.

Throughout this chapter, we will assume that the valuation of a bidder when she does not get any item is zero; that is, for each bidder i, $v_i(\emptyset) = 0$.

Example 4.1 Suppose there are three items, A, B, and C, and Alice is a bidder. Her valuation function could be:

$v_{\text{Alice}}(\{A\}) = 10,$ $\quad v_{\text{Alice}}(\{A,C\}) = 25,$

$v_{\text{Alice}}(\{B\}) = 15,$ $\quad v_{\text{Alice}}(\{B,C\}) = 20,$

$v_{\text{Alice}}(\{C\}) = 12,$ $\quad v_{\text{Alice}}(\{A,B,C\}) = 30,$

$v_{\text{Alice}}(\{A,B\}) = 20.$

Notice that we have to specify a valuation for each possible combination. Here, $v_{\text{Alice}}(\{A\}) = 10$ means that Alice values having A (and only this item) as 10: she is willing to spend up to 10 to get just A. The fact that $v_{\text{Alice}}(\{A,B\}) = 20$ and $v_{\text{Alice}}(\{A,C\}) = 25$ means that she attaches more value to having both A and C than to having both A and B. She is willing to spend up to 20 to get A and B, and up to 25 to get A and C.

Example 4.2 Again, we consider three items, A, B, and C, but this time Alice's valuations are

$v_{\text{Alice}}(\{A\}) = 20,$ $\quad v_{\text{Alice}}(\{A,C\}) = 0,$

$v_{\text{Alice}}(\{B\}) = 15,$ $\quad v_{\text{Alice}}(\{B,C\}) = 0,$

$v_{\text{Alice}}(\{C\}) = 1,$ $\quad v_{\text{Alice}}(\{A,B,C\}) = 0,$

$v_{\text{Alice}}(\{A,B\}) = 0.$

In this case, we can see that Alice values A more than she values B, which is valued more than C. Having $v_{\text{Alice}}(\{A,B\}) = 0$ means that she attaches no value to having both A and B. That is, those valuations depict the case of a bidder who wants only one item.

The two previous examples illustrate the richness of the approach. The valuation function can describe any situation in terms of a bidder's preferences. We could have a situation where a bidder prefers to have as many items as possible, but if she does not get all the items, she prefers some combinations over others. This case is depicted in example 4.1. Example 4.2 is the case of a buyer who has preferences among the items but only wants one. Any combination of two or more items is not desirable for Alice. This could be the case for a coach of a team who needs one player with a specific skill but does not want two players with that skill.

The Vickrey-Clarke-Groves Auction

The formal way to describe the valuation of bidder i is simply to say that it is a function $v_i(\cdot)$ that gives, for each possible set S of items, a number $v_i(S)$. That is, $v_i(S)$ is the amount of money that bidder i is willing to spend to get all the items in the set S (and none of the items that are not in S). In example 4.1, if $S = \{A, C\}$, then we have $v_{\text{Alice}}(S) = 25$.

4.3 The VCG Auction

The VCG auction works as follows:

Step 1 Each bidder submits a bid function, a function that specifies how much she bids for each possible combination of items.

Step 2 The auctioneer computes an assignment, which basically says who gets what. The assignment that the auctioneer computes is called the optimal assignment. We explain in detail in section 4.3.1 what this assignment is.

Step 3 The auctioneer computes for each bidder the price of the items she obtains under the optimal assignment.

Step 1 is the simplest, and there is not much to comment about except that it is like the Vickrey auctions in that bids are submitted simultaneously. So the VCG auction is a sealed-bid auction.

Remark 4.1 The Vickrey-Clarke-Groves *auction* that is studied in this chapter is actually a special case of what economists call the Vickrey-Clarke-Groves *mechanism* (as we explain in appendix B, a mechanism is a more general problem than an auction).

Edward Clarke and Theodore Groves have each separately proposed similar *mechanisms* to solve the problem of contributions in public good problems.[1] When considering auctions over multiple *identical* items (with private values), the mechanism defined by Clarke and Groves becomes identical to the auction defined by Vickrey in his seminal contribution. So, in this chapter we present an auction that is a more general version of the (original) Vickrey auction (because the multiple items need not be identical) but that is a special case of the Clarke-Groves mechanism (because it is an auction).

4.3.1 Computing the Optimal Assignment

An assignment is a function μ that sets, for each bidder, which items she gets. So, $\mu(i)$ is the assignment of bidder i under the assignment μ.

Given a set of bidders and a set of items, there are many assignments; for instance, if there are two bidders, say Alice and Bob, and two items, say A and B. The possible assignments are depicted in table 4.1.

Table 4.1
Possible assignments with two bidders and two items

Assignment	Alice	Bob
μ_1	A, B	nothing
μ_2	nothing	A, B
μ_3	A	B
μ_4	B	A
μ_5	nothing	nothing
μ_6	A	nothing
μ_7	B	nothing
μ_8	nothing	A
μ_9	nothing	B

Under assignment μ_3, for instance, Alice gets A (so $\mu_3(\text{Alice}) = \{A\}$), and Bob gets item B (so $\mu_3(\text{Bob}) = \{B\}$). Notice that in an assignment it is not possible for two bidders to get the same item.

To each assignment there corresponds a **social value**: the sum of the bidders' valuations, evaluated at the assignment. In other words, if we consider an assignment μ, we look at the set of items each bidder receives under μ, and then we look at his or her valuation for that set. Adding those values for each bidder gives the social value of the assignment μ.

Formally, if μ is an assignment, the social value is

$$\sum_{i \in N} v_i(\mu(i)).$$

Example 4.3 Consider again the case of Alice and Bob, and suppose that their valuations are given by the following table.

	Alice	Bob
A	10	5
B	5	3
A, B	11	6

So, Alice values having just item A at 10 but having both A and B at 11.

There are four possible assignments (we ignore the trivial assignments where not all the items are allocated).

The social value of assignment μ_1 is 11 because under that assignment Alice gets both A and B, and her valuation for both items is 11 (recall that we assumed that a bidder's valuation when being assigned nothing is 0).

Assignment	Alice	Bob	Alice's valuation	Bob's valuation	Social value (total)
μ_1	A, B		11	0	11
μ_2		A, B	0	6	6
μ_3	A	B	10	3	13
μ_4	B	A	5	5	10

An auction consists, among other things, in choosing a rule to decide which bidder gets which item. In all the simple auctions we saw in chapter 2, the rule was such that the bidder with the highest valuation is the one bidding the highest and obtaining the item. (In auction theory, we usually say that the auction consists of selling the item to the bidder that values it the most.) The value that the winner attaches to the item she gets is then split between the winner and the seller. If v is the valuation of the winner and p the price, then the winner's net payoffs is $v-p$ and the seller's payoff is p. The sum of these two payoff is $v-p+p=v$. So selling to the bidder with the highest valuation amounts to maximizing the surplus between the seller and the bidders, and thus it allows the seller to sell the item at the highest possible price.

The VCG auction follows that principle in that we want to award the items to the bidders that value them the most. Since there are many different items and each bidder can have several of them, the assignment that maximizes the social value will be the assignment that maximizes the total surplus. Hence, the assignment of interest for the auctioneer is the assignment that maximizes the social value, called the **optimal assignment**, which we denote by μ^*. So, μ^* is such that

$$\sum_{i \in N} v_i(\mu^*(i)) \geq \sum_{i \in N} v_i(\mu(i)) \quad \text{for any other assignment } \mu.$$

In example 4.3, the assignment that maximizes the social value is assignment μ_3, where Alice gets A and Bob gets B, and the social value is 13. In more detail, we have

$$\sum_i v_i(\mu^*(i)) = v_{\text{Alice}}(\mu^*(\text{Alice})) + v_{\text{Bob}}(\mu^*(\text{Bob}))$$
$$= v_{\text{Alice}}(\mu_3(\text{Alice})) + v_{\text{Bob}}(\mu_3(\text{Bob})) \quad \text{(because } \mu^* = \mu_3\text{)}$$
$$= v_{\text{Alice}}(\{A\}) + v_{\text{Bob}}(\{B\})$$
$$= 10 + 3 = 13.$$

In the VCG auction, the auctioneer computes the optimal assignment by using the bids submitted by the bidders.

4.3.2 Calculating Prices

Once we have the optimal assignment μ^*, we can start computing the price that each bidder will pay for her assignment. The general idea is that each bidder will pay the externality she imposes on the other bidders. That is, we compute for each bidder i the difference between

- the social value of the other bidders at the optimal assignment and
- the maximum social value of the other bidders when bidder i is not present in the auction.

Formally, we proceed as follows. Let i be a bidder, and let μ^* be the optimal assignment. The first thing we have to compute is the social value of *the other bidders* at the optimal assignment μ^*. This is simply the amount of the maximum social value from which we subtract the value that i obtains at the assignment μ^*, or equivalently,

$$\sum_{j \neq i} v_j(\mu^*(j)) = v_1(\mu^*(1)) + v_2(\mu^*(2)) + \cdots + v_{i-1}(\mu^*(i-1))$$
$$+ v_{i+1}(\mu^*(i+1)) + \cdots + v_n(\mu^*(n)).$$

Note that the sum is made across all bidders except bidder i.

Next, we consider the same auction, with the same bids, but without bidder i. For this "auxiliary" auction, we look at the highest social value. This implies that we have to identify the assignment that will maximize the social value.

Formally, we have to solve the following problem:

$$\max_{\mu} \sum_{j \neq i} v_j(\mu(j)).$$

Here, the subscript μ to max means that we consider the assignment μ that maximizes the sum (which is taken over all bidders except bidder i). That is, we want the highest social value, but we will not consider i's assignment when computing the sum. If we include bidder i in the sum, then the solution is the optimal assignment μ^*.

The price for bidder i is then the difference between the social value of the other bidders when i is not present and the social value of the other bidders when i is present:

$$p_i = \underbrace{\max \sum_{j \neq i} v_j(\mu(j))}_{\text{others' maximal social value } \textit{without} \text{ bidder } i} - \underbrace{\sum_{j \neq i} v_j(\mu^*(j))}_{\substack{\text{others' social value} \\ \textit{with} \text{ bidder } i \\ \text{under } \mu^*}}.$$

If we consider Alice, the social value at the optimal assignment is 13, but the social value at that assignment of all bidders except her (i.e., just Bob) is equal to 3.

If Alice is not present, then there is only Bob, and the assignment that gives the highest social value is the one that assigns him both A and B. In this case, Bob obtains a valuation equal to 6.

The difference between the highest social value of Bob when Alice is not present and the social value at the optimal assignment of Bob is $6-3=3$. So the price that Alice has to pay for obtaining A (her assignment under μ^*) is 3.

We can also compute the price Bob has to pay to get $B = \mu^*(\text{Bob})$. At the assignment μ^*, Alice obtains a value of 10. If Bob is not present, then Alice can obtain a valuation equal to 11 (being assigned both A and B), so the difference is $11-10=1$. Therefore the price that Bob has to pay to get item B is 1.

One may argue that the prices Alice and Bob have to pay in example 4.3 are relatively low. Alice only pays 3 for an object she values at 10, and Bob pays only 1. Why are the prices so low? Section 4.4 gives an answer to that question: those "low" prices are what give the bidders an incentive to submit their true valuations.

4.4 Incentives under the VCG Auction

In the previous sections, everything was presented (and calculated with the example) as if the bidders submitted their valuation functions truthfully. But in fact nothing prevents them from submitting untruthful valuation functions. It turns out that bidders can only have lower payoffs by doing so. That is, submitting one's true valuation function is a dominant strategy in the VCG auction.

Result 4.1 Bidding truthfully is a dominant strategy in the VCG auction.

Proof Suppose that bidder i bids $\hat{v}_i(\cdot)$ instead of $v_i(\cdot)$, her true valuation. We will show that the net payoff of bidder i when bidding $v_i(\cdot)$ is higher than when bidding $\hat{v}_i(\cdot)$.

To proceed, we need to introduce some notation that will ease the exposition. We denote by $\hat{\mu}$ the assignment that maximizes the social value when bidder i is present with bidder i submitting $\hat{v}_i$ and all the other bidders not changing their bids. That is, $\hat{\mu}$ is the assignment such that

$$v_1(\hat{\mu}) + v_2(\hat{\mu}) + \cdots + v_{i-1}(\hat{\mu}) + \hat{v}_i(\hat{\mu}) + v_{i+1}(\hat{\mu}) + \cdots + v_n(\hat{\mu})$$

is the highest.

The social value across all bidders but bidder i that is obtained when bidder i submits $\hat{v}_i$ will be denoted by $\widehat{V}_{N\setminus\{i\}}$. That is, $\widehat{V}_{N\setminus\{i\}} = \sum_{j \neq i} v_j(\hat{\mu}(j))$.

We also write $V_{N\setminus\{i\}}$ for the maximum social value that the other bidders can get when i is not present in the auction, $V_{N\setminus\{i\}} = \max_\mu \sum_{j \neq i} v_j(\mu(j))$. It is important to note that $V_{N\setminus\{i\}}$ does not depend on whether i bids $\hat{v}_i$, or v_i as it is computed assuming i is not present in the auction.

Finally, we denote by $V^*_{N\setminus\{i\}}$ the social value that obtains the optimal assignment μ^* across all bidders except bidder i (but with bidder i present). That is, $V^*_{N\setminus\{i\}} = \sum_{j \neq i} v_j(\mu^*(j))$.

Let

$U_i = $ net payoff when bidding truthfully, $v_i(\cdot) = \underbrace{v_i(\mu^*(i))}_{\text{valuation of the assignment when submitting } v_i} - \underbrace{\left(V_{N\setminus\{i\}} - V^*_{N\setminus\{i\}}\right)}_{\text{price paid when submitting } v_i}$

and

$\widehat{U}_i = $ net payoff when bidding $\hat{v}_i(\cdot) = \underbrace{v_i(\hat{\mu}(i))}_{\text{valuation of the assignment when submitting } \hat{v}_i} - \underbrace{\left(V_{N\setminus\{i\}} - \widehat{V}_{N\setminus\{i\}}\right)}_{\text{price paid when submitting } \hat{v}_i}.$

To prove the result, we just have to compute $U_i - \widehat{U}_i$ and show that it is always positive:

$$U_i - \widehat{U}_i = \left[v_i(\mu^*(i)) - \left(V_{N\setminus\{i\}} - V^*_{N\setminus\{i\}}\right)\right] - \left[v_i(\hat{\mu}(i)) - \left(V_{N\setminus\{i\}} - \widehat{V}_{N\setminus\{i\}}\right)\right]$$

$$= \underbrace{\left[v_i(\mu^*(i)) + V^*_{N\setminus\{i\}}\right]}_{\sum_j v_j(\mu^*(j))} - \underbrace{\left[v_i(\hat{\mu}(i)) + \widehat{V}_{N\setminus\{i\}}\right]}_{\sum_j v_j(\hat{\mu}(j))}.$$

Recall that μ^* is the assignment that maximizes the social value. So,

$$\sum_j v_j(\mu^*(j)) - \sum_j v_j(\hat{\mu}(j)) \geq 0 \quad \Leftrightarrow \quad U_i - \widehat{U}_i \geq 0.$$

That is, bidder i's highest net payoff is when bidding truthfully. ∎

4.5 Relation with the Vickrey Auction

When there is only one item to be sold, the VCG auction is simply the Vickrey auction. To see this, let i be the bidder with the highest valuation and let j be the bidder with the second-highest valuation.

The Vickrey-Clarke-Groves Auction

Clearly, the assignment μ^* that maximizes the social value is the one that gives the item to be sold to the bidder with the highest valuation, and the other bidders get nothing.

The valuation of bidder j is denoted by v_j. When bidder i is present, she is the one getting the item. In this case, bidder j's payoff is 0, and all other bidders except bidder i also get a payoff equal to 0. So the social value across all bidders except bidder i when bidder i is present is 0.

However, if bidder i is not present, then bidder j gets the item and enjoys a payoff equal to v_j. The other bidders get nothing and thus have a payoff equal to 0. So the social value when bidder i is not present is equal to $v_j + 0 + 0 + \cdots + 0 = v_j$.

In the VCG auction, the price paid by a bidder is the externality she imposes on the others, so bidder i pays $v_j - 0 = v_j$. That is, the winner in a VCG auction is the bidder with the highest valuation, but the price she has to pay is equal to the second-highest valuation. This is exactly what happens in a Vickrey auction.

4.6 The Complexity of the VCG Auction

Apart from Internet ads (see section 5.5), the VCG auction is rarely used in real life. Why? A crucial step in the VCG auction is to find an assignment that maximizes the social value. This implies that we have to compare all possible assignments, and the total number of assignments can be a gigantic number. While in certain cases it is possible to effectively reduce the number of assignments one has to consider, in most cases such a reduction will not significantly reduce this figure.

Perhaps more worrisome is that we would also need to ask bidders to submit valuations over a large number of possibilities. To see how daunting the task can be, suppose there are only 10 items. In this case, a bidder would have to submit $2^{10} - 1 = 1023$ different valuations, one for each possible combination of items.[2]

If the bidder wants, say, just 2 items out of the 10, then she can simply say that for all combinations that have 3 or more items her valuation is 0. But she still has to submit valuations for all possible combinations that consist of only 1 item (there are 10 cases), and all combinations that consist of only 2 items (there are 45 such combinations). That is, the bidder in this case would have to submit 55 valuations.

For the auctioneer, the task is even greater. Suppose there are 10 bidders and 10 items, and suppose that the auctioneer also knows that each bidder only wants 1 item. There are in this case $10! = 3,628,800$ different assignments. To compute the price for each bidder, the auctioneer has to consider again $10! = 3,628,800$ different assignments, but she has to do that 10 times, one time for each bidder. It follows that to find the optimal assignment and the prices for each bidder, the auctioneer must consider $10! + 10 \times 10! = 39,916,800$ combinations.

These calculations are just for 10 bidders and 10 items. If we multiply the number of items and the number of bidders by 2, the total number of combinations the auctioneer has to consider is *not* twice the number of combinations for the case 10 bidders–10 items but rather is equal (approximately) to 5.10×10^{19}. In other words, the *size* of the problem grows exponentially. For small numbers, a computer can find a solution relatively quickly. But as the number of cases grows, the time a computer needs to find a solution becomes very long, so that it may become impossible (unless we use all the computers on Earth for several thousand millennia!). Computer scientists say that this problem is *NP complete* (which means that we do not know how to find a solution in "reasonable time").

So, the VCG auction is a relatively simple auction mechanism that has very good incentive properties, but that is of little help if we want to design an auction for a practical problem. However, the VCG auction is used as a benchmark for auction theorists when considering alternative (and more feasible) auction formats. We will also see in chapter 5 how the VCG auction is used.

5 Keyword Auctions

5.1 Introduction

One of the most common auctions today is the one for keywords on search engines. When a user enters a search term (a "query"), she obtains a page with the (theoretically) most relevant links that match the query. For most queries, the first links that are displayed are not the most relevant links but "**sponsored links**," links that some advertisers paid to be displayed above the search result. Figure 5.1 shows an example of a search. The first four links on the result page (in the black frame) are ads. The advertisers pay the search engine to be displayed before the search results. This placement is important because it is a fact that users tend to click more on the first links. The first link receives on average more clicks than the second link, the second link receives more clicks than the third link, and so on. Also, each time a user clicks on a sponsored link, the corresponding advertiser pays the search engine some amount (which does not happen if the user clicks on a nonsponsored link). The links that are below the black frame are not ads; they are called "**organic results**."

What works with queries on search engines also works on any other website. Today nearly all the ads we can see on any website are the outcome of an auction. Whether that website is facebook.com, nytimes.com, or a recipe blog, the system used to decide which ads are displayed on your screen can be traced back to the auction mechanism that was pioneered by search engines. The purpose of this chapter is to analyze how we got there and how the allocation of sponsored links is made.

Advertising on the Internet is a huge market. It represents about 90%–95% of Google's revenues. For Facebook or Yahoo!, the ratio of the revenue that comes from advertising has a similar ratio. Some keywords are very cheap, but others are relatively expensive. Some advertisers are willing to pay more than $50 per click![1]

Advertising on the Internet raises several challenges that are not present for the other forms of media. First, in contrast with, for instance, printed ads in magazines

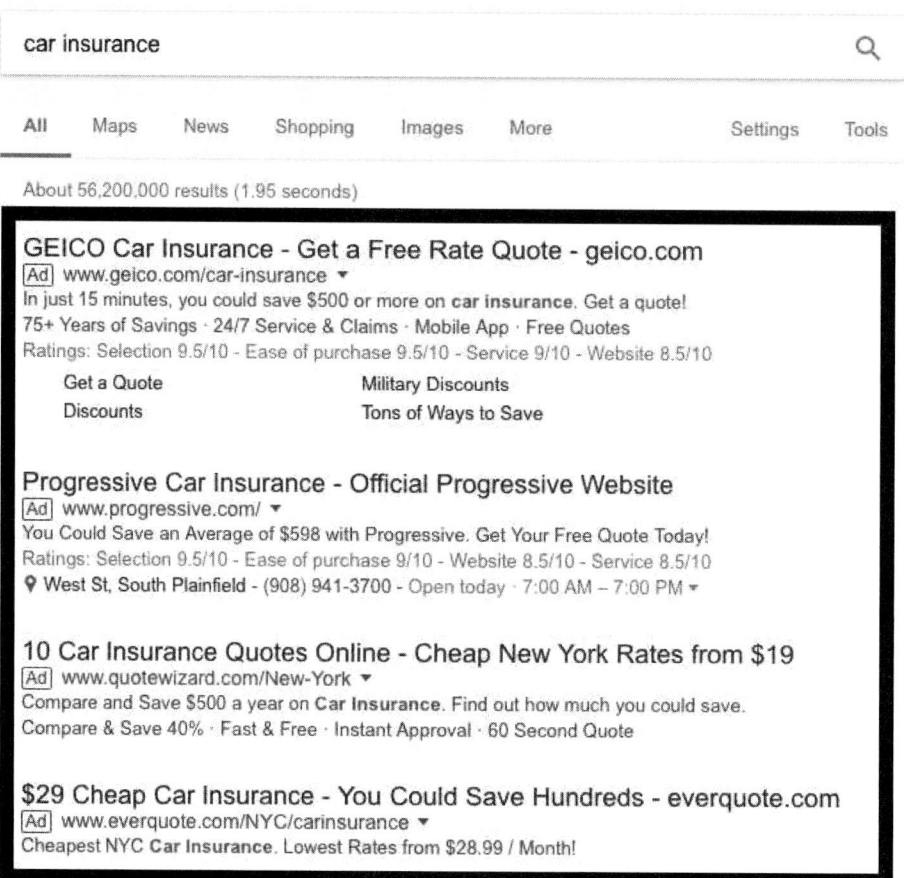

Figure 5.1.
A search with sponsored links. If we click on one of the four links inside the black frame, then the corresponding advertiser has to pay the search engine some amount. Clicking on www.allstate.com below the black frame is free for Allstate. It is not a sponsored link.

or commercials on TV, the situation can change almost continuously, and advertisers can have strategies that change very rapidly. The fact that the whole system is based on computers allows considering that each request made by a user on the Internet is a different setting (and hence a different auction). The pace of communications on the Internet permits advertisers to adjust their strategies very quickly. If the model set by search engines has loopholes, it is very likely that advertisers will find them very quickly. In other words, advertising on the Internet must use a well-defined model and be robust to all kinds of manipulations by the advertisers.

Advertising on the Internet also raises a new question in terms of pricing. With the cookies and other tracking technologies, it is now possible for third parties to check whether a user performed a transaction on a website. Suppose, for instance, that some company A is putting an ad on website B. It is possible for B to know whether a user who clicked on the link pointing to A's website has performed a transaction. This is not the case for traditional media such as magazines, radio, or TV. If I buy something in a store (or on the website of that store), it is difficult if not impossible for the seller to know whether I came to the store because I saw their ad in a magazine or for another reason. Therefore, in the world of advertising on the Internet, advertisers' best option would be to pay the search engine only when a user makes a transaction on their website (and that user comes from the ad displayed on the search engine's page). Search engines, on the contrary, would prefer to be paid each time they display an ad, independent of whether the user clicks on it (and if the user clicks on the ad, independent of whether he made a transaction).

The industry has settled for now on an intermediate solution: advertisers pay each time a user clicks on the ad. That solution is called **pay per click** (PPC).

5.2 Running Billions and Billions of Auctions

In the following sections, we will describe and analyze the auction format that most websites are using now (including the bidding process and how the "winners" are chosen and the prices they have to pay), but without much ado we can say that each time a user enters a search query, there is an auction. So if there are 100 users making a search query, then we will have 100 auctions.

Take some keyword, say "insurance." There are a number of advertisers who would like to have a link to their website displayed on top of the search results whenever a user searches for "insurance." All those advertisers will bid to get displayed. To this end, they submit their bids to the search engine. The bidding process has become very sophisticated. Nowadays bidders can add several clauses to their bids. For instance, a bidder can say that it bids only in the evening from 6 p.m. to 8 p.m. A bidder can also bid only if the search is made from a specific region (a pizza stand in New York has no interest in bidding for a "pizza" query by someone from Los Angeles). A bidder can also set a maximum number of clicks or amount paid, saying that as soon as its expenses (or number of clicks) attain some number, then the advertiser drops out of the auction. But, for the sake of simplicity, we assume that bidders submit "simple" bids.

So each time there is a user searching for the keyword "insurance," the search engine retrieves all the bids made by advertisers for that keyword. Then the outcome of the auction is computed, and immediately after this is done, the search engine builds the web page that is sent to the user's browser, with the auction winners' links on the top of the page and the search results immediately after them. If there is another user searching for the keyword "insurance" immediately after, then the search engine will run another auction, again retrieving the bids of the advertisers. Those bids may be the same or different bids.

5.3 The Origins

At the beginning, the advertising model for websites was very similar to the one for billboards or pages in magazines: the advertiser paid for some number of displays, where prices were negotiated or "imposed" by the websites.[2]

The first service to run an auction with a PPC format was GoTo in 1998 (renamed Overture in 2001, and bought by Yahoo! in 2003). The keyword auction became successful very quickly because it was fast and it ran in real time (advertisers could adjust their bids at any moment). Yahoo! and MSN were the main and biggest clients of Overture. However, the fact that bids could change very quickly made the system very unstable, and a new solution had to be found.

The main problem with the Overture auction format was that it was a first-price auction: bidders were ranked according to their bid but had to pay their bid each time a user clicked on the sponsored link. But with a first-price auction there may not be an equilibrium, a "stable" situation where nobody wants to change their bid. The following example illustrates the point.

Example 5.1 Suppose that there are three bidders (1, 2, and 3), with values per click of $10, $4, and $2, respectively. There are only spots for two sponsored links, and the first spot yields many more clicks than the second spot (so that any bidder prefers the first spot to the second spot). Let b_1, b_2, and b_3 be the bids of bidders 1, 2, and 3, respectively.

Notice that for bidders 1 and 2 it is sufficient to bid more than $2 to ensure a spot. If $b_1, b_2 > 2$, then there is a bidding war between bidders 1 and 2, until we reach $4. At this point, only bidder 1 can afford to bid more than $4. So bidder 2 can only get the second spot. It is then sufficient for bidder 2 to bid $2.01 (just enough to prevent bidder 3 from entering) instead of $4. But then bidder 1 can adjust its bid and bid $2.02, and the bidding war starts again.

Keyword Auctions

Another issue with the first-price auction was related to the fact that at the beginning advertisers had different speeds in the auction. Some advertisers were able to adjust their bids very frequently, while others were slower, often with a real person placing the bids instead of bidding software.[3] Such discrepancies in terms of speed increased the risk that bids would remain at a low level for a long time. That was of course very beneficial for the advertisers, but it generated low revenues for the search engines.

5.3.1 Payoff Flows

When computing advertisers' payoffs in an Internet ad auction, we need to take into account the click frequency. The reason for doing this is that advertisers do not update their bids for each auction (i.e., for each user search query). Instead, advertisers usually set a bidding strategy that consists of how much they want to bid and a stopping rule, after which the advertiser is no longer participating in the auction. The stopping rule can be, for instance, a specific date, a total amount spent on the campaign, or a combination of both (e.g., bid for the next three weeks, but stop before that date if the total spending on the campaign reaches $100,000). Put differently, an ad auction can be understood as a repeated auction (one auction for each search query that matches the keyword) instead of a unique auction.

To take into account the click frequency, we first have to set a unit of time (e.g., an hour, a day, a week). For example, suppose that an advertiser values the click at $4 (the advertiser is willing to spend $4 to have a user clicking on a link pointing to its website). If the link has a frequency of 200 clicks per hour, then in one hour the "revenue" to the advertiser is $200 \times \$4 = \$800$. If the advertiser only pays $3 per click to the search engine, then in one hour the advertiser pays $600. It follows that the "net profit" for the advertiser is $200.

Suppose now that there is another position, for which the advertiser pays only $1 per click but with a frequency of 50 clicks per hour. In this case, the net payoff *per click* is $\$4 - \$1 = \$3$. We could think that this position is better for the advertiser. But with only 50 clicks/hour, this gives a net payoff of $50 \times (\$4 - \$1) = \$150$ per hour, which is less than for the other position. In the extreme case, what is the point of being willing to pay a very little amount per click if the ad is positioned in such a way that no user ever clicks on it?

So, in the context of sponsored links, we have to work with *flows of payoffs*: payoffs over some amount of time. The time unit we choose is irrelevant; it could be one hour, one day, or one week. What is important is to compare the various positions in terms of net payoff over time, and not net payoff per click. The fact that we now have to add the click frequencies in the analysis will have profound effects on the analysis of the auction mechanism.

5.4 The Google Model: Generalized Second-Price Auction

The solution that emerged very quickly and for which Google was a pioneer is what is now called the **generalized second-price auction (GSP auction)**. In this auction, each advertiser places a bid. The bids are then ranked, and the sponsored links are attributed in the following way:

- the highest bidder is shown first,
- the second-highest bidder is shown second,
- the third-highest bidder is shown third,
- etc.

In general, search engines only display three or four sponsored links on the main result page, but the GSP auction can be described (and studied) for any number of sponsored links.[4]

The price system in the GSP auction is still PPC, but this time the prices paid by the advertisers are based on a scheme similar to that of the Vickrey auction. That is, the price for the highest bidder is the second-highest bid, the price for the second-highest bidder is the third-highest bid, and so on. The general statement is thus simply that, for each $k \geq 1$, the kth highest bidder pays the bid (when the user clicks) of the $(k+1)$th bidder.

5.4.1 Quality Scores

Around 2005, Google modified the rules of its GSP auction, subtly changing how it would rank the bidders and calculate the price. Since then, Google's modification has been adopted by its competitors as well. This section is devoted to explaining this modification. For the sake of simplicity, in this section we refer to Yahoo! as the "traditional" GSP auction we have described in section 5.3 and refer to Google as the "modified" GSP auction, which we now describe.

The GSP auction, as we have just defined it, seems to be an effective way to rank advertisers and determine the price they pay for each click. There is an issue, however, with this auction: it does not take into account the click frequency. Click frequency is crucial because it will determine advertisers' payments and, more importantly, the search engine's revenue. A key (but not surprising) observation made by Google was that the clickthrough rate depends not only on the rank of the sponsored link but may also depend on who the advertiser is. For instance, ads from highly visible brands are more likely to be clicked than ads from lesser-known ones, irrespective of their ranks on the search result page. To see this, suppose that there are two advertisers, A and B, competing for one spot. Advertiser A bids \$1 per click and generates 1000 clicks per day, while advertiser B bids \$50 per click but generates only 10 clicks per day. In the GSP auction, B would be

Table 5.1
Quality scores depend on the user's identity

Visited site	User is	
	a man	a woman
Hanes	High score	Low score
Victoria's Secret	Low score	High score
Craig's List	Very low score	Very low score

ranked first, although the search engine would prefer to display the other advertiser. So it needs to find a way to *rank A above B* in spite of its having a lower bid.

Equally important for the search engine is whether the ad is relevant for the user. A young, wealthy, single, urban person is not necessarily attracted by the same car models as parents with kids living in a suburb. So an ad for a minivan is relevant for the latter but not for the former, and the other way around for a coupe.

Another factor that can also matter is the *user experience* (UX) brought by the advertiser. UX is one of the key elements that can explain why some websites thrive and others fail. While many elements in UX are subjective and difficult to appraise objectively, there are nevertheless a number of features that can be measured precisely, such as the amount of data to be transferred to load a website or the number of elements (including ads) displayed on a website.

Google thus combines for each advertiser its clickthrough rate, the ad's relevance, the user experience, and some other factors to determine what is known as the **quality score**. Note that the quality score depends on the user: two different users searching for the same keywords may generate different quality scores for the same advertiser. Consider for example a search for the keyword "underwear" (but not a few days before February 14 or the end of the year holiday season!). Table 5.1 shows how the quality score varies depending on who is the person searching for that keyword.

Once the search engine has calculated the quality scores of each bidder, it proceeds as follows:

Step 1 Calculate for each bidder its **final score**:

Final score = Bid × quality score.

Step 2 Rank bidders according to their final scores.

Step 3 Calculate the price paid by each advertiser. The price paid by the advertiser ranked k is the *lowest* price p such that

p × quality score > final score of $(k+1)$th bidder.

Example 5.2 There are three bidders, Pim, Pam, and Poum, who compete for two spots. The bids are $b_{Pim} = \$6$, $b_{Pam} = \$4$, and $b_{Poum} = \$2$, and the quality scores are $QS_{Pim} = 2$, $QS_{Pam} = 4$, and $QS_{Poum} = 1$.

With those quality scores and bids, Pim's final score is 12, Pam's final score is 16, and Poum's final score is 2. So Pam is ranked first, Pim is second, and Poum is third.

Pam just needs a final score higher than 12 to win against Pim. As long as she bids at least \$3.01, she is ranked first. So the price for Pam is \$3.01 per click.

Pim just needs a final score higher than 2 to win against Poum, which is obtained if he bids \$1.01. Thus the price for Poum is \$1.01 per click.

For Poum, since his ad is not displayed, he pays nothing.

When advertisers are identical across all dimensions other than their value per click, Google's and Yahoo!'s auctions are identical. For the rest of this chapter, we will analyze the GSP auction in "Yahoo! style." We will assume that bidders are ranked according to their bids and not their scores.

5.4.2 Truthtelling under the GSP Auction

Truthtelling turns out not to be a dominant strategy under the GSP auction. This implies that the GSP auction is not a straightforward generalization of the Vickrey (or English) auction (for which truthtelling is a dominant strategy).

Example 5.3 Consider three bidders competing for two spots. The values for bidders 1, 2, and 3 are, respectively, $v_1 = \$10$, $v_2 = \$4$, and $v_3 = \$2$. Suppose that the click frequency of the first spot is 200 clicks/hour and that of the second spot is 199 clicks/hour.

We consider first the case when bidders bid truthfully, where each bidder bids its valuation. It follows then that the first spot is awarded to bidder 1 and the second spot to bidder 2. Bidder 3 loses the auction. The price paid by bidder 1 is then \$4/*click*, and for bidder 2 it is equal to \$2/*click*. So the payoff for bidder 1 is $200 \times (\$10 - \$4) = \$1200$.

Suppose now that bidder 1 bids \$3 instead of \$10. In this case, it is bidder 2 who is awarded the first spot, and bidder 1 gets the second spot. The price/click paid by bidder 1 then becomes \$2. So the payoff (taking into account the click frequency) becomes $199 \times (\$10 - \$2) = \$1592$. Clearly, bidding \$3 is more profitable for bidder 1 than bidding its valuation, even if it is now relegated to the second position. So bidding one's valuation cannot be a dominant strategy.

5.4.3 Equilibrium under the GSP Auction

The absence of dominant strategies in the GSP auction complicates the analysis. We have to characterize the equilibrium of the GSP auction. This is bad news because the GSP auction turns out to be a very complicated game, for we must

Keyword Auctions

view the GSP auction as a repeated auction. From a game theory point of view, this means that we are considering here an infinitely repeated game with asymmetric information (i.e., each bidder may not know the valuations of her opponents). The bad news is that the analysis of such games is one of the most notoriously difficult areas in game theory... so if we want to go any further, we have to make some simplifying assumptions so as to make the model tractable. The analysis of the GSP auction has been pioneered by Benjamin Edelman, Michael Ostrovsky, and Michael Schwarz.[5]

5.4.4 Assumptions about the Long-Run Equilibria

The first assumption we will make is that after some time we can assume that each advertiser knows the valuations of the other advertisers. This is not too heroic an assumption in the case of Internet advertising. Indeed, recall that advertisers can adjust their bids *very* frequently, and it is perfectly reasonable to assume that advertisers try several bidding strategies in order to find the "right" bid. Even if they lose some money during those trials, it will weigh little in comparison to the stream of payoffs they will receive once they have reached a steady strategy (which will last for much more time than the experiments they would have made).

The second assumption relates to the Nash equilibrium. Since bids can be changed at any time, it seems reasonable to assume that stable bids must be such that no advertiser wants to change her bid. That is, bids must form a Nash equilibrium of the simultaneous-move, one-shot game.[6] This assumption implies that given the bids of all other advertisers, no advertiser would like to modify its bid.

5.4.5 Refinement: Envy-Free Equilibrium

The first two assumptions developed in section 5.4.4 will greatly simplify our analysis, but they are not sufficient for forming a meaningful analysis.

What we need is to capture the dynamic aspect of the auction. If we just retain the assumption that the bids must form a Nash equilibrium of the one-shot game, we lose the fact that we are considering a repeated auction. The idea is to consider simple strategies that advertisers can consider when anticipating the reactions of their opponents. The type of strategy we will consider consists of trying to force out the advertiser who occupies the position immediately above. Suppose that a bidder, say i, bids $\$b$ per click and is awarded the kth position. Let $\$b'$ be the bid of the advertiser, say j, who is ranked just above, meaning the one who obtains the $(k-1)$th position.

Notice that if bidder i slightly raises its bid (so that the ranking of bidders is not affected), then its payoff does not change: the ranking is unchanged, and bidder i's price is still the bid of the bidder below. But if bidder i modifies her bid, this will change bidder *j*'s payoff. It will decrease because now bidder *j* will have to pay slightly more per click. It may be the case then that bidder *j* can be better off changing his bid, in such a way that the relative rankings of bidders *i* and *j* are now reversed: with the new bids by bidder *i* and bidder *j*, it is now bidder *i* who is above bidder *j*. Since after some time bidders know each others' valuations, bidder *i* can predict what would be her final net payoff if bidder *j* reacts in such a way. If, in the end, bidder *i* is better off, then changing her bid (so as to trigger a reaction by *j*) is worth it. Otherwise, bidder *i* does not change her bid and remains ranked below bidder *j*. If no advertiser wants to change its bid so as to trigger a reaction by the bidder ranked just above and ends up swapping its position with this bidder, then we say that the advertisers' bids are **locally envy-free**.

Example 5.4 Consider four bidders competing for three spots. The bids and the outcome of the GSP auction are depicted in the following table.

Ranking	Bidder	Valuation	Bid	Price paid	Click frequency
1	Alice	$15	$15	$10	200
2	Bob	$14	$10	$7	100
3	Carol	$7	$7	$3	50
4	Denis	$3	$3	$0	0

The net payoffs for Alice and Bob are

$u_{\text{Alice}} = 200 \times (\$15 - \$10) = \$1000,$

$u_{\text{Bob}} = 100 \times (\$14 - \$7) = \$700.$

Suppose now that Bob raises his bid to 13. This does not affect the ranking or the payoffs of Bob, Carol, or Denis. But Alice's net payoff becomes

$u'_{\text{Alice}} = 200 \times (\$15 - \$13) = \$400.$

In this case, Alice is better off if she changes her bid to 9. By doing so, the result of the auction becomes

Ranking	Bidder	Valuation	Bid	Price paid	Click frequency
1	Bob	$14	$13	$9	200
2	Alice	$15	$9	$7	100
3	Carol	$7	$7	$3	50
4	Denis	$3	$3	$0	0

Keyword Auctions

With these new bids (13 for Bob and 9 for Alice), the net payoffs for Alice and Bob become

$u''_{\text{Alice}} = 100 \times (\$15 - \$7) = \$800,$

$u''_{\text{Bob}} = 200 \times (\$14 - \$9) = \$1000.$

It now suffices to compare the situations for Bob. At the beginning, his net payoff was equal to $700, but at the end (after changing his bid to $13 and the reaction of Alice) his final payoff is $1000. So the initial bids of $15, $12, $7, and $3 (for Alice, Bob, Carol, and Denis, respectively) do not constitute a locally envy-free equilibrium.

We can write the condition for envy-freeness more formally. To this end, let

α_k = number of clicks per period for position k;

v_i = valuation per click for advertiser i;

$g(k)$ = identity of the bidder in position k;

p_k = payment per period for advertiser in position k (i.e., the bid of the bidder in position $k+1$ multiplied by the click frequency of position k).

Definition 5.1 An equilibrium of the simultaneous-move game induced by the GSP auction is *locally envy-free* if a bidder cannot improve her payoff by exchanging bids with the bidder one position above her; that is, if for each position k, with $k > 1$,

$$\alpha_k v_{g(k)} - p_k \geq \alpha_{k-1} v_{g(k)} - p_{k-1}. \tag{5.1}$$

Equation (5.1) reads as follows. The left-hand side is the net payoff of the bidder who is assigned to the kth position (advertiser $g(k)$). The right-hand side is the payoff *that same bidder* would get if she were assigned the position just above her (i.e., position $k-1$) and made the same payment as the current bidder assigned to the $(k-1)$th position.

It turns out that if an equilibrium is locally envy-free in the GSP auction, then the outcome (who gets what position and at what price) is what economists call a **stable assignment**, where an assignment in this context consists of indicating for each position the advertiser that will be displayed and the price per click (or payment per period) that the advertiser will pay.

Definition 5.2 An assignment is **stable** if:

1. For each position, the price per click is nonnegative,
2. No advertiser pays more per click than her valuation,

3. There does not exist an advertiser i and a position k that can **block** the assignment:
   - Advertiser i is not assigned to position k.
   - $\alpha_k v_i > (\alpha_{\mu(i)} v_i - p_{\mu(i)}) + p_k$, where $\mu(i)$ is the position of advertiser i.

The key part in the definition of a stable assignment is the third requirement. It says that the total surplus that can be shared between an advertiser and a position not assigned to each other cannot exceed the payoff they obtain under the current assignment. In the definition, advertiser i and position k are not assigned to each other. Here $(\alpha_{\mu(i)} v_i - p_{\mu(i)})$ is advertiser i's payoff with the current assignment (called μ), and p_k is position k's net payoff. If the total surplus they can get together, $\alpha_k v_i$, is greater than the sum of their current payoffs, then we say that i and k **block** the assignment; that is, the assignment cannot be stable.

Suppose that, for each position on the search result page, Google asks some employees to manage the position with the goal of getting as much revenue as possible. That is, there is an employee who is in charge of the first position, another employee who is in charge of the second position, and so on. In this situation, each employee has to find an advertiser to be displayed at that position.

Consider position k (Mr. k at Google) and the advertiser i. Position k has a click frequency of α_k, and advertiser i is willing to pay v_i per click. Hence, the value of being at position k for advertiser i is $\alpha_k \times v_i$. The problem for Mr. k and advertiser i is to split this value. For instance, if advertiser i pays nothing, then Mr. k's revenue is 0 and advertiser i's revenue is $\alpha_k \times v_i$. More generally, if the payment to position k is p_k, then

advertiser i is left with $\alpha_k \times v_i - p_k$,

and

Mr. k's payoff is p_k.

Suppose that the price advertiser i pays is very small. Then Mr. k has very little revenue. For simplicity, assume that the price is 0. Suppose that advertiser j is currently talking with Mr. $k+1$, the employee in charge of the $(k+1)$th position, but is contemplating a very bad deal for advertiser j. Let us assume that the price advertiser j would have to pay is equal to the whole revenue, $\alpha_{k+1} \times v_j$. So, advertiser j is left with a net payoff equal to zero. Clearly, Mr. $k+1$ does not want to renegotiate with advertiser j, as any other price would decrease Mr. $k+1$'s payoff.

But in this case advertiser j and Mr. k may have an opportunity to get a better deal. It suffices for advertiser j to pay a small amount to Mr. k to be better off,

Keyword Auctions

and Mr. k would also be better off because it is better to get a little from advertiser j than nothing from advertiser i. In this case, we say that the assignment "Mr. k with i" and "Mr. $k+1$ with j" is not stable, because with the prices paid by i and j, both Mr. k and advertiser j would prefer to be assigned together than with their respective partners.

An assignment will be stable if the prices paid by the advertisers are such that no pair (position – advertiser) that are not together under the assignment can find a price such that they would prefer to be together.

Example 5.5 Consider the first and the second positions and two advertisers, Alice and Bob. The click frequency of the first position is 200 and that of the second position is 150. The valuation per click for Alice is 15, and Bob's valuation is 13.

First consider the following assignment.

Position	Advertiser	Price	Click frequency	Valuation
1st	Alice	$6	200	$15
2nd	Bob	$5	150	$13

In this assignment, Alice has the first position and pays $6/click, and Bob has the second position and pays $5/click.

The value to be shared under this assignment between the first position and Alice is $200 \times \$15 = \$3000$. From this, there is $200 \times \$6 = \$1200$ for the first position and $1800 left for Alice (that is her net payoff).

The value to be shared under this assignment between the second position and Bob is $150 \times \$13 = \$1950$. From this, there is $150 \times \$5 = \$750$ for the second position and $1200 left for Bob.

The total value that Bob and position 1 can generate together is $200 \times \$13 = \$2600$. Notice that if position 1 stays with Alice and Bob stays with position 2, the total net payoff of advertiser Bob and position 1 together is $1200 + $1200 = $2400. This is less than what they can generate together ($2600).

For instance, if position 1 and Bob agree on a price of $6.5, then the net payoff for position 1 is $200 \times \$6.5 = \$1300$, and the net income for Bob is $200 \times (\$13 - \$6.5) = \$1300$. Both Bob and position 1 would be better off under this deal, so the initial assignment is not stable.

Consider, for instance, the following assignment (with new prices).

Position	Advertiser	Price	Click frequency	Valuation
1st	Alice	$7	200	$15
2nd	Bob	$4.9	150	$13

In this case, the value to be shared between the first position and Alice is still $3000, and it is $1950 for the second position and Bob.

The payoff of Alice is 200 × ($15 − $7) = $1600, and for position 1 it is $3000 − $1600 = $1400. For Bob, the net payoff is 150 × ($13 − $4.9) = $1215, and position 2's payoff is $735.

If position 1 and Bob could find a deal that made them both better off, we should have:

Position 1's payoff with Bob > Position 1's payoff with Alice,

Bob's payoff with position 1 > Bob's payoff with position 2.

Notice that the net payoff of position 1 with Bob added to Bob's net payoff with position 1 is simply the total valuation to be shared (accounting for the frequency), 200 × $13 = $2600.

If we sum these two inequalities, we then must have that the total valuation that Bob and position 1 have together must be greater than the sum of their net payoffs under the current assignment, $1400 + $1215 = $2615. Since together they can have at most $2600 to be shared, and since the sum of their current payoffs is greater than this, it must be that whatever price they choose, either Bob or position 1 will have a lower net payoff. In other words, it is not profitable for Bob and position 1 to block the current assignment. It can be verified that the same occurs for Alice and position 2, so this assignment is stable.

We now have two different concepts at hand: envy-free equilibrium and stable assignments. It turns out that there is a close relationship between envy-free equilibrium and stability.

Result 5.1 (Edelman, Ostrovsky, and Schwarz) The outcome of any locally envy-free equilibrium in the GSP auction is a stable assignment.

Furthermore, if there are more advertisers than positions, then any stable assignment is the outcome of a locally envy-free equilibrium.

This result is important because the stability of an assignment is something that can be easily checked by an advertiser. Each advertiser knows the click rate of each position (it was one of our assumptions) and the valuations of the other bidders, so it is easy for each advertiser to see if there is scope for a better deal with a position other than the one it was assigned.

We now provide a proof of the first part of result 5.1. This proof is interesting because it sheds light on the relation between envy-freeness and stability. Suppose we have an envy-free equilibrium and that $p_1, p_2, \ldots p_n$ are the payments received by positions $1, 2, \ldots, n$. For notational simplicity, we assume that in our equilibrium advertiser k is matched to position k for all k.

Keyword Auctions

The only thing we have to do is show that for any advertiser k and position h that are *not* assigned together, the following inequality holds:

$$\alpha_k v_k - p_k \geq \alpha_h v_k - p_h. \tag{5.2}$$

Why is this inequality enough? Position h's net payoff is p_h. If this position and advertiser k were assigned together, then the total surplus to be shared between position h and advertiser k would be $\alpha_h v_k$. Therefore, advertiser k's **maximum payoff** she can hope to receive if position h agrees to be assigned to her (instead of the advertiser she is currently assigned to) is given by

$$\alpha_h v_k - p_h.$$

If equation (5.2) holds, then we have:

k's payoff at position $k >$ maximum payoff k can get with position h,

which implies that position k and advertiser h do not block the assignment.

So we need to show that equation (5.2) holds for any pair of position and advertiser that are not assigned to each other under the envy-free equilibrium. We distinguish between three cases.

Case 1 advertiser k and position $h > k$

One of the assumptions we made for the equilibria of the repeated GSP auction is that bidders' bids also constitute a Nash equilibrium of the one-shot game. This means that any deviation (i.e., new bid) such that advertiser k ends up at position h is not profitable for advertiser k:

$$\underbrace{\alpha_k v_k - p_k}_{\substack{\text{equilibrium payoff}\\\text{of advertiser }k}} \geq \underbrace{\alpha_h v_k - \hat{p}_h}_{\substack{\text{payoff of advertiser }k\\\text{if it deviates}\\\text{and gets ranked number }h}}. \tag{5.3}$$

In equation (5.3), $\hat{p}_h$ is the payment that position h receives if advertiser k deviates in such a way that it ends up being ranked number h. Advertiser k can obtain h's position by bidding b such that $b_{h'+1} < b < b_{h'}$. (By bidding such an amount b, all advertisers currently ranked $k+1, k+2, \ldots, h$ move up one position.)

Notice, however, that the payment to position h depends on α_h (this does not change) and the bid of the advertiser at position $h+1$. The advertiser in that position is the same whether advertiser k deviates or not, so we have

$$\underbrace{\hat{p}_h}_{\substack{\text{what advertiser }k\\\text{would pay if it deviates}}} = \underbrace{p_h}_{\text{what advertiser }h\text{ pays}}. \tag{5.4}$$

So we can rewrite equation (5.3), and we get

$$\alpha_k v_k - p_k > \alpha_h v_k - p_h,$$

which is what we wanted.

Case 2 advertiser k and position $k-1$

Rewriting equation (5.2), replacing h by $k-1$, gives

$$\alpha_k v_k - p_k \geq \alpha_{k-1} v_k - p_{k-1}.$$

But that is precisely the condition of envy-freeness! So this case is de facto proved.

Case 3 advertiser k and position $m < k - 1$.

This is the most delicate case. Unlike case 1, if bidder k deviates by taking the (now) higher position m, the payment received by position m will change. In other words, k's net payoff when deviating will *not* be

$$\alpha_m v_k - p_m.$$

Put differently, equation (5.4) does not hold here: we have $\hat{p}_m \neq p_m$. If k deviates, the value of p_m changes, and thus whether equation (5.2) holds is not obvious. To see this, suppose that at equilibrium the advertisers are (in this order) Alice, Bob, Carol, and Denis. Suppose that Denis deviates to take Alice's position (i.e., Denis now bids more than Alice), so now Alice is ranked second. The payment for position 1 *before* Denis's deviation was determined by Bob's bid (along with the click frequency of the first position). After Denis deviates, the payment to the first position is determined by Alice's bid.

To proceed, we first show that in an envy-free equilibrium we necessarily have an **assortative assignment**, which means that bidders are ranked by their valuations, namely $v_k \geq v_{k+1}$ for all k. Notice that the equilibrium condition (where nobody wants to move one position down, discussed in case 1) implies that

$$\alpha_k v_k - p_k \geq \alpha_{k+1} v_k - p_{k+1}. \tag{5.5}$$

The envy-freeness condition says that no advertiser wants to move one position up:

$$\alpha_{k+1} v_{k+1} - p_{k+1} \geq \alpha_k v_{k+1} - p_k. \tag{5.6}$$

Adding equations (5.5) and (5.6) yields

$$\alpha_k v_k + \alpha_{k+1} v_{k+1} \geq \alpha_{k+1} v_k + \alpha_k v_{k+1} \iff v_k(\alpha_k - \alpha_{k+1}) \geq v_{k+1}(\alpha_k - \alpha_{k+1}).$$

Keyword Auctions

Since $\alpha_k > \alpha_{k+1}$ (a higher position means more clicks), we have

$$v_k \geq v_{k+1},$$

which confirms that in equilibrium we have an assortative assignment.

Now, since the equilibrium is locally envy-free, we have

$$\alpha_k v_k - p_k \geq \alpha_{k-1} v_k - p_{k-1},$$
$$\alpha_{k-1} v_{k-1} - p_{k-1} \geq \alpha_{k-2} v_{k-1} - p_{k-2},$$
$$\alpha_{k-2} v_{k-2} - p_{k-2} \geq \alpha_{k-3} v_{k-2} - p_{k-3},$$
$$\vdots$$
$$\alpha_{m+2} v_{m+2} - p_{m+2} \geq \alpha_{m+1} v_{m+2} - p_{m+1},$$
$$\alpha_{m+1} v_{m+1} - p_{m+1} \geq \alpha_m v_{m+1} - p_m.$$

Since $\alpha_h < \alpha_{h-1}$, for any $h > 1$, we have

$$\underbrace{\alpha_h v_h - p_h \geq \alpha_{h-1} v_h - p_{h-1}}_{\text{envy-freeness condition}} \quad \Rightarrow \quad v_h \leq \frac{p_{h-1} - p_h}{\alpha_{h-1} - \alpha_h}. \tag{5.7}$$

Now, $v_k < v_h$ for any $h < k$ implies that we can replace v_h by v_k in the second inequality of equation (5.7), and we get for any $h = 2, \ldots, k-1$,

$$v_k \leq \frac{p_{h-1} - p_h}{\alpha_{h-1} - \alpha_h} \quad \Rightarrow \quad \alpha_h v_k - p_h \geq \alpha_{h-1} v_k - p_{h-1}. \tag{5.8}$$

So we can rewrite the inequalities we had before replacing v_j by v_k for $j = m, \ldots, k-1$,

$$\alpha_k v_k - p_k \geq \alpha_{k-1} v_k - p_{k-1},$$
$$\alpha_{k-1} v_k - p_{k-1} \geq \alpha_{k-2} v_k - p_{k-2},$$
$$\alpha_{k-2} v_k - p_{k-2} \geq \alpha_{k-3} v_k - p_{k-3},$$
$$\vdots$$
$$\alpha_{m+2} v_k - p_{m+2} \geq \alpha_{m+1} v_k - p_{m+1},$$
$$\alpha_{m+1} v_k - p_{m+1} \geq \alpha_m v_k - p_m.$$

We now add all these inequalities. Observe that the left-hand side of the first, second, third, ... inequalities is the same as the right-hand side of the second, third, ... inequalities. We thus obtain

$$\alpha_k v_k - p_k \geq \alpha_m v_k - p_m.$$

This is equation (5.2) where h is replaced by m, which is what was needed, so the proof is complete.

5.4.6 The Generalized English Auction

The Vickrey auction is a one-shot, simultaneous-move game that is equivalent to the English auction. Would it be possible to generalize the English auction so that the GSP auction would be its one-shot, simultaneous-move version? The most intuitive auction we can design that would capture the main property of the GSP auction (that the price that each bidder pays is the bid of the bidder just below her) would be the following:

- There is a clock showing the current price, which increases over time.
- The price starts at 0, and at the beginning all the advertisers are present in the auction.
- Each advertiser can drop out at any time. The bid of an advertiser is equal to the price on the clock at the time she dropped out.
- The auction ends when the next-to-last advertiser drops out; namely, as soon as there is only one advertiser left.

With the rules, the outcome of the auction is the following:

- The last advertiser that is present in the auction (all the other advertisers having dropped out) is ranked first.
- All the other advertisers are ranked according to the time (or price) at which they dropped out. If advertiser i dropped out at a higher price than advertiser j, then i is ranked above j.
- Each advertiser pays the bid of the advertiser ranked just below her.

This auction looks very much like the English auction, so we can call it the *generalized English auction*. One would think that the outcome of such an auction would be equivalent to the GSP auction. As we will see, the generalized English auction is not equivalent to the GSP auction. Before establishing this, we first need to characterize the equilibria of the generalized English auction.

Result 5.2 (Edelman, Ostrovsky, and Schwarz) There is a unique (perfect Bayesian) equilibrium of the generalized English auction, where an advertiser with valuation v drops out at the price

$$p^* = v - \frac{\alpha_k}{\alpha_{k-1}}(v - b_{k+1}), \qquad (5.9)$$

where k is the number of advertisers remaining in the auction (including herself).

The preceding formula reads as follows. Take an advertiser, and suppose that when that advertiser dropped out there were k advertisers left in the auction (including herself). Let v be the valuation of that advertiser. Because she dropped out when there were k advertisers left, she will be ranked kth in the auction. So this advertiser will pay the bid of the advertiser ranked just below, the advertiser who will be ranked the $(k+1)$th in the outcome of the auction. That price is then b_{k+1}, the price on the clock when the advertiser ranked $k+1$ dropped out.

Call A the advertiser who is ranked just below—the advertiser ranked $(k+1)$th. Notice that once A has dropped out, all the remaining bidders know the time and price at which A dropped out (which is b_{k+1}), and thus they all know that the first one to drop out after A will pay the price b_{k+1}. So a bidder with valuation v just "waits" until the price that is displayed satisfies the equality in equality (5.9). As soon as the price on the clock is p^*, the bidder drops out.

The intuition behind result 5.2 is the following. Suppose there are k bidders remaining (including me), and the price at which the last advertiser dropped out is b_{k+1} (so it is the price that the next bidder who drops out will have to pay). If I am the first to drop out after the bidder who bid b_{k+1}, my net payoff for the kth position will be

$$\alpha_k \times (v - b_{k+1}). \tag{5.10}$$

Indeed, the click frequency of the kth position that I will get is α_k, which I multiply by my net payoff per click (my valuation minus the price per click, $v - b_{k+1}$).

Suppose instead that I wait for another bidder to drop out, and then I drop out just after that bidder. Let p be the price at which that bidder stops. So that bidder gets the kth position bidding p and I get the $(k-1)$th position, and I will pay p per click. So my net payoff will be

$$\alpha_{k-1} \times (v - p). \tag{5.11}$$

Since $\alpha_{k-1} > \alpha_k$ (the first positions have a higher click frequency), if p is very close to b_{k+1}, then the payoff for waiting to be the $(k-1)$th bidder—given in equation (5.11)—is greater than the payoff for dropping out now—given in equation (5.10). Note, however, that as p increases, my payoff is getting smaller. So there is some p^* such that I am just indifferent between waiting (and being the $(k-1)$th bidder) and dropping out (and being the kth bidder). That is, p^* is such that

$$\alpha_i \times (v - b_{k+1}) = \alpha_{k-1} \times (v - p^*).$$

Solving this equation for p gives the formula in equation (5.9).

If I do not drop out when the price is p^*, then I take the risk that someone will drop out before me, but since it will be for a price $\hat{p} > p^*$, my payoff for being the

$(k-1)$th bidder (equation (5.11)) will be lower than my payoff had I dropped out at a price between b_{k+1} and $\hat{p}$ (equation (5.10)).

The problem with this result is that the payoffs advertisers obtain at the equilibrium are not those of the GSP auction (with truthful bidding) but rather are equal to the payoffs of the VCG auction. So in fact the generalized English auction is the sequential version of the VCG auction.

To see this, suppose there are three bidders competing for three spots. Let v_1, v_2, and v_3 be the values of bidders 1, 2, and 3, respectively. The click frequency of the first position is α_1 and that of the second position is α_2. Assume that $v_1 > v_2 > v_3$.

To make the point, we consider only the price that bidder 2 is paying. We first consider the price for bidder 2 with the VCG auction. If bidder 2 is not present in the auction, then the assignment that maximizes the total valuation is the first position for bidder 1 and the second position for bidder 3, generating a total value of

$$\alpha_1 \times v_1 + \alpha_2 \times v_3.$$

If bidder 2 is present, the assignment that maximizes the social value is position k for bidder k ($k=1,2$). In this case, the social value for bidders 1 and 3 (all bidders except me) is

$$\alpha_1 \times v_1 + \alpha_3 \times v_3.$$

The difference is then

$$\alpha_1 \times v_1 + \alpha_2 \times v_3 - (\alpha_1 \times v_1 + \alpha_2 \times v_3) = v_3 \times (\alpha_2 - \alpha_3). \tag{5.12}$$

This price is what bidder 2 pays for each click when in the second position, if we run a VCG auction. Notice that this price is obtained using the valuation per period of time (the one used to define the frequency). This means that the total net payoff flow of bidder 2 under the VCG auction is

$$\alpha_2 \times v_2 - v_3(\alpha_2 - \alpha_3). \tag{5.13}$$

Now consider the pricing under the generalized English auction. In this case, using equation (5.9), bidder 3 will drop out when the price is

$$p = v_3 - \frac{\alpha_3}{\alpha_2}(v_3 - 0). \tag{5.14}$$

Note that the 0 at the end is the bid of a hypothetical fourth bidder. But since there is no such bidder, that bid is zero. In the generalized English auction, we obtain that bidder 2 will be the first bidder to drop out after bidder 3. Indeed, suppose that bidder 2 drops out before bidder 3, so bidder 2 will drop out when the price is

$$p = v_2 - \frac{\alpha_3}{\alpha_2}(v_2 - 0). \tag{5.15}$$

Comparing equations (5.14) and (5.15), we can see that the price at which bidder 3 drops out (and is the first one to do so) is lower than the price at which bidder 2 would drop out (being the first one to do so) if

$$v_3 - \frac{\alpha_3}{\alpha_2}(v_3 - 0) < p = v_2 - \frac{\alpha_3}{\alpha_2}(v_2 - 0) \quad \Leftrightarrow \quad \frac{\alpha_3}{\alpha_2} < 1.$$

Since $\alpha_3 < \alpha_2$, the preceding equation is always satisfied; in other words, bidder 3 drops out before bidder 2.

The bid of bidder 3, the price at which bidder 3 drops out, given by equation (5.14), is a *price per click*. This is the price that bidder 2 will pay. To compute the price per unit of time, we have to multiply it by the frequency of the second position, so the price per unit of time for bidder 2 is

$$\alpha_2 \times \left(v_3 - \frac{\alpha_3}{\alpha_2}(v_3 - 0)\right) = v_3 \times (\alpha_2 - \alpha_3). \tag{5.16}$$

Clearly, the price per unit of time under the VCG auction (equation (5.12)) and the price per unit of time under the generalized English auction (equation (5.16)) are identical. So the generalized English auction can be understood as the sequential version of the VCG auction when considering keyword auctions.

5.5 The Facebook Model: VCG for Internet Ads

The generalized second-price auction introduced by Google is not the only way ad space is sold on the Internet. Today, the VCG auction is seen by many people as a serious alternative, and Facebook is a prominent example. In fact, the VCG auction, or simple Vickrey auction, is an attractive solution for *display* ads; that is, ads that have a graphical content (as opposed to a result of a search query, which is just a title and a web address). To understand the pros and cons of each auction format, we first compare (in a simple framework) the two auctions, and then we discuss the motivation for using the VCG auction.

5.5.1 Comparing VCG and GSP Auctions

The purpose of this section is to give an analytical characterization of the seller's revenue. So we have a set of advertisers, indexed $i = 1, 2, \ldots, n$, and each advertiser i has a valuation per click v_i. There are several ad slots, and each slot $s = 1, \ldots, S$

comes with clickthrough rate α_s, with $\alpha_1 > \alpha_2 > \cdots \alpha_S$ (i.e., the first slot has a higher clickthrough rate than the second slot, etc.). We have already seen that for the GSP auction we can restrict our analysis to (locally) envy-free equilibria.

We now simplify the notation by assuming (without loss of generality) that the equilibrium under the GSP is such that advertiser 1 is assigned the first position, advertiser 2 is assigned position 2, and so on. The envy-freeness condition written for advertiser $i+1$ is $v_{i+1}\alpha_{i+1} - p_{i+1} \geq \alpha_i v_{i+1} - p_i$, which can be rewritten as

$$p_i \geq v_{i+1}(\alpha_i - \alpha_{i+1}) + p_{i+1}. \tag{5.17}$$

Equation (5.17) simply says that the *revenue for position i*, equal to p_i, must be at least equal to $v_{i+1}(\alpha_i - \alpha_{i+1}) + p_{i+1}$.

If we have only three slots (and thus four bidders), the minimum revenues for each position must then, under equilibrium condition (5.17), yield

$$p_1 \geq v_2(\alpha_1 - \alpha_2) + p_2, \tag{5.18}$$

$$p_2 \geq v_3(\alpha_2 - \alpha_3) + p_3, \tag{5.19}$$

$$p_3 \geq v_4\alpha_3. \tag{5.20}$$

The left-hand side of the last equation is only $v_4\alpha_3$, because the fourth bidder obtains no slot and pays nothing ($p_4 = 0$).[7] If we sum these three equations, we obtain

$$p_1 \geq v_2(\alpha_1 - \alpha_2) + v_3(\alpha_2 - \alpha_3) + v_4\alpha_3. \tag{5.21}$$

Summing equations (5.19) and (5.20), we obtain

$$p_2 \geq v_3(\alpha_2 - \alpha_3) + v_4\alpha_3. \tag{5.22}$$

So equations (5.21), (5.22), and (5.20) are the *lower bound* (in any locally envy-free equilibrium) of the revenues of positions 1, 2, and 3, respectively.

We now calculate the seller's revenue with the VCG auction. Remember that for the VCG auction it is a dominant strategy for the advertisers to bid their valuations. So we can assume that for each bidder i the bid is v_i, his valuation. With v_1, v_2, v_3, and v_4 being the valuations of the advertisers, we assume that $v_1 > v_2 > v_3 > v_4$. It is easy to see that in this case the total social valuation is maximized when giving position 1 (with the highest frequency, α_1) to advertiser 1, position 2 to advertiser 2, and so on. If advertiser 1 participates in the auction, then advertiser 2 (who has the second-highest valuation) takes the second position and advertiser 3 takes the third position. So the total social valuation (excluding advertiser 1) is

$$v_2\alpha_2 + v_3\alpha_3. \tag{5.23}$$

If advertiser 1 does not participate in the auction, then advertisers 2 and 3 obtain positions 1 and 2, respectively, and advertiser 4 (who gets no position if advertiser 1 is present) now gets position 3. So the total valuation for all bidders (except bidder 1) now becomes

$$v_2\alpha_1 + v_3\alpha_2 + v_4\alpha_3. \tag{5.24}$$

The difference between formulas (5.23) and (5.24) is the price that advertiser 1 has to pay, the revenue for position 1,

$$p_1 = v_2(\alpha_1 - \alpha_2) + v_3(\alpha_2 - \alpha_3) + v_4\alpha_3. \tag{5.25}$$

For advertiser 2, the total valuation of the other bidders when advertiser 2 is present is

$$v_1\alpha_1 + v_3\alpha_3,$$

and the total valuation of the other bidders when advertiser 2 is *not* present (so advertiser 1 keeps the first position but advertisers 3 and 4 take the second and third positions, respectively) is

$$v_1\alpha_1 + v_3\alpha_2 + v_4\alpha_3,$$

and thus the revenue of the second position is obtained by taking the difference,

$$p_2 = v_3(\alpha_2 - \alpha_3) + v_4\alpha_3. \tag{5.26}$$

Finally, for the third position, we obtain the following revenue (the only difference is that advertiser 4 takes the third position when advertiser 3 is not present):

$$p_3\alpha_3 = v_4\alpha_3. \tag{5.27}$$

We can now compare, for each position, the lower bound of the revenue with the GSP auction versus the revenue with the VCG auction. That is, compare:

- equation (5.21) with equation (5.25);
- equation (5.22) with equation (5.26); and
- equation (5.20) with equation (5.27).

Clearly, for each position, the *lowest* equilibrium revenue for the seller with the GSP auction is equal to the revenue under the VCG auction, so we have the following result.

Result 5.3 The revenue of the search engine with the GSP auction is at least as high as that with the VCG auction.

5.5.2 The Rationale for VCG

In section 5.5.1, we just saw that the seller (Google, Facebook, or any website) would have higher revenues with the GSP auction than with the VCG auction. So, why do Facebook and other websites use the VCG auction? As we intuited at the beginning of section 5.5, the VCG auction turns out to be a popular choice for *display ads*. In fact, Google's ad placement program, called AdSense, utilizes the VCG auction.[8]

The first and major difference between search queries and display ads is that for web pages that have display ads the advertiser does not have precise knowledge of what the user is interested in. From the cookies stored in our computer and the content of the page, the advertiser does know what we like, but this information is less precise than with a search query. Indeed, with a query, the user is *actively signaling* to the advertiser what she is looking for. This implies that with display ads advertisers have more uncertainty about how much a click is worth. With the VCG auction, it is a dominant strategy to bid one's (true) valuation. Facebook's contention is that with the VCG auction advertisers will spend more resources and energy figuring out exactly what their valuations are and make their ads better, instead of trying to game the GSP auction and identify the right bidding strategy. So the auction used by Facebook makes life simpler for the advertiser, which may entice bidders to prefer campaigning on Facebook rather than on a search engine.

There are two other motivations for using the VCG auction. First, display ads often allow the possibility for an advertiser to display *several ads* on the same page. If a website has several ad spaces, some advertisers may want to bid for only one space, and other advertisers may bid for multiple displays. This type of situation is the domain of combinatorial auctions like the VCG auction, which further motivates the use of the VCG auction instead of the GSP auction. A second motivation, and perhaps one of the most important for Facebook, is that by lowering its revenue for each auction, Facebook is trying to maintain a long-term relationship with the advertisers. The principle is not new. If the clients (the advertisers) have good deals, they have good returns, and they are more likely to come back and spend more.

At the end of chapter 4, we explained that one of the reasons why the VCG auction is not often used in practice is that it can be computationally intractable because of the number of combinations the bidders and the auctioneer have to consider. For Facebook, this is not really an issue. On a web page, the number of ads is limited, drastically reducing the number of computations that Facebook has to run to compute the optimal assignments. Facebook also reduces the number of bidders by considering only those that seem relevant for the user. Add to this the extraordinary computing power that firms like Facebook or Google have at their disposal, and the complexity of running a VCG auction to display a few ads on a screen becomes something manageable.

6 Spectrum Auctions

Perhaps one of the most fascinating applications of auction theory is the implementation of new auction mechanisms to allocate licenses to cell phone carriers. With the advent of cell phone technology in the 1990s, regulators across the globe had to set up mechanisms to allocate licenses. Designing allocation mechanisms turned out to be complex, mostly because the auction literature did not have a ready-to-use mechanism. In this chapter, we will study the various allocation mechanisms used by regulatory agencies, discussing why some designs were successful and others were not.

6.1 How Can Spectrum Be Allocated?

Before we consider the problem of allocating *cell phone licenses*, it is important to discuss first the problem of allocating licenses in general. There are many ways to proceed, but the main ones are

- a lottery;
- a beauty contest; and
- an auction.

6.1.1 Lotteries

One way to allocate licenses or any other object could be to run a lottery. Any interested party would just have to sign up, perhaps paying some participation fee. This is, for instance, what the U.S. Department of Homeland Security is doing to allocate 50,000 green cards each year. The use of a lottery has several advantages, the main one being that it is a fair process in the sense that all participants have equal chances. Another positive aspect of a lottery is that it makes the use of bribes or favoritism less likely.

But a major issue with lotteries is that licenses may not be awarded to the people who would make the best use of them. This is important because a license is not an object that is awarded for personal use. A license holder, whether it is for cell phone services or operating a bridge or a tunnel, uses the license to provide a service to customers. It is thus important for regulatory agencies to allocate licenses in a way that maximizes the likelihood that they will be awarded to the entities that will provide the best services for the users. Because the use of a lottery leaves virtually no control over who gets a license, it is, in the case of spectrum, not the right approach. From an economic perspective, a lottery is not an efficient mechanism. The individual being awarded the license may not be the one who values it the most.

6.1.2 Beauty Contests

Another solution to allocate licenses is to use the so-called **beauty contests**. This procedure consists of asking the companies wishing to obtain a license to present a project, and then a committee selects the winning projects. This is the procedure that is often used to select large architectural projects. It is obvious that a beauty contest solves many of the shortcomings of a lottery, the main one being the guarantee that only "good" candidates are selected. In a beauty contest, regulatory agencies usually lay out a list of requirements that candidates must satisfy, thereby allowing for better screening than with a simple random draw. One can argue that nothing prevents the regulators from also publishing a similar list of requirements in the case of a lottery. However, the beauty contest allows for a selection based on criteria that are more difficult to explain, if not subjective. Selection committees in a beauty contest often contain experts who can help policy makers distinguish between bad and good projects. Often, the arguments by those experts rely on their experience, intuition, and other criteria that are more difficult to list explicitly.

For complex and important projects such as cell phone licenses, bridges, railroads, or other public facilities, the use of experts who can evaluate candidates using their experience and knowledge seems for many people the right solution. One caveat of this approach is that it is more difficult to make the selection process transparent. At the time candidates submit their projects, they may be unable to know the exact criteria that will be used to select their projects. More worrisome is that beauty contests can easily lead to favoritism and corruption. Since the ultimate decision will be based on criteria that are not (or could not be) explicitly given, misconduct by committee members becomes more difficult to prove (and thus to prevent).

Another issue with beauty contests, and perhaps one of the most important problems, is that governments and regulatory agencies do not usually know well how businesses operate, and this knowledge may be paramount for gauging candidacies accurately. To mitigate this problem, regulators often impose a list of requirements that the winners must satisfy not only at the time the project is awarded but also for the next five, ten, or even twenty years. For instance, in the case of cell phone services, the government can impose a minimum infrastructure investment (e.g., the number of radio towers to be deployed). But such stipulations have a drawback, as they require from the government a monitoring effort that is often difficult to realize.

Important firms often favor beauty contests and lobby regulators for implementing this type of selection process. One reason for this is that it may be easier for them to lobby and effectively pressure selection committees. But perhaps the main reason is that beauty contests are usually cheap for firms, whereas auctions, as we will see, can be relatively expensive for the winners. An argument that is often used to justify the use of beauty contests is that, because licenses are cheaper, firms will be able to propose lower prices to consumers. This argument is flawed, however. A license is a fixed cost, so it does not affect the profit maximization decisions of the firms. A firm choosing the price that maximizes the profit function will choose the same price if it maximizes the same profit function minus the license cost. This is so because the price of a license does not depend on the number of clients or the price firms charge them.

6.1.3 Why Run Auctions?

The use of auctions to allocate licenses offers several advantages that make it a more attractive solution than a lottery or a beauty contest, especially when we consider cell phone licenses. To understand why, it is best to go back in time to the early 1990s, when cell phone technology started to become ready for the general public.

When a company launches a new version of a product, it can have a relatively precise idea of how consumers will react. Past experience can be analyzed to make predictions about the likelihood of success of the new product. If, on the contrary, a company launches a completely new product, obtaining estimates about consumer demand becomes much more difficult. In the case of cell phone licenses, potential carriers in the early 1990s did not have precise knowledge of how valuable a license would be (which would depend, among other things, on the public's reaction). Cell phones for the general public was a new market in the early 1990s. Note that the difficulty in calculating the value of a license is not a problem confined to auctions. Beauty contests often (if not always) require candidates to pay a fee, which can be understood as the price of a license. Even if companies obtain their licenses at a low price, they still have to figure out how much to invest, what price they should charge to consumers, and so on. All these decisions depend on how valuable the new market is.

As we have seen throughout the previous chapters, auctions are *price discovery mechanisms*. That is, they are ideal tools for agencies, governments, or simply sellers to discover how much bidders value the items to be allocated. If well designed, auctions can be the perfect tools to extract information from bidders by giving bidders incentives to reveal their valuations. This contrasts with beauty contests, where little information is extracted from the bidders about how much they value a license, leaving to the regulator alone the task of estimating the value of a license. Even if carriers like AT&T or GTE (the ancestor of Verizon) have an imprecise knowledge of the value of a license, there is no doubt that their estimates are much better than those of the regulator.

Another attractive argument for governments is that auctions can be a valuable source of income. We will see in this chapter that successful auction designs permit governments to collect substantial amounts of money. From an economic standpoint, government revenue obtained from the allocation of licenses is better than revenue obtained from taxes (taxes create distortions that can alter both the demand and the supply, lowering consumers' incomes and producers' profits).

6.2 Issues

Designing auctions for spectrum is not an easy task. There are many problems that must be considered, such as defining the content of a license, estimating the right number of licenses to be awarded, or simply the design of the auction itself. In this section, we first discuss some general issues related to the problem of auctioning licenses and then review some more specific concerns or issues that influenced the auction designs used since the 1990s to allocate licenses.

6.2.1 General Issues

Cell phone communication is done through radio waves, the same kind of signal used by wifi, FM radio, or aerial TV, or by using fiber optics. The first task of the regulator is to define the spectrum that is available for cell phone communications, meaning the range of frequencies that will be used by the phones. Once the spectrum has been decided, the first issue when designing a spectrum auction is to decide whether carriers bid for a

- **block of frequencies**, where the total spectrum is divided into several blocks (e.g., the first block between 698.2 MHz and 716.2 MHz, a second block between 777.2 MHz and 792.2 MHz), and a carrier has a license if it manages to win, say, at least two blocks, or

- **licenses**, where to each license is assigned a set of frequencies and bidders bid on licenses.

These two ways of defining what carriers bid for are substantially different. When bidders bid for a block of frequencies, the number of licenses can be endogenous. We will have to wait until the end of the auction to know how many carriers obtained a license. This contrasts with the auction, where bidders bid directly for a license. In this case, the number of carriers can be known as soon as the auction starts (we just need to know the number of bidders).

Defining the object of the auction (frequencies or licenses) is obviously linked to the question of the ideal market structure (i.e., the number of cell phone carriers). On the one side, we want a sufficiently high number of carriers. The more carriers we have, the more they will compete on price and quality, which is good for the consumer. This is a classic argument in economics. The extreme case is when there is a monopolist, where only one firm has a license, which typically entails lowering incentives to invest and provide high-quality service to customers. Also, a monopolist charges consumers a price higher than the competitive price. However, on the other hand, too many competitors is not socially desirable either. Cell phone communications require large investments, which may not be possible when the market is too fragmented.

Determining the "right" number of licenses is not an easy problem, and it was even more difficult back in the 1990s when the technology was new. The lack of information about how valuable a license would be (which is linked to the problem of the number of licenses) posed a major problem when designing an auction. In the classic auctions we have seen, it was assumed that bidders know their valuations, which allows the design of robust auction mechanisms like the VCG auction. When bidders have only imprecise knowledge of their valuations (partly because they do not know precisely the adoption rate of the new technology, how elastic the demand is, and other factors), it is difficult to conceive an auction mechanism where bidders would bid optimally. Another complication brought by spectrum auctions is that bidders' valuations are likely to be correlated. If the cell phone market is promising (i.e., valuable) for, say, AT&T, it is likely to be also for its competitors. As it turns out, auctions with interdependent valuations are much more complex to analyze than when valuations are independent.

6.2.2 Collusion, Demand Reduction, and Entry

One of the main objectives when auctioning licenses (or spectrum) for cell phone services is to avoid having a unique winner. If this is the case, then consumers do not have any choice, a clearly undesirable outcome. So we need several winners, which implies that we need a *multiunit auction*. There are several items to be sold, allowing several winners (e.g., as in the VCG auction in chapter 4).

The problem is that collusion is much easier to sustain when there are several items to be sold. It suffices that the colluding parties agree to bid on different items. By avoiding competing with each other in the bidding game, bidders can make final prices lower. One advantage with this strategy is that it is difficult to distinguish from *tacit collusion*. Firms being accused of collusion in that case can simply defend themselves by saying that they were just taking advantage of the bidding behavior of their competitors. Avoiding bidding on the same items in order to lower prices is known as **demand reduction**. Note that this phenomenon is more likely to be observed when we run dynamic auctions such as an English or a Dutch auction, as it requires colluding bidders to observe the bids of their partners.

Example 6.1 To see how demand reduction works, consider the case of two bidders, Jane and Fred, and two objects, A and B. Both Jane and Fred value either object at \$100, and they value having both objects at \$200.

The auction we consider is the following. Each object is sold, at the same time, through an English outcry auction (which we will describe in more detail in section 6.3). So in this auction bidders have to announce a price for each of the objects they are bidding for. For simplicity, we assume that the minimum price increment for any bid is \$1. This means that if during the auction the current bid for an object is, say, \$26, then if a bidder wants to place a bid for that object, the bid must be at least $\$26 + \$1 = \$27$.

An example of a strategy that brings demand reduction is the following:

- If I am not the highest bidder on any object, I bid the minimum price increment for the lowest-priced item.

For instance, suppose that the current highest bid for object A is \$50 and it is \$60 for object B. These two bids are made by the other bidder. Then my bid consists of bidding only for the lowest-priced object, A. The bid I will submit is \$50 + the minimum bid increment, \$51.

- If I am the highest bidder for one of the objects, then I do not submit any additional bid.

It is not difficult to see that in this case Jane and Fred can arrange to sustain low final prices for these two items. If the current bids are, say, \$20 and \$25 for objects A and B, respectively, and both bids are made by Jane, then with the strategies we just described, we have the following:

- Fred bids for object A (the cheapest) and bids the minimum allowed bid that is above \$20. So he bids \$21.
- Jane does not bid further.

Now the current bids are \$21 and \$25 for objects A and B, respectively. Each bidder is the highest bidder for one of the items, so they stop bidding, and the auction ends with these prices.

Spectrum Auctions

Without any explicit or tacit agreement on such strategies, Jane and Fred would engage in a bidding war until the price of $100 is reached. To see this, note that if Jane is not getting object A and Fred is getting it at a price $p < \$100$, then we have:

- If Jane refrains from bidding, then she obtains a net payoff for object A equal to 0.
- If Jane bids $p + \$1$, then she can obtain a net payoff for object A equal to $100 - (p+1) \geq 0$.

So it is better for Jane to bid above Fred as long as the current bid is less than $100. The same occurs with object B. So without coordination between Jane and Fred, the auction ends with each item being sold at $100.

Another desirable objective when selling licenses is to **promote entry**, promote the arrival of new carriers in the market. One obvious way to achieve this is to offer more licenses than there are incumbents. There are two reasons for this. First, having additional carriers will increase competition, which encourages better and more affordable services for customers. Of course, as we intuited earlier, we may not want too many carriers. But there is another, perhaps less obvious reason for promoting entry: it increases competition *during* the auction. If there are more licenses to be offered than there are incumbents, then it is likely that there will be several new competitors for the additional licenses. This means that there will be more bidders than there will be licenses to be sold, and thus collusion among bidders will be more difficult to sustain. In example 6.1, Jane's and Fred's strategies are less likely to be successful if there is a third bidder. If there are only, say, two bidders for two licenses, it is easy for them to implement a demand reduction strategy. Having more bidders than objects is a good way to minimize the risk of having demand reduction.

6.2.3 Maximum Revenue

Another objective (and perhaps the main one) for a seller when running an auction is to maximize revenue. This expression has several meanings. First, there is the obvious interpretation: the government would like to raise as much money as it can. This money can serve to finance some programs and/or lower taxes. Whoever sells some object on eBay would like to sell it at the highest possible price. So maximizing revenue is a legitimate objective in an auction.

There is another, more subtle interpretation. Maximizing revenue in most auctions is tantamount to selling the object to the bidder that values it the most, the bidder that has the highest profitability for the market.[1] If the market is sufficiently competitive, meaning if there are enough cell phone operators, then bidders with high valuations will be those more able to compete on quality and price. In other words, selling the licenses to the operators with the highest valuations is a guarantee that, through competitive forces, consumers will have the best experience at the best price.

6.2.4 The Exposure Problem

In an auction such as the spectrum auction, bidders' valuations are likely to depend on each other. In other words, we have an auction with interdependent or common values. This is not surprising. If the cell phone market looks profitable for a cell phone operator, it is also likely to look profitable for its competitors, because they compete for the same customers.

Also, for a large country like the United States, the auction will typically consist of dividing the country into several regions and offering several licenses for each region but no nationwide license.[2] The reason for doing this is related to the problem of obtaining a competitive market once the licenses are sold. For a large country like the United States, the number of firms able to offer a nationwide cell phone service is relatively small, and thus not likely to be as competitive as one might wish. By offering only regional licenses, the regulator is making it easier for small or medium-sized firms to enter the auction and compete. In the United States, there are about 80 regional carriers (in 2016, when including Alaska, Hawaii, and Puerto Rico), while there are only four nationwide operators (AT&T, Verizon, T-Mobile, and Sprint).[3] Each national carrier thus holds many licenses (one for each region), and many other carriers also have several licenses. An auction spectrum for wireless communication is thus an auction of multiple objects, where bidders can acquire several objects.

How do we design an auction for many objects when we allow bidders to buy several of them? A quick answer would be to use the VCG auction (see section 4.3) or any *combinatorial auction*; that is, any auction that allows bidders to bid on *combinations* of items instead of submitting a bid for each item separately. On paper, an auction like the VCG auction indeed looks like the ideal auction mechanism for such a problem. But in practice, especially for an auction like the U.S. spectrum auction, the VCG auction is virtually unfeasible. The reason lies in the number of licenses, and thus in the number of combinations of (regional) licenses the bidders have to consider. We can have, for instance, a bidder that is interested in having a license for a big metropolitan area, hesitating between having a license for, say, Washington, D.C., and Philadelphia or Philadelphia and New York City. If that bidder can only afford one of these two options (and not the whole package Washington-Philadelphia-New York), and provided there are several licenses for sale in each of these three cities, the number of combinations our regional bidder would have to consider could be high, making the bidding process extremely difficult. But there is a bigger difficulty. Even if one assumes that bidders are able to compute their bids for all the combinations of licenses they are interested in, we still have the problem of the auctioneer having to compute the VCG allocation and prices, and this number can be extremely large, making the computation nearly impossible to perform in a reasonable amount of time. In other words, a combinatorial auction, although apparently the best approach for designing a spectrum auction, is not viable.

One way to eliminate the problem of computing the assignment of licenses is to run parallel auctions, one for each license. This can be done, for instance, using English auctions. But doing so creates a problem because now it will be more difficult for bidders to express preferences among *combinations* of items. Being unable to express such preferences is known as the **exposure problem**.

An example of the exposure problem is the following. Consider a firm that wants to be a small regional carrier for some region that offers, say, three licenses. In a VCG auction, that firm would submit, for instance, a positive bid for each license but a zero bid for any combination of two or more licenses. If we run parallel auctions, a natural strategy for our firm would be to bid for each of the three licenses, with the hope of obtaining one of them. The problem with this is that the firm is taking the risk of obtaining two or three licenses. When running parallel auctions, it is difficult for our firm to bid in a way that guarantees it will win *at most* one license.

A similar problem can occur when a carrier sees two licenses, say licenses A and B, as complements. That bidder is willing to pay, say, 100 for A and 100 for B if it has both of them, but only, say, 50 for having only A or B. So that bidder is willing to bid up to 100 for A and up to 100 for B only if it is a winner for both licenses. If it is only the winner for either A or B, it only wants to bid up to 50. When running parallel auctions, that carrier can end up being the highest bidder on one of the licenses at a price higher than 50 and losing the other license. A combinatorial auction eliminates the risk of having such situations.

6.2.5 Winner's Curse

An auction like the spectrum auction differs from the auctions we have studied in the previous chapters in an important aspect: bidders do not have a precise idea of their valuations. This was obviously the case in the 1990s when we had the first generation of cell phones. This was a new technology, and it was difficult to make a clear prediction about how valuable the market would become. This uncertainty was legitimate, for history is full of examples of new technologies that did not catch on. For instance, neither the laser disc (the ancestor of the DVD, introduced at the end of the 1970s) nor the digital audio tape (introduced in 1987) managed to attract enough demand to be successful products.

When bidders are uncertain about their valuations, a classic exercise is to estimate a lower and an upper bound of the valuation, calculate the expected valuation, and use these estimates to determine one's bid. A problem can occur when the bid (and the price paid by the winner) will be above the lowest possible valuation. In this case, the bidder takes a risk, because the *realization* of the valuation (which may be known only after several years) can turn out to be *lower* than the price paid by the bidder. Paying for an object that turns out to be worth less than the price is known as the **winner's curse**. A milder version of the winner's curse is when the realized value is less than expected though still higher than the price paid by the bidder. In this case, the bidder's net gain is still positive, although it is less than the gain the bidder anticipated.

The winner's curse is particularly relevant with common value auctions. In such auctions, the value of the object is roughly the same for all bidders (but they do not know it; they only have an estimate of it when they bid). Since bidders make their estimates independently of each other, those estimates will differ. However, if there are enough bidders, then the average estimate is likely to be close to the correct value. Therefore the highest bidder necessarily overestimates the valuation of the object and thus pays more than the valuation. It is in that sense that the winner is *cursed*. The winner's curse is not confined to the problem of selling cell phone licenses. Auctions for oil drilling permits constitute a classic example of auctions with a winner's curse problem.

One way to mitigate the winner's curse is to have a dynamic auction such as an English auction. This could be an English outcry auction or an English auction where the identity of the active bidders is public throughout the bidding process. The intuition is the following. In such dynamic auctions, observing one's competitors' bids helps to refine one's valuation. Clearly, this learning process can only occur if bidders observe, at least partially, the bidding strategies of their opponents. This is why we specified that, ideally, bidders should be able to observe at any time the identities of the active bidders, should we run an English auction. By learning the other bidders' strategies, a bidder can update its estimates about the valuations of the other bidders and thus make a more precise estimate of its own valuation.

To sum up, in the presence of common values, when bidders have imperfect knowledge of their valuations, the winner's curse can be mitigated by having a dynamic auction where bidders can observe their opponents' bids.

6.3 The Simultaneous Ascending Auction

For a complex problem like the sale of cell phone licenses (i.e., selling many related objects), one of the most successful auction mechanisms that has been proposed is the **simultaneous ascending auction**. This auction was first proposed by Paul Milgrom, Robert Wilson, and Preston McAfee when they were designing the 1994 U.S. spectrum auction. Since then, this auction mechanism has been used in other situations, such as energy markets. There are many different implementations of the simultaneous ascending auction, but, roughly speaking, they all consist of running different ascending auctions in parallel.

The simultaneous ascending auction is particularly adapted to situations where the problem of price discovery (related to the winner's curse) is more severe than the exposure problem, although experience has shown that in many situations this auction mechanism does a sufficiently good job at mitigating the exposure problem. If, however, the items for sale are such that complementarities are strong (e.g., I want item A only if I also have item B), then a combinatorial auction is perhaps more suitable.

One can view the simultaneous ascending auction as a natural extension of the English auction when there are several goods. Note that here the "extension" refers to the way the auction proceeds, not the outcome of the auction (with respect to the optimal strategies and the outcome, the natural extension would be the VCG auction of chapter 4). One particular aspect of this auction is that each of the ascending auctions (one for each item) is linked in the sense that the auction ends when no bidder is willing to raise the bid on *any* of the items. So, if there are, say, two items for sale, A and B, the auction ends when bidding has stopped for both A and B. If after a while nobody bids on item B but the bidding process for item A is still going on, then the auction is not closed.

A clear advantage of having an ascending auction is that, as the bidding process goes on, bidders increase their knowledge about the valuations of the items and are thus better able to assess what the final prices would be. Put differently, the simultaneous ascending auction is clearly aimed at enhancing price discovery, minimizing the winner's curse. If there is a substantial risk of having the winner's curse, bidders tend to adopt cautious strategies, lowering their bids. Hence, a by-product of the enhanced price discovery is that, because the risk of the winner's curse is lowered, bidders are able to bid more aggressively, raising more revenue for the seller.

With respect to the exposure problem, the auction rules often allow bidders to withdraw their bids. This permits bidders to further refine their demands. For instance, a bidder who views two items as being complements (she wants both or none, but not just one of them) can withdraw her bid for one of the items if she sees that she will not be able to win the other item. Peter Cramton further argues that the price discovery property of this auction can also mitigate the exposure problem. By better learning the value of several items, a bidder can realize that some combinations are less valuable than expected (and thus feel less desire to bid on combinations of items).[4]

In an auction like the simultaneous ascending auction, bid withdrawal has to be severely constrained. Often, a bidder must pay a penalty when withdrawing her bid. The reason is the following. Since the auction is essentially designed to allow bidders to discover the true value of the items, a bidder may be tempted to first bid on several items to force the other bidders to bid (and thus reveal their valuations' estimates) and then withdraw her bid. This bidder would then learn the bids of the others while sending a mixed message to her opponents about her valuations. Such free-riding behavior is clearly a threat to the proper functioning of the auction and may even restrain the bidders from bidding because they have doubts about whether the opponents' bids are sincere or shill bids.

6.4 Case Studies

Since the 1990s, many countries have run spectrum auctions, and whenever there is a new generation of cell phone technology (2G, 3G, LTE, etc.), or some substantial range of spectrum has been freed, there is another opportunity to run an auction. While nowadays most countries use a simultaneous ascending auction, this has not always been the case. Also, since many details are left unspecified in the general description of this auction mechanism, there are also many versions of this auction, each one with a slight variation. Among all those auctions, some have been real successes, while others did not work as expected (but much has been learned from these failures).

6.4.1 The U.S. 1994 PCS Broadband Auction

It is difficult to discuss real-life application of the simultaneous ascending auction without mentioning the 1994 U.S. spectrum auction, the first of its kind. Without further ado, this auction was successful. It is in fact quite possible that this auction mechanism would not have been further studied, refined, and used had this auction been a failure.

6.4.1.1 Description In this auction, the Federal Communications Commission (FCC, the U.S. regulator in charge of wireless communications) sold 99 PCS licenses. There were at that time two competing technologies:

- *PCS (Personal Communication Service)*: Communications are done in the "1900 MHz band." Phones have low power, and the radius for transmission is about 1 mile.
- *Cellular*: Operates in the 700 MHz band. Phones use more power, and the communication radius is bigger (about 8 miles).

For the licenses, the country was divided into 48 regions (also called *Major Trading Areas*), offering two types of licenses for each region (called types A and B). For New York City, Los Angeles, and Washington, D.C., only one type of license was offered for sale (the other type having been awarded earlier). Through that, the FCC was selling auction *licenses*, with each license being assigned a block of spectrum. So the number of licenses was known in advance (the other possibility would have been selling blocks of spectrum, with an endogenous number of licenses—see section 6.2.1).

The auction lasted a bit more than three months, and the FCC raised $7 billion (adjusting for inflation, that is a bit more than $11 billion in 2016 dollars). There were initially 30 bidders, with 18 bidders winning at least one license.

6.4.1.2 Auction rules One key aspect in the design of the auction was the bidding process. The auction proceeded by rounds, and during each round bidders had to announce their bids. In order to avoid bidders staying idle and just observing the bids of their opponents (so as to learn their valuations), it was decided that in order to be eligible for a license a bidder had to remain *sufficiently active* in every round. A bidder's activity was defined as follows. In any given round, a bidder's activity level is the sum of

- the population coverage of the licenses for which she was bidding in the *previous round* and
- the population coverage of the licenses for which she was bidding in the *current round*.

For instance, suppose there are three regions, A, B, and C, and their population coverages are, say, 10 million, 1 million, and 3 million. In the previous round, I put in a bid for a license in regions A, B, and C, but in the current round I only bid for a license in regions A and B. Then my activity level is $(10+1+3)+(10+1) = 24$.

The auction was divided into different phases.

- *Phase I*. A bidder having an activity level equal to x cannot, in any subsequent round, reach an activity level higher than $3x$.

So a bidder having an activity level in some round of, say, 24 million could not place bids in any later round that would bring his activity level to more than $3 \times 24 = 72$ million.

- *Phase II*. A bidder having an activity level equal to x cannot, in any subsequent round, reach an activity level higher than $\frac{3}{2}x$.
- *Phase III*. A bidder having an activity level equal to x cannot, in any subsequent round, reach an activity level higher than x.

It is easy to see that putting an upper bound on the activity level is aimed precisely at limiting free-riding, observing the bids of one's opponent (and thus learning about their valuations) but not bidding too much in order to avoid revealing one's valuations. So a bidder bidding on only a few licenses at the beginning of the auction has a low activity level and could not bid on many licenses later on.

On top of controlling the activity level, the bids had to satisfy a minimum increment level. Also, it was decided to make this auction as transparent as possible. After each round, all the bids, as well as the identities of the bidders, were disclosed. This policy had a risk, as some small bidders could drop out quickly if they saw they would not be able to obtain a license, which would reduce competition during the auction. However, making the auction as transparent as possible also allowed bidders to have better knowledge about the values of the licenses, reducing the risk of the winner's curse.

6.4.1.3 How did it go? The first indicator of success is the revenue. With $7 billion, the total revenue exceeded industry and government estimates. There are reasons to believe this level of revenue resulted from the use of a simultaneous ascending auction. As we have explained, with this auction mechanism, bidders can refine their estimates about the values of the licenses, minimizing the risk of the winner's curse, which in turn allows bidders to bid more aggressively.

Economists often look at the efficiency to assess market mechanisms. For an auction, that would be whether the revenue was maximized, which entails having the licenses sold to the bidders with the highest valuations. An inefficiency would thus occur when a firm manages to obtain a license at a price lower than the market price. In the case of the 1994 spectrum auction, such an exercise is difficult because we do not observe firms' true valuations, but an analysis of postauction transactions permits us to perform an approximate efficiency test. Observing bidders reselling their licenses after the auction would suggest that the auction did not produce an efficient outcome. Since a bidder would only resell its license at a price higher than the price paid in the auction, observing resales would mean that the licenses were awarded at a price lower than what they are worth.

In the case of the 1994 PCS auction, there was no resale except for GTE (General Telephone & Electric Corporation, the ancestor of Verizon). This move by GTE is, fortunately for the auction designers and the FCC, not because of a lack of efficiency. GTE resold its license at the price it paid at the auction, and the sale was most probably the result of a corporate strategy shift toward cellular technology instead of PCS technology.

The analysis of the price paid for each license also sheds some light on the success of the auction. First, the auction generated what we can call *market prices*. By this we mean that similar licenses (e.g., licenses covering populations similar in size, geography, and other socioeconomic indicators) were sold at similar prices.

Second, bidders in the auction managed to buy adjacent bands in the spectrum. Between each spectrum block, there is a buffer band that is meant to reduce interference. For instance, my block could be, say, 1850–1870 MHz, and the next block could be, say, 1875–1995 MHz. The band between these two blocks, 1870–1875 MHz, is left unused so that there is no interference between my communications and those of the carrier that has the 1875–1995 MHz block. Buying licenses that have adjacent bands allows carriers to use the buffer between the blocks, thus increasing their capacities.

Third, bidders with nationwide interests managed to get a license in each region and to obtain the same band across regions. This is important because in spite of the presence of bidders with a nationwide interest, there was no nationwide license for sale. Those bidders had to *construct* a nationwide license on their own. There was some fear that in such a large auction that would not have been possible.

Finally, bidders managed to obtain licenses with local synergies. Bidders having almost secured a license for a region were bidding more aggressively for neighboring licenses, and often managed to obtain that second license. It is, for instance, more interesting to have one license for Illinois and another one for Wisconsin than one for Illinois and one for Arizona.

6.4.2 Mistakes

With the success of the 1994 U.S. spectrum auction, many countries adopted this auction mechanism. One issue is that the auction has to be tailored each time to fit the needs of a country's regulator and the characteristics of the market. For example, countries differ in size and population, so the "ideal" number of licenses may vary from one country to another.

A famous example of an auction failure is the 1999 German auction. Although the German regulator used a simultaneous ascending auction, several details made the auction go awry. For this auction, the spectrum was divided into ten blocks. However, contrary to the 1994 U.S. spectrum auction, the items for sale were not licenses but instead blocks of spectrum. It was decided that an operator should obtain at least two blocks to have a license. So that means that there could be as few as two licenses and a maximum of five licenses. The German regulator hoped that allowing carriers to have different numbers of licenses would permit the entry of both large and small operators.

There was in Germany at that time two large, dominant firms: Mannesmann and T-Mobil (a subsidiary of Deutsche Telekom). What is important to note is that both T-Mobil and Mannesmann had at that time a good idea of each other's valuations as well as the valuations of the smaller firms.

The auction opened at 10:15 a.m. on October 28, 1999, and minimum bids were set at DM1 million (about $500,000 in 1999, which is roughly $720,000 in 2016 dollars). Bidders had 30 minutes to post a bid, and for the subsequent rounds a bid for a block had to be at least 10% of the highest bid from the previous round.

At 10:23 a.m., 8 minutes after the beginning of the auction (i.e., 22 minutes before the first bids were due), Mannesmann submitted the following bids:

- DM 36.36 million for blocks 1 to 5.
- DM 40.00 million for blocks 6 to 9.
- DM 56.00 million for block 10 (a bigger block).

In the second round, T-Mobil submitted the following bids:

- DM 40.01 million for blocks 1 to 5.
- Nothing for blocks 6 to 10.

In the third round, there was no further bidding, and the auction ended. What happened is that Mannesmann figured out that the smaller bidders could not bid over DM 40 million, so any block priced at more than that would only be feasible for Mannesmann or T-Mobil. One can see that the "weird" bid of DM 36.36 million carries a message: 10% higher than that is DM 40 million. So Mannesmann's message to T-Mobil was essentially saying: "I take the blocks 6 to 10 and I leave you with the first 5 blocks." T-Mobil understood the message and bid accordingly. In the end, the revenue of the German auction was DM 416 million (about €200 million), a small number compared to the €55 billion that was raised a year later for the 3G spectrum auction.

The reasons why the German auction design was flawed right from the beginning are the following. First, the maximum number of licenses was just equal to the number of incumbents. This gives less incentive for entry and thus lowers competition during the auction. Second, the number of licenses was endogenous, which gave incentives for large bidders to outbid the smaller ones. By managing to have quickly only one bidder left for each item for sale, T-Mobil and Mannesmann managed to swiftly end the auction. Indeed, a simultaneous ascending auction works like an English auction: the auction stops as soon as the number of bidders is equal to the number of items for sale (provided each bidder is the top bidder for a license). The design of the auction thus gave a strong incentive for large bidders to coordinate.

The Dutch spectrum auction (July 2000) was another example of a bad design. For that auction, five licenses were put up for sale. The problem was that there were five incumbent operators, which gave very little incentive for new entrants to participate in the auction. To make the problem worse, the Dutch regulators authorized large operators from other markets to make deals with incumbents. In the end, there was only one small entrant (Versatel) that competed with the incumbents. The auction failed definitely when an incumbent (Telfort) threatened to sue Versatel if it continued to bid. The Dutch government chose not to fine that incumbent, essentially because it would have ended the auction and it was not clear how to substantially fine that incumbent. Once the number of bidders was equal to the number of licenses, the auction was virtually done before it had even started. In the end, the Dutch government only raised €3 billion, much less than it expected.

Having sufficient bidders is thus crucial for an auction mechanism to be successful. With too few bidders, especially when the number of bidders is equal to the number of licenses (or the number of licenses + 1), it is relatively easy for the biggest bidders to grab all the licenses at a modest price. One solution, proposed by Paul Klemperer, is to run a so-called **Anglo-Dutch auction**.[5] In this auction, we first run an English auction, but we stop as soon as there are $n + 1$ bidders left, where n is the number of licenses or items to be sold. This phase has the same purpose as the simultaneous ascending auction; that is, to facilitate price discovery and minimize the risk of the winner's curse. In the second phase (the "Dutch" part), we run a sealed-bid first-price auction where winners pay their bids and the reserve price is the price at which the first phase stopped.

In the case of the 2000 Dutch spectrum auction, there are reasons to believe that such a design would have worked better. With five bidders for five licenses, the auction would have proceeded right away with the sealed-bid first-price auction. The small bidder, Versatel, would likely have bid more aggressively than in the ascending auction the Dutch government ran. This is because in an ascending (outcry) auction, bidders always start with low bids, and by observing the bidding behavior of their opponents, small bidders may be discouraged from bidding further (they anticipate that they will lose the auction). In contrast, a sealed-bid first-price auction reveals less information to the bidders as to the valuations of their opponents, and thus a bidder is more likely to believe that she has one of the highest valuations. This argument can be extended to potential entrants: by giving small bidders more chances in a sealed-bid first-price auction, the auction could have attracted more bidders.

7 Financial Markets

7.1 Introduction

Financial markets rely heavily on auctions, despite their often being presented from an aggregate point of view, with a traditional demand and supply analysis. All of us have seen in the media about the price of a stock going up or down and changes in the interest rate of the bond of some country. Commentators generally focus on explaining the change in price, but most people have little knowledge about how these prices are determined in practice. The answer is very easy: through an auction! But the answer is even simpler: the auction format that is used is, in general, very simple. The main difference between financial markets and traditional auctions (such as eBay or Internet ads) is that in many instances financial markets have bidders on both sides of the market: bidders who want to sell and bidders who want to buy. Such auctions are called **double auctions**. In spite of this simplicity, there is still much to learn by studying such auctions, especially when considering, for instance, stock exchanges since the development of electronic trading.

We start this chapter with the simplest case of a double auction, the sale of Treasury bonds, and then we consider (slightly) more complex double auctions.

7.2 Treasury Auctions

Treasury bonds are loans sold by a government to finance its activities. They are initially sold by the Treasury (the "primary market"), but buyers can sell them later to other investors (the "secondary market"). We first outline the context of Treasury auctions and then describe how these auctions work.

7.2.1 Outline
There are several types of such loans, which depend on the "maturity," when payment by the government is due.

- **Treasury bills**: Their maturity is one year or less. Interest is paid at the end. Bills are usually sold at a discount from their face value. The full amount is paid back when the bill matures.
- **Treasury notes**: Their maturity is two to ten years. Interest is paid annually or semiannually.
- **Treasury bonds**: These are similar to Treasury notes, but with a maturity of more than ten years.

In the United States, about two-thirds of the nation's debt is held in Treasury bills, notes, and bonds. When the Treasury wants to raise some money (for a bond, bill, or note), there are two variables to be determined:

- the amount to be raised, and
- the interest offered to investors (the yield).

To raise money, the U.S. Treasury runs... an auction. This auction will determine the quantity to be sold (the amount being raised) and the "price" of the bonds, notes, or bills. Since the "items" being sold consist of money, the price is simply the interest being offered.

U.S. Treasury auctions have very clear rules, which among other things specify categories of buyers. There are two categories:

- **Primary dealers**: These are banks or securities broker-dealers that can trade directly with the Federal Reserve. They are required to participate actively in U.S. Treasury auctions. There are about 20 primary dealers.
- **Other buyers**: This includes any other institutions (e.g., pension funds, investment funds, insurance companies) or individuals (you or me).

When the Treasury wants to raise some money, a public announcement is made containing the following information:

- amount the Treasury is selling;
- auction date;
- issue date;
- original issue date (if a reopening);
- maturity date;
- terms and conditions of the contract;
- definition of customers eligible to participate;
- noncompetitive and competitive bidding closing times; and
- any additional information relevant for the investors.

Financial Markets

Table 7.1
Deadlines for bidding

	Bills	Notes, bonds
Noncompetitive bids	11:00 a.m.	12:00 p.m.
Competitive bids	11:30 a.m.	1:00 p.m.

Most of the information we just described is relatively standard and expected. What is "new" here is the notion of "competitive" and "noncompetitive" bids.

Noncompetitive bids are bids that just consist of a *quantity* to be bought. So a bidder submitting a noncompetitive bid is just telling how much money she wants to invest in a U.S. Treasury security, but she is not stating the price at which she wants to buy it. Noncompetitive bidders are usually small investors or individuals. They are given priority service, which means that noncompetitive bidders are guaranteed to receive securities for a value corresponding to their bid. Also, noncompetitive bids cannot exceed $5 million per auction.

So, for example, if a bidder submits a noncompetitive bid for a Treasury note of $2.5 million, that bidder knows before the auction ends that at the end of the auction she will hold Treasury notes that are worth $2.5 million. What that bidder does not know before the end of the auction is the interest rate of the notes.

Competitive bids are more complex, for they consist of a *quantity* to be bought (as for noncompetitive bids) but also an *interest rate*, the "price" at which the investor is willing to buy the security.[1]

The deadlines for submitting bids differ, depending on the type of security and the type of bid. Table 7.1 indicates the closing times (on the day of the auction).

7.2.2 How Treasury Auctions Work

Treasury auctions are different from the traditional auctions (e.g., English or Dutch auctions) because bidders' bids consist of two numbers: a price (the interest they ask) and a quantity (how much money they want to invest). But we have already seen auctions where bidders submit "complex" bids, such as the VCG auction.

In the VCG auction, bidders signal how many items they want simply by submitting a bid of zero on the combinations they do not want. In this case, we could do the same thing. For instance, a bidder who wants to invest, say, $2,000,000 with an interest rate of, say, 2.5% would be seen as a bidder submitting the following bid function:

$$\text{bid} = \begin{cases} 2.5\% & \text{if I get \$2,000,000} \\ \infty\% & \text{if I get any other positive amount} \end{cases}$$

If the bidder gets any other amount, then she requests to be paid an infinite interest rate. Clearly, the Treasury will never want to charge such an interest rate (when the auction is announced, there is a maximum interest rate the Treasury will agree to pay). So that bidder is sure never to be allocated an amount different from $2,000,000 or $0. Such a bid would be like a bid in a VCG auction: a price the bidder is willing "to pay" for any possible combination of items. Here the items are dollars, and the "price" is the interest the investor wants to charge. But doing so would be too complicated. It is easier to see things from the perspective of the seller: the Treasury. The Treasury has a specific amount of money it wants to raise, say, for instance, $1 billion. So the Treasury is simply looking for the interest it needs to pay so that there are enough buyers who together want $1 billion of securities.

Also, the bid we just described is not correct. When a bidder submits a bid, say $2,000,000 at 2.5%, the Treasury will understand that the bidder wants *up to* $2,000,000 of securities at a maximum rate of 2.5%. That is, if there is only $1,500,000 available, then that bidder will accept investing only $1,500,000 at an interest rate of 2.5%. Similarly, that bidder is also willing to accept investing $2,000,000 at any rate higher than 2.5%.

The Treasury proceeds as follows.

1. Subtract the total noncompetitive bids from the amount the Treasury seeks to raise. We do that because, as we stipulated earlier, noncompetitive bidders are guaranteed to be served.

2. The amount remaining is left for the competitive bidders. For them, we first rank them according to the interest rate asked by the bidders, starting with the lowest. Then we allocate to the competitive bidders the amount they want to get, starting with the bidder with the lowest interest rate. Note that each time we allocate some amount to a bidder, the amount that is left to be raised decreases and the interest rate requested increases. We do that until there is no money left to be allocated.

3. At some point, we will eventually reach a point where there is some amount left to be allocated but it is less than what is requested by the next bidders in our ranking who have not been allocated any amount. In this case, there is a tie. If there is only one bidder at the next interest rate, then that bidder gets only what is left. If there are several bidders requesting the same interest rate, then we divide the amount that is left to be assigned proportionally between those bidders, according to the amount they want to raise.

We now consider an example to illustrate how the auction works. Suppose that the Treasury is seeking to raise $11 billion. For our purposes, it does not matter if it is in bonds, bills, or notes. For this auction, the total amount of noncompetitive

Table 7.2
Competitive bids

Name	Yield	Amount
Bidder 1	2.998%	$3.5 billion
Bidder 2	2.999%	$2.5 billion
Bidder 3	3.000%	$3.0 billion
Bidder 4	3.000%	$5.0 billion
Bidder 5	3.001%	$2.0 billion
Bidder 6	3.002%	$1.0 billion

bids is $1 billion. So that means that the competitive offering, that is, the total amount that is left to competitive bidders, is $10 billion.

Suppose that the bids (that we already ordered according to the requested interest rate) are those given in table 7.2. The Treasury starts with bidder 1 and provides the requested amount, $3.5 billion. So there is $10 − $3.5 = $6.5 billion left to be allocated.

Next, we consider bidder 2, who requested $2.5 billion. This is less than what is available ($6.5 billion), so bidder 2 gets the amount she requested. So now there is only $6.5 − $2.5 = $4 billion to be allocated.

Now we have to consider the next bidder, bidder 3, whose bid is $3 billion at 3.000%. But bidder 3 is not the only one requesting a yield of 3.000%. In the Treasury auctions, bidders cannot be discriminated, so we need to consider all the bidders who bid a yield of 3.000%. That is, we need to consider bidder 3 and bidder 4.

The total amount requested by bidder 3 and bidder 4 is $3 billion + $5 billion = $8 billion. This is more than what is available ($4 billion), so there is a tie. As we explained earlier, those bidders will be allocated an amount that is proportional to what they asked.

The total amount asked by bidder 3 and bidder 4 is $8 billion. Out of those $8 billion, there are $3 billion asked by bidder 3, which is

$$\frac{3}{3+5} = \frac{3}{8} = 37.5\%.$$

So bidder 3 will get 37.5% of the remaining $4 billion, which is 0.375 × $4 billion = $1.5 billion.

As for bidder 4, his share is 3.125 + 1.125:

$$\frac{5}{3+5} = \frac{5}{8} = 62.5\%.$$

Table 7.3
Final outcome of the Treasury auction

Name	Yield	Amount asked (billions $)	Amount awarded (billions $)	Rate awarded
Bidder 1	2.998%	$3.5	$3.5	3.000%
Bidder 2	2.999%	$2.5	$2.5	3.000%
Bidder 3	3.000%	$3.0	$1.5	3.000%
Bidder 4	3.000%	$5.0	$2.5	3.000%
Bidder 5	3.001%	$2.0	$0	N/A
Bidder 6	3.002%	$1.0	$0	N/A

So bidder 4 will get 62.5% of the remaining $4 billion, which is 0.625 × $4 billion = $2.5 billion.

Now that bidder 3 and bidder 4 are allocated $1.5 and $2.5 billion there is $4 billion − $1.5 billion − $2.5 billion = $0 to be allocated. So the auction stops.

But what about the "price" of the auction? In other words, what about the interest rate (the yield) that the Treasury will pay to the investors? Are investors getting different yields? The answer is no. They all get the same yield/price. An auction where all (winning) bidders get the same price is called a **uniform-price auction**.

For an auction such as a Treasury auction, investors get the last yield that was considered. In our example, it is the yield corresponding to the bids of bidder 3 and bidder 4, 3.000%. This means that all the bidders who have been allocated some money will receive an interest rate of 3.000%. That rate is called the **drop-out rate**. This is also the interest rate that the noncompetitive bidders will get. The final outcome of our auction is given in table 7.3.

So in our example only bidders 1, 2, 3, and 4 managed to invest, and the interest paid by the Treasury, the stop-out rate, is 3.000%. This is the rate that is announced to the media.

7.2.3 Analysis

The Treasury auction works a little bit like a kth-price auction. In chapter 2, we saw the second-price auction, which is an auction in which the winner pays the *second*-highest bid. In a kth-price auction, the price paid by the winner (or the winners, if there are several of them) is the kth-highest bid. In our example, the auction turns out to be a *third*-price auction because the "price" (the yield) is the bid of the third-highest bidder (here bidder 3 and bidder 4, who turn out to bid the same yield).

Note that bidder 1 and bidder 2 are "happy" in the sense that they manage to invest at a higher interest rate than they were willing to accept (so the return on their investment will be higher than what they requested, 2.998% and 2.999%, respectively). In a second-price auction, or any kth-price auction (with $k > 1$), the same happens: the winner or winners end up paying a price that is lower than their bid.

So the Treasury auction looks like a kth-price auction. But there is a caveat. When we talk about the kth-price auction, we usually assume that the value of k is announced *before* the auction takes place. For instance, if we do a second-price auction, bidders know, before bidding, that the price paid by the winner will be the second-highest bid. In a Treasury auction, things are different because we cannot know in advance the value of k.

With the bids we considered in our example, $k = 3$. So we may want to say that it is a third-price auction. But had bidder 2 bid, say, $7 billion, then the drop-out rate would have been 2.999% (and bidder 2 would get awarded only $6.5 billion). In this case, it would be a second-price auction. Similarly, if bidder 4 would have bid a yield of 3.001%, then the drop-out rate would have been 3.001%, in which case the auction would have been a fourth-price auction.

Note that, theoretically, the auction could also end up being a first-price auction. This can happen if the highest bidder wants to invest at least as much as what is left after noncompetitive bidders are served ($10 billion). But again we cannot know in advance if it will turn out to be a first-price auction. Also, in practice it cannot happen, because the Treasury usually sets a maximum amount a single bidder can bid that is significantly lower than the total amount the Treasury wants to raise, so there are at least two competitive bidders.

A final comment is about the "behavior" of the noncompetitive bidders and that of the Treasury. Noncompetitive bidders only bid an amount; they do not indicate the yield they want. So their bid can be understood as an amount they want to invest at "whatever interest." In other words, noncompetitive bids consist of sending an *order* that says "I want to buy that amount," but since the price is not specified in the order, it is implicitly assumed that the price will be the price given by the *market*. Such bids (or orders) are called *market orders*. (We will study market orders in detail in chapter 8.) The same happens with the Treasury. It wants to raise some amount, but it is not asking for a specific yield. It is as if the Treasury is saying, "I want that amount of money, at whatever price (yield) the market will propose." Here again, the Treasury's request is similar to a market order: it will accept the price given by the market. In section 7.3, we will see what happens when the "bids" from both sides of the market specify a quantity and a price.

7.3 Double Auctions

Double auctions are auctions in which we have bidders from both sides of the market. They are used when we have several individuals or institutions that possess some good to sell and we also have several buyers who want to buy that good.

The goods or items to be sold through a double auction are commodities; that is, buyers do not care who the seller is.

In the Treasury auction, we saw that there were parties who could submit a "market order," a bid that only indicates the quantity they want to buy, with the implicit assumption that they will accept whatever price the market is proposing. Double auctions can also handle such bidders. In fact, the Treasury auction we saw in section 7.2.2 is a double auction, with the particularity that there is only one actor on the selling side, and it is using a market order.

In a double auction, a bid that is not a market order consists of a price and a quantity. We also need to distinguish between a bid from a buyer and a bid from a seller. To this end, the bid of a seller is usually called an *ask*, and that of a buyer is simply a *bid*.

For a seller, placing an ask of, say, price = 10 and quantity = 50 will mean that the seller is willing to sell *any* quantity *up to* 50 at a price of *at least* 10. If the seller manages to sell, say, 30 units at a price of 12, this is fine for her. Similarly, selling only 20 units at a price of 10 is fine. But any transaction where that seller is asked to sell 51 units or more will not be accepted, even if the price is well above 10.[2]

But the interpretation of the seller's bid has to be more precise. If the price is higher than 10, then our seller's supply is 50. However, if the price is *exactly* 10, then our seller's supply is any quantity between 0 and 50. The reason is the following. The price 10 is understood as the seller's valuation, also called the reservation value. That is, she values the good to be exchanged at 10. Recall that in an auction we defined the payoff of a bidder, if she wins the auction, to be *valuation − price*. For a seller, it is the reverse: *price − valuation*. So, for our seller, the net payoff per unit, if she sells, is *price* − 10. If she does not sell, then the payoff is 0. So, for any price higher than 10, the seller makes a positive net profit if she sells, and thus her supply is 50. However, if the price is equal to 10, then her net profit is zero and she is thus indifferent between selling and not selling. This entails that the supply for that seller is

$$\text{supply} = \begin{cases} 50 & \text{if } price > 10 \\ \text{any quantity between 0 and 50} & \text{if } price = 10 \\ 0 & \text{if } price < 10 \end{cases}$$

For a buyer, the interpretation of a bid is similar. If a buyer sends a bid of price = 12 and quantity = 30, then that buyer is willing to buy any quantity up to 30 units as long as the price is at most 12. If the price is more than 12, then the buyer does not make any transaction. The demand for that buyer is constructed like the supply of the seller. If the price is strictly less than 12, then the buyer wants to buy 30 units. At a price of 12, she is indifferent between buying any quantity

Financial Markets

Table 7.4
Bids

Trader	Bid	Quantity	Min cumulative demand	Max cumulative demand
Alice	market	4	4	4
Bob	$8	6	4	10
Carol	$6	4	10	14
Denis	$4	4	14	18

between 0 and 30, and for a price higher than 12, her demand is zero:

$$\text{demand} = \begin{cases} 30 & \text{if } \text{price} < 12 \\ \text{any quantity between 0 and 30} & \text{if } \text{price} = 12 \\ 0 & \text{if } \text{price} > 12 \end{cases}$$

We are now ready to see how the double auction works. It will be easier to explain it through an example. As we will see, it is very close to the traditional supply-demand analysis. Our objective in this auction will be to find an equilibrium price, a price where the number of units the buyer wants to buy is equal to the number of units the seller wants to sell. We consider in our example four buyers, Alice, Bob, Carol, and Denis. Their bids are described in table 7.4.

Here Alice sends a market order stating that she wants to buy 4 units at any price. Bob's bid is putting a price limit of $8. It is the maximum price Bob accepts to pay, and he wants at most 6 units. The bids of Carol and Denis can be described similarly.

The last column is the "cumulative demand." It starts with the highest price, ∞. Indeed, Alice sending a market order is tantamount to saying that she will accept *any* price. When the price goes down to $8, Bob's bid kicks in. At that price, the *maximum total demand* is 10: the 4 units that Alice wants to buy and the 6 units that Bob wants to buy. But since Bob's demand at that price is any quantity between 0 and 4, the *minimum total demand* is 4:4 + 0. So the total demand at a price of $8 is any number between 4 and 10.

If the price goes down to $6, Carol's demand is added. The total demand is now 14 = 4 for Alice + 6 for Bob + 4 for Carol. In fact, and this will be important, for a price of $6, the total demand is any quantity between 10 and 14. The difference, 4, is Carol's demand. Recall that at a price of $6 Carol is indeed indifferent between buying any quantity between 0 and 4. The asks for the sellers are given in table 7.5.

The cumulative demand (dotted lines) and supply (solid lines) are depicted in figure 7.1. For any price that is more than $8, only Alice wants to buy. So that

Table 7.5
Asks

Trader	Ask	Quantity	Min cumulative supply	Max cumulative supply
Gina	market	3	3	3
Henry	$4	5	3	8
John	$6	4	8	12
Mary	$10	4	12	16

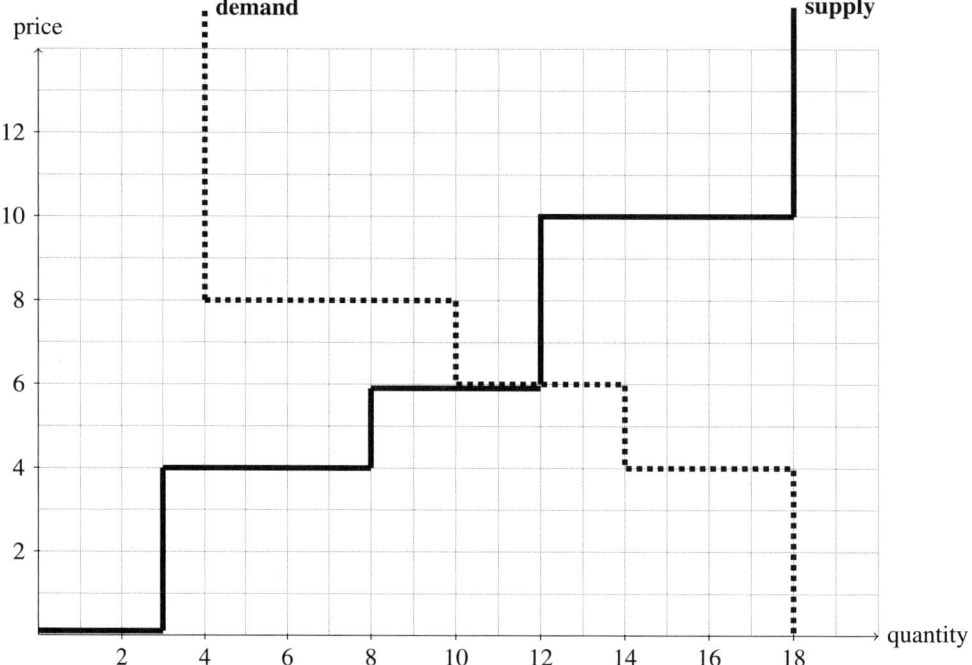

Figure 7.1.
Cumulative demand and supply

is why the demand goes to infinity. When the price drops to $8, the cumulative demand is any number between 4 and 10, so that is why the demand is horizontal. At a price of $4, the total demand is between 14 and 18 units. For any price below $4, the total demand is 18. So that is why the demand ends with a vertical line. The rest of the demand "curve" and the supply "curve" are constructed similarly.

From figure 7.1 we see that the demand and the supply cross at $6. So this will be the final price, which is the same for all individuals. That is, the double auction is a uniform-price auction.

We now have to determine which buyers are served and which sellers sell. Obviously, the buyers who want to buy at a price lower than $6 will not be served; in other words, Denis will not buy anything. Similarly, the sellers who want to sell at a higher price (here Mary) will not sell their units.

The allocation rules are the following:

- The buyers whose bids are strictly more than $6 are served. Here they are Alice and Bob. The total quantity they obtain is 10 (4 for Alice and 6 for Bob).
- The sellers who ask strictly less than $6 are served. Here they are Gina and Henry. The total quantity they sell is 8 (3 for Gina and 5 for Henry).

Now we calculate what happens to Carol and John. At $6, the total demand is 14 and the total supply is 12. Notice first that we have to take the smaller number of these two, so the "realized" demand will be only 12. This means that all the sellers, except Mary, will sell all their units, and the remaining buyer (Carol) will not have all the units she was willing to take: she was willing to buy up to 4 units, but she will receive only $12 - 10 = 2$, where 12 is the total supply at $6 and 10 is what Alice and Bob already obtained.

The double auction has a very nice property in that the surplus is maximized, where the surplus is simply the net payoff of each agent. That is, for a buyer, the surplus is his valuation minus the price he pays if he obtains the good, and for a seller it is the price she obtains (if she sells) minus her valuation.

It is now easy to see that the Treasury is in fact a double auction. It is a special case in that there is only one seller and that seller submits a market order.

7.4 Initial Public Offering

Most if not all companies start as *private companies*: the company's equities are privately held and not listed by some stock exchange. Shares of a private company can be sold, but in that case the seller can control who she is selling the shares to. Companies that do well often try to go *public*. In this case, the shares of the company (or a part thereof) are tradable on a stock exchange, and anyone can buy those stocks.

The process of going from private to public is known as an **initial public offering** (IPO) and basically consists of putting shares of the company up for sale. The money raised in this process goes directly to the company itself and to the current shareholders (often the founders of the company and early investors) who chose to sell all or a portion of their shares. The money going directly to the company (with the issuance of new shares) is often called the *primary offering*, and the sale of shares by current shareholders is called the *secondary offering*. Usually both the primary and the secondary offerings are part of the same process. An IPO is thus the opportunity for a company to raise (substantial) capital, increase and diversify the composition of its shareholders, or also reward existing employees who receive equity participation such as stock options.

7.4.1 Allocation and Pricing through Contracts

A common method for an IPO is to use one or more investment banks that work as intermediaries. Those banks, called *underwriters*, are in charge of selling the shares to investors and the retail clients of the underwriters. If you buy some shares offered at an IPO, it is likely that you actually buy those shares from your broker, who got them from one of the underwriters (or a client thereof). In many cases, the underwriters (together with the company's shareholders) set a price target at which the shares will be sold. That price can be firmly set before the underwriters approach investors or can be compiled during the negotiations between the underwriters and the investors.

One key question when considering an IPO is the price at which the stock should be sold. A price set too high brings the risk that not enough investors will buy the stock, and consequently the company may not raise sufficient funds (and early investors not be sufficiently rewarded). A price set too high will also make the price fall during the first days of trading. A famous example of a price set too high is Facebook's IPO. The IPO was held on May 18, 2012, and the offering price was $38. After the IPO, when Facebook shares started to be traded on stock exchanges, the price of a share dropped significantly, and by the end of August, three and a half months after the IPO, the price of the shares was just slightly above $18. In other words, Facebook was overpriced during its IPO.

Because of the uncertainty about the demand, shares are often underpriced during an IPO. One advantage of doing so is that it can generate interest in the company's stock. It is not rare to have some IPOs with a price so low that on the opening day of trading the share price increases significantly. An extreme case is the IPO of the social networking website theglobe.com in 1998. The share price for the IPO was set at $9, and on the first trading day the price spiked to $97 (before falling to $63.50 at closing).

7.4.2 Auctions for IPOs

As the examples of Facebook and theglobe.com make clear, determining the right price for an IPO is not an easy task, even though the stakes are high. For this reason, a number of companies opt for using an auction to discover the "right" price for their stocks. One of the pioneers in doing so is Google (though it was not the first).

The usual way to proceed in that case is to run a uniform-price auction like the Treasury auctions (see section 7.2.2 for the details). That is, the issuer (the company) announces the number of shares it wants to sell, and bidders propose a price and a quantity. Bids are ranked according to the price, and shares are sold starting with the highest bidders.

Running an auction for an IPO is an efficient price-discovery mechanism in the sense that it minimizes the risk that the price of the shares will drop during the first days of trading. This does not prevent the price from going up, however. As for the Treasury bonds or bills, an IPO is a situation where bidders' valuations are not private; that is, buyers' valuations are correlated. This makes sense because most investors are solely interested in the profitability of a company and the future prospects concerning the price of its stock. If the price of the IPO is higher than expected, investors might revise their beliefs about their valuations, which can in turn boost the demand during the first trading days of the newly public company.

Investment banks and other actors in financial markets often show resistance about the use of an auction for an IPO. One of the reasons is that an auction promotes *equal access*: only the bids matter when allocating shares during an IPO. In the traditional IPO method (i.e., not an auction), the allocation of shares is often the result of private negotiations between the underwriters and their clients, so that favorable treatments are easier to implement.

8 Trading

Most security markets use double auctions to determine the prices and quantities to be traded. That is, traders send orders to buy or sell securities, and a double auction is used to determine trades. There is a notable difference, however, in the way the auction is run. In the double auction model we saw in chapter 7, we assumed that all buyers and all sellers were present *at the same time* in the auction. In trading markets this is not the case, for quotations (i.e., the price determinations) are made throughout the day, and buyers' bids or asks are processed sequentially as they arrive at the exchange.

8.1 Stock Markets

To be precise, stock markets run a double auction each time a trader sends a buy or sell order. As in the double auction model, buyers submit bids and sellers submit asks. But they also specify the type of order they send, which somehow specifies *when* the order has to be executed. The reason is the following. Suppose that, for some reason, all traders submit their orders at the same time.[1] We will see in section 8.3 that this is impossible, but for now imagine it can happen. It could be that the supply and the demand do not meet; that is, there is no equilibrium price. In that case, no order will be executed. Note that, for this to happen, it must be that no trader sent a *market order*. Indeed, if one trader did so, say a seller, then there must be a buyer who could match that order.[2] We need to specify what to do next. Are the orders canceled? What if later there is another trader who submits an order that could match an order previously submitted? Since the traders are the "owners" of their orders, they have to tell the exchange (i.e., the institution running the auction, such as the NYSE or the NASDAQ) what to do with their orders. By default, exchanges will assume that the order "stays" until it can be matched with another order.

We can distinguish two types of orders:

- **Market orders**: Such orders are similar to the ones we have seen for double auctions. If it is a buy order, then it means that the buyer wants to buy a certain quantity at whatever price is possible. In this case, exchanges will choose the lowest price for the buyer. For instance, if there is a buy at the market order for 1 share of a stock and there are two selling orders, one at $10 and the other at $15, then the buyer will buy the security at $10.

If there is no match possible (in this case, it means there are no orders from the opposite side of the market), then the order remains valid until it can be executed (or is canceled by the trader who submitted that order).

- **Limit orders**: These orders are the "nonmarket" orders we saw in section 7.3. That is, they specify a maximum price if it is a bid or a minimum price if it is an ask, and a quantity.

If the order cannot be executed then that order remains valid until it can be executed (or is canceled by the trader who submitted that order).

Orders that are not executed or only partially executed are stored in the **limit order book**. So the highest bid in the limit order book is necessarily lower than the lowest ask. The difference between the lowest ask and the highest bid is known as the **bid-ask spread**. If an order arrives, say a bid, such that it is higher than the ask, then we say that the order **crosses the spread**. However, if the bid is below the lowest ask, then the order is stored in the limit order book.

What happens when the spread is crossed? If an order crosses the spread, then there is room for a transaction. The main question is at which price the transaction will be made. In general, the price is the one set by the order that arrived first. For example, suppose that the first order of the day is an ask, say for $10.00. There is no bid, so that order is stored in the limit order book. Later, a bid arrives for, say, $11.00. The bid is higher than the ask, so there is a transaction, and the price is $10.00. Similarly, if the first order were a bid for, say, $15.00, and later an ask arrives for, say, $14.00, then the transaction is made at $15.00. We say that the trader who placed an order that is stored in the book is called a **liquidity provider**, and a trader whose order crosses the spread is called a **liquidity taker**.

In practice, there are many types of orders, although all are derived from market orders or limit orders. First, note that a market order can be understood as a limit order when setting the bid to ∞ or the ask to 0. To see this, consider, for instance, a buyer sending a limit order with an infinite bid, so that order necessarily crosses the spread (assuming there is an ask in the limit order book). If the buyer had instead sent a limit order but at a price above the lowest bid, the effect would be the same: the order crosses the spread, and the transaction is performed.

Other common types of orders are:

- *immediate-or-cancel*: Orders have to be executed as soon as the exchange receives them. If they cannot be executed, they are immediately canceled.
- *stop-loss order*: Orders are triggered only when the price reaches a certain level.
- *good-till-canceled*: The order remains active (i.e., stays in the limit order book if not executed) until a certain date, after which the order is automatically canceled.

There are three questions that remain and that are extremely important for professional traders.

- *What if there are several possible transactions?*

Suppose the limit order book contains three orders, with the asks being $10.00, $11.00, and $12.00, and suppose that a buy, order arrives with a bid of $13.00. Clearly, the spread is crossed. But at what price will the transaction be done? The price will be the most favorable for the liquidity taker (here the buyer). So a seller who wants to maximize the probability of being part of a transaction should set a low price, and if it is a buyer, a high price should be set.

To understand why it is the liquidity taker that takes the best price, imagine a farmer's market where there are two stands selling, say, apples. One stand belongs to Bob and the other to Denis. Bob has six apples he wants to sell, and the price he asks is $12 per apple, whereas Denis has only three apples and asks $11. Then Erik arrives and wants to buy five apples and is willing to pay up to $14 per apple. Erik sees the two stands. Obviously, he will buy first the apples from Denis, because they are cheaper. Denis has only three apples to sell, so Erik buys them all. But he still needs two more apples. Then he goes to Bob and buys only two apples, at a price of $12.

- *What if the quantities in the bid and the ask do not match?*

In this case, the transaction is made at the lowest quantity. The order with the highest quantity is then "rewritten" with the same price (bid if it is a buy order, ask if it is a sell order), and the quantity is now the original quantity–the quantity exchanged.

Suppose, for instance, that there is an ask in the limit order book for $10.00, with a quantity of 500. (For simplicity, assume this is the only ask.) Suppose that there is a buy order arriving with a bid of $10.01, but the requested quantity is only 300. Clearly, the spread is crossed. So the buyer and the seller engage in a transaction, but the quantity exchanged is only 300. So now the buyer's order is fully executed but the seller's order is only partially executed. That order is updated and now becomes an order for selling at $10.00 but with a quantity of $500 - 300 = 200$.

- *What if two orders have the same price?*

In this case, priority is given to the *earliest* order. Suppose, for instance, that in the limit order book there are two orders, with their time of arrival given in the first column (for simplicity, assume these are the only asks).

Time	Trader	Action	Type	Price	Quantity
10:00 a.m.	A	sell	limit order	$20	3
10:01 a.m.	B	sell	limit order	$20	5

Suppose that at 10:02 a.m. a buy order arrives, for a quantity of 2, and the bid is $20.01. The orders from traders A and B compete with each other, but because A's order arrived first, it has priority. So trader A is served, while B is not. The buyer only wants 2 units, so A's order is only partially executed. So after this transaction the limit order book becomes the following.

Time	Trader	Action	Type	Price	Quantity
10:00 a.m.	A	sell	limit order	$20	1
10:01 a.m.	B	sell	limit order	$20	5

As we have just said, if an order can be executed, then it is always made at the most favorable price for the liquidity taker. We now consider an example to illustrate how this auction works. Since we will assume that traders will submit their orders at different times, we will have to specify when they are submitted. For simplicity, we will assume that there is no lag between the moment a trader submits an order and the moment it is received by the exchange (we will have more on this in section 8.3).

We start by assuming that no order has been submitted yet, so the limit order book is empty. At 10:00 a.m., Alice submits the first order of the day as follows.

Time	Trader	Action	Type	Price	Quantity
10:00 a.m.	Alice	buy	limit order	$10	4

So Alice wants to buy 4 units at a maximum price of $10. Since there are no other orders and her order is a limit order, it is stored in the limit order book.

Later, we have the following order.

Time	Trader	Action	Type	Price	Quantity
10:30 a.m.	Bob	sell	limit order	$12	6

The price asked by Bob (who wants to sell) is higher than what Alice is willing to pay. So Bob's order is not executed and is thus stored in the limit order book. At this moment, the bid-ask spread is $2.00, the difference between the lowest ask (from Bob) and the highest bid (from Alice). Next we have the following order.

Time	Trader	Action	Type	Price	Quantity
10:40 a.m.	Carol	sell	immediate-or-cancel	$14	5

Carol wants to sell, but her ask price is higher than the highest price proposed by a buyer, so her order is not executed. However, unlike Bob's order, Carol's order is not stored in the limit order book.

Next come the following orders.

Time	Trader	Action	Type	Price	Quantity
10:50 a.m.	Denis	sell	limit order	$11	3
11:00 a.m.	Erik	buy	limit order	$14	5

When Denis's order arrives, the highest price for a buyer is still Alice's, which is below Denis's ask price. So Denis's order is not executed. Ten minutes later, Erik's order arrives. He wants to buy and is requesting a price that can be matched by two sellers: Bob and Denis. If we were to construct the total demand (from Alice's and Erik's orders) and the supply (from Bob's and Denis's orders), they would meet.

As we said, orders are always executed at the most favorable price. But favorable for whom? For Erik, Denis, or Bob? As we explained, the transaction will be made for the price that is the most favorable for the liquidity taker, Erik.

Erik's order will be executed, buying 3 units from Denis at a price of $11 and 2 units from Bob at $12. In the end, Denis's order will be deleted from the limit order book, and Bob's order will be updated by changing the requested quantity from 6 to 4 (he sold 2 units to Erik). So at 11:00 a.m. the limit order book is as follows.

Time	Trader	Action	Type	Price	Quantity
10:00 a.m.	Alice	buy	limit order	$10	4
10:30 a.m.	Bob	sell	limit order	$12	4

Note that because not all orders arrive at the same time, the price will evolve throughout the day. The procedure we just described is a **continuous time double auction**. It is a double auction, but it is run each time a new order arrives. Once an order has arrived, it is either executed (and the orders involved in the execution are updated) or it is stored in the limit order book (if it is not an immediate-or-cancel order).

We can see that it is the trader who arrives "late" who has a surplus from the trade. In our example, it is Erik. He has a surplus of $14 − $11 = $3 for each of the first 3 units and of $14 − $12 = $2 for each of the next 2 units. Had Erik's order arrived just before Bob's, it would have been Bob who obtained a surplus.

So we may think that it is in the interest of a trader to send his order as late as possible. But this may not be the best strategy. To see this, suppose that Erik waits and instead sends his order at 11:30 a.m. It could be that there is another order arriving, say, at 11:15 a.m., which is a bid for 3 units at $12. This matches Denis's order perfectly. So at 11:16 a.m. there are only Alice's and Bob's orders in the limit order book. Then, at 11:30 a.m. Erik's order arrives. It matches Bob's order, but the total surplus for Erik is smaller than when he sends his order at 11:00 a.m.

Similarly, for Bob it would be better for him to send his order after Erik's. That way Bob would obtain a surplus. So traders have to play a complicated game: they not only have to determine what quantity they want to buy and at what price but also have to determine *when* they send their order. A seller wants to send her order *before* the other sellers (her order has more chances to be executed because it arrives earlier) but also wants to send her order *after* the buyers (because in that case the price would be the bid). A key feature that makes the game even more complex is that, while the limit order book is public (i.e., traders can see the orders it contains), traders cannot know which orders will arrive in the future. This is one of the reasons why traders in real life try to use sophisticated strategies to anticipate the orders that are about to arrive.

8.2 Opening and Closing

Today, most exchanges operate with electronic orders, and buyers and sellers meet via the limit order book. But that was not always the case. Before markets went electronic, they were run by "market dealers" or "specialists," real people receiving the orders from the buyers and sellers and facilitating the determination of the price (and taking a fee for doing so). Because they were run by people, exchanges obviously needed to establish an opening time and a closing time for each trading day. The usual time frame that was adopted was opening at 9:30 a.m. and closing at 4:00 p.m.

When exchange markets went electronic, the question of having closing and opening times became less important because computers do not sleep. Also, having electronic trading means that a trader in Japan can send orders to New York when it is nighttime in the United States. So, would it not be better to have the market operating 24 hours a day?

To answer that question, we need to acknowledge that having clear opening and closing times (with a market open roughly a quarter of a day: 6 hours and 30 minutes) has a huge effect beyond simply facilitating the lives of people working at an exchange: it makes the market "thick." Market thickness refers to the notion of having a market with sufficient buyers and sellers. This notion is vague because there is no exact formula to determine how thick a market should be. When a market is thick, there are enough people from both sides of the market to make transactions attractive, so it attracts buyers. The same happens if we want to buy, say, furniture at a flea market. Obviously, we will prefer to go to a large flea market with many sellers, as it will increase the likelihood that we will find a nice opportunity. The same occurs for sellers: they prefer to go to a market with many buyers.

Today, markets like the NASDAQ or the NYSE still have opening and closing times, but they allow trading before and after market hours. But there is a huge difference between market hours (9:30 a.m.– 4:00 p.m.) and the "premarket trading hours" and "aftermarket trading hours": there are many more orders and transactions during market hours than during premarket and aftermarket hours. The reason is the same as we outlined with our example of flea markets. The market hours play the role of a coordinating device that helps to maximize market thickness.

So, how are orders handled during premarket and aftermarket trading hours? There are no particular differences; exchanges use limit order books. But since there are fewer orders and active traders outside market hours, the problem of price discovery is more challenging, especially in the premarket trading hours. In a market, each trader will arrive with some belief about the value of a good or stock she wants to purchase. A crucial aspect of markets like stock markets is that participants will revise their beliefs by observing the orders of the other agents. But if activity is low, there is not much to learn.

Another issue that makes the problem even more difficult is that exchanges want to see a smooth opening. At 9:30 a.m., when markets open, there is suddenly a spike in the number of orders being sent. If there is much uncertainty (i.e., there is much to learn) about the value of a particular stock, traders will send and revise their orders frenetically, and we may end up with congested markets that are unable to operate (like a traffic jam on a highway). Also, if markets are not open before 9:30 a.m., it will be likely that we will have hundreds (if not thousands or more) of trades at different prices in a short time interval. This will create high volatility, something few people like.

Since there are precise market hours, we need a "price consolidation mechanism" that will help prices to converge toward a unique price. Ideally, this convergence should be achieved just before the market opens. The way to do that is to use a double auction, like the one we saw in section 7.3. During that auction, traders revise their bids, but no transactions are made. The only question is when we will end the auction.

Ending the auction right before the market opens is not a solution. It amounts to having a hard deadline. We have seen with the comparison of the eBay and Amazon auctions that with hard deadlines we end up with sniping (see section 3.4.1). In other words, hard deadlines lead to market manipulation. Another type of possible manipulation is the following. Consider a trader who would like to buy a stock at a low price, say $4. If the trader anticipates that the final price will converge to a much higher price, say $9, he may be discouraged and decide not to send his order. But there is a way out: discouraging the other traders. To do so, our trader sends two orders: a market order and an order at a high price (for instance, $15). This may discourage other buyers (who will anticipate that the final price will be high), and thus they may cancel their orders. At the last moment, our trader cancels the $15 order, which may result in a low price.

So we need to find a solution that can eliminate the scope for market manipulation. Clearly, most of the problems come from the presence of a deadline: at 9:30 a.m., the continuous time quotation starts. Obviously, this implies that a solution à la Amazon (i.e., extending the deadline if there is an order at the last moment) is not possible either: at 9:30 a.m., the market opens with the trading procedure we saw in section 8.1.

One solution is to have the end time of the double auction be random. This is, for instance, what the London Stock Exchange, the Tel Aviv Exchange, Euronext Amsterdam, and Xetra (Deutsche Börse), among others, are doing. When the end time of the auction is random, traders cannot manipulate the market (e.g., via sniping). Traders thus have fewer incentives to send fake orders (that they would later cancel), and thus we can expect that the price determined by the auction will be close to the "real" price.

The NASDAQ and NYSE do not use a random stopping time. The solution adopted by the NYSE consists of correcting the "market imbalance," which is done by considering a "reference price" for the security (a stock, a future, a bond, etc.). The reference price is the price that, given the buy and sell orders received so far, maximizes the number of trades. If there are several such prices, then the NYSE uses the price that is the closest to the price of the security at the closing time of the previous day. With that reference price, we look at the total demand and the total supply; that is, the total number of units buyers would like to buy at that price and the total number of units the seller would like to sell at that price. If the demand is higher, we have a *buy imbalance*, and if it is the supply that is higher, we have a *sell imbalance*. Before the market opens, the NYSE will disclose the market imbalance for each security. From 8:30 to 9:00 a.m., the balance is published every five minutes. Then from 9:00 to 9:20 a.m. it is published every minute, and from 9:20 until the opening (or sometimes before), the imbalance is published every 15 seconds. At 9:28 a.m., the NYSE also publishes (together with the imbalance) an indicative equilibrium price (like the one we calculated in section 7.3), so the strategy adopted by the NYSE consists of gradually disclosing information about market conditions, at an increasing frequency. That way trades can adjust, and the price obtained at the end of the auction, at 9:30 a.m., will be close to the price traders would expect from the continuous trading model (the one that uses the limit order book).

The solution adopted by NASDAQ, called the *opening cross*, is different. The purpose of NASDAQ's solution is to merge the double auction with the regular limit order book trading method. To this end, until 9:28 a.m., traders can submit, modify, or cancel their orders (traders can start sending their orders at 4:00 a.m.). NASDAQ then uses the orders it received to calculate a tentative price, using a 10% threshold in order to calculate the opening price (i.e., the price that will eventually be published when the market opens). For instance, if a buyer wants to buy at, say, $100 and a seller wants to sell at, say, $110, we first calculate the midpoint, which is in this case $105. Ten percent of this midpoint is $10.50, which is subtracted from the buyer's price (we obtain $89.50) and added to the seller's price (we obtain $120.50). We then have a price range of $89.50–$120.50, which indicates that the opening price will be between $89.50 and $120.50. This information is updated every five seconds. Since traders will modify their orders (until 9:28 a.m.), this price range will evolve, and in general the range will decrease over time. At 9:28 a.m., NASDAQ starts accepting only orders that correct the imbalance, and it publishes the imbalance information every five seconds. Finally, at 9:30 a.m., all pending orders are considered for execution as NASDAQ merges the *opening book* (the orders received for the preopening) and the *continuous book* (the order received for the regular trading hours).

8.3 High-Frequency Trading

In recent years, there has been growing interest (if not complaint) in *high-frequency trading*. High-frequency trading refers to the speed at which traders can make transactions or, more precisely, the speed at which traders can send (or cancel) orders. Sometimes people argued that high-frequency traders make several thousand (if not million) transactions per second. In fact, it is more about *how fast* orders reach the exchanges.

8.3.1 Market Structure

To understand what high frequency entails, we first need to understand how financial markets are structured. To keep things simple, we focus on the U.S. financial markets. There are in the United States two types of exchanges: the *national securities exchanges* and *dark pools*.

Table 8.1
U.S. national securities exchanges

	Exchange	Market share
NASDAQ group	NASDAQ (Q)	13.96%
	NASDAQ BX (B)	2.82%
	NASDAQ PSX (X)	0.93%
NYSE group	NYSE (N)	12.21%
	NYSE Arca (P)	9.76%
	NYSE MKT (A)	0.20%
BATS group	EDGX (K)	6.50%
	BATS BZX (Z)	6.02%
	BATS BYX (Y)	4.63%
	EDGA (J)	2.37%
	Chicago Stock Exchange —CHX (M)	0.38%
	IEX (V)	2.14%

Source: http://www.bats.com/us/equities/market_statistics/.

The national securities exchanges contain the most famous exchanges. A list of them is maintained by the U.S. Securities and Exchange Commission (SEC). The list as of March 2017 is given in table 8.1, together with the market shares of each exchange. We can see that several exchanges belong to the same companies.

The official exchanges listed in table 8.1 must follow very strict regulations set by the SEC about how trades are performed. But these exchanges are not the only ones where investors can buy and sell securities. Apart from these exchanges, there are also exchanges that are less regulated, which are often referred to as "dark pools." Nonofficial exchanges that operate in a way similar to the official exchanges (i.e., with orders being sent electronically) are known as *electronic communication networks* (ECNs).[3] Most recent exchanges, such as Bats or IEX, used to be ECNs before becoming national security exchanges. Exchanges are considered private venues where investors can buy and sell securities. For the general public, there is not much difference between an official exchange and a dark pool except that exchanges (and orders) in a dark pool are kept private. It is worth noting, however, that dark pools were pioneers in the transition to electronic trading. The first electronic dark pools appeared in the late 1980s to early 1990s, while it was only 10 or 15 years later that the official exchanges switched to entirely electronic trading.

The NASDAQ and NYSE switched to electronic trading in the middle of the last decade, and most of today's trading at the NYSE no longer happens on Wall Street in Manhattan, New York. The New York Stock Exchange (NYSE) is in fact located in Mahwah, NJ, and is simply a giant data center.[4] The transition to fully electronic markets permitted investors to trade securities on different exchanges. For instance, Facebook is listed on NASDAQ (the Nasdaq Stock Market), but an investor can sell or buy Facebook shares on the New York Stock Exchange, on the Chicago Stock Exchange, or any other exchange (including dark pools). All that is needed is to have someone from the opposite side of the market with whom our investor can make a transaction. Investors will generally choose a venue based on the conditions offered by the different exchanges. For instance, some exchanges require the liquidity taker (Erik in our example in section 8.1) to pay a fee for each transaction, while other exchanges ask the liquidity provider to pay the fee. How many sell or buy orders are in the limit order book (called the *depth* of the book) may also influence the choice of where to send an order.

Trading the same security (e.g., a Facebook share) at different places can create a problem (or an opportunity) in that the price of the security can vary from one place to another. For instance, a Facebook share can be at $115.03 on NASDAQ, while in Chicago the price of a Facebook share is $115.84. In finance, having two different prices for the same item is called an *arbitrage opportunity*. It is easy to see that an investor can easily make some profit *with no risk* when there is an opportunity of arbitrage. In our case, an investor can buy Facebook shares on NASDAQ and resell them in Chicago, making a profit of $115.84 − $115.03 = $0.81 per share. It may not be a lot, but if multiplied by thousands (or millions) of shares it can represent a substantial amount of money.

8.3.2 Market Regulation

To prevent such arbitrage opportunities, along with the switch to electronic trading the SEC implemented a new regulation in 2005–2007, the **Regulation National Market System**, often referred to as the "Reg NMS" by professionals. The new regulation was designed, as the SEC puts it, to implement "a series of initiatives designed to modernize and strengthen the National Market System for equity securities."

One of the main purposes of the Reg NMS is to ensure that investors are treated in a fair way across the exchanges, and one of the ways to achieve this goal is by encouraging competition between exchanges. As we have already commented, the price of a security may vary across exchanges. In order to avoid heterogenous prices, the Reg NMS requires exchanges to "share" the bids and asks they received so as to constitute the **national best bid and offer** (NBBO). The NBBO lists, for each security, the lowest ask among all exchanges (the "best offer") and the highest bid (the "best bid"). If an exchange is about to execute a trade, say, with a seller being a liquidity taker, then the exchange must check whether the bid for that trade it has in its book is at least as high as the best bid in the NBBO. If the best bid in the NBBO is higher (and thus corresponds to a bid at another exchange), then the exchange has to *reroute* the seller's order to the exchange that has the highest bid. That way, buyers and sellers are assured of obtaining the best deal for their quote.

Example 8.1   NASDAQ has in its limit order book a bid for 10 shares of company XYZ at a price of $10.

The current highest bid is at Bats, for 20 shares at a price of $10.05.

NASDAQ just received a sell order for 5 shares of XYZ at a price of $9.95, so that order crosses the spread. If NASDAQ fills this order, the seller will sell its 5 shares at a price of $9.95. But since this is not the best national bid, NASDAQ has to send the order to Bats, where the seller will sell its 5 shares at a price of $10.05.

There is no doubt that the regulation is well intentioned. The seller in example 8.1 is clearly better off with his order being rerouted. The problem is that this regulation makes sense when there is no latency between exchanges; that is, when communication between the exchanges (NASDAQ and Bats in the example) is *instantaneous*. By instantaneous we mean here that it takes *zero* seconds:

- for NASDAQ to check whether the highest bid it has is the national highest bid;
- for NASDAQ to send the order to Bats; and
- for Bats to receive and process the order.

When we write zero seconds, we mean 0 seconds. Not 0.001 seconds, 0.000001 seconds, or 0.00000000000001 seconds. Even though computers and optical fibers (or microwave dishes) are fast, a 0 second delay is physically impossible to achieve. This creates a problem because by the time the order to sell 5 shares of XYZ has arrived at Bats, the price may have changed (even in the NBBO), and our investor may end up selling its shares at a price lower than the initial bid on NASDAQ. To avoid such troubles, many exchanges offer traders variations of the limit order that prevent orders from being rerouted. A famous type of order that achieves this is the so-called *hide not slide* order introduced by Direct Edge. The Reg NMS rule that says that orders must be rerouted if the investor can get a better deal only applies to *displayed* orders, orders that are in the limit order book and that are shown to all traders and investors. The hide not slide order is simply an order that is not displayed; it is kept secret by the exchange. So only the trader who sent that order and the exchange that received it know the order exists. The "not slide" part is simply telling that the exchange must not reroute that order.

Another complication, which could be (theoretically) avoided, is that high-frequency traders usually have much faster access to the exchanges than the exchanges have between themselves, and with this access high-frequency traders can *anticipate* the bids and asks displayed in the NBBO before it is actually published.

8.3.3 Surfing on the Latency

When markets are fluid, meaning that market participants can easily place orders and there are sufficient buyers and sellers, arbitrage opportunities quickly disappear. In our example, buying shares on NASDAQ will increase the price of the Facebook shares, and selling those shares in Chicago will decrease the price displayed in Chicago. Eventually the two prices will converge to the same unique price. Since the prices between the two exchanges will eventually converge, the investors who are the fastest are the ones who reap the most benefits from an arbitrage opportunity. This is precisely what high-frequency traders are after: being the first to arrive.

To understand how speed is pricey for traders (and to get an idea of the magnitude of the lag time they are considering), the story of Spread Networks is an interesting one. In 2010, Spread Networks completed the construction of a communication line between Chicago and New York. The line was remarkable in that it was a *line*: the cable put in place by Spread Networks was nearly a straight line. Before that, cables between Chicago and New York were zigzagging along railroads, mountains, rivers, and suburbs. Having a straight line reduced the time it took for a round trip between New York and Chicago from 16 milliseconds to 13 milliseconds. To get an idea of how fast this is, it is helpful to compare that to the blink of an eye, which takes about 400 milliseconds. Spread Networks limited the number of firms that could use its cable. This created scarcity and allowed Spread Networks to charge several million dollars per year to whoever would use their cable. Trading firms had no choice but to pay for access to the cable. Indeed, the potential loss of being late was far larger than the benefit of using the cable. The estimated cost of Spread Networks's cable is about $300 million. The irony of the story is that a few years later other companies installed a network of microwave dishes, which allow even faster communication, pushing the limit to slightly above 8 milliseconds.

To see how arbitrage opportunities can easily arise in everyday markets, we report an analysis made by Eric Budish, Peter Cramton, and John Shim.[5] Their study focused on the prices of the two largest financial instruments that track the S&P 500 index, namely the "SPDR S&P 500 exchange traded fund" (henceforth SPY, its ticker symbol) and the "S&P 500 E-Mini futures contract" (henceforth ES).[6] Since both securities track the same index (i.e., a portofolio of stocks), one would

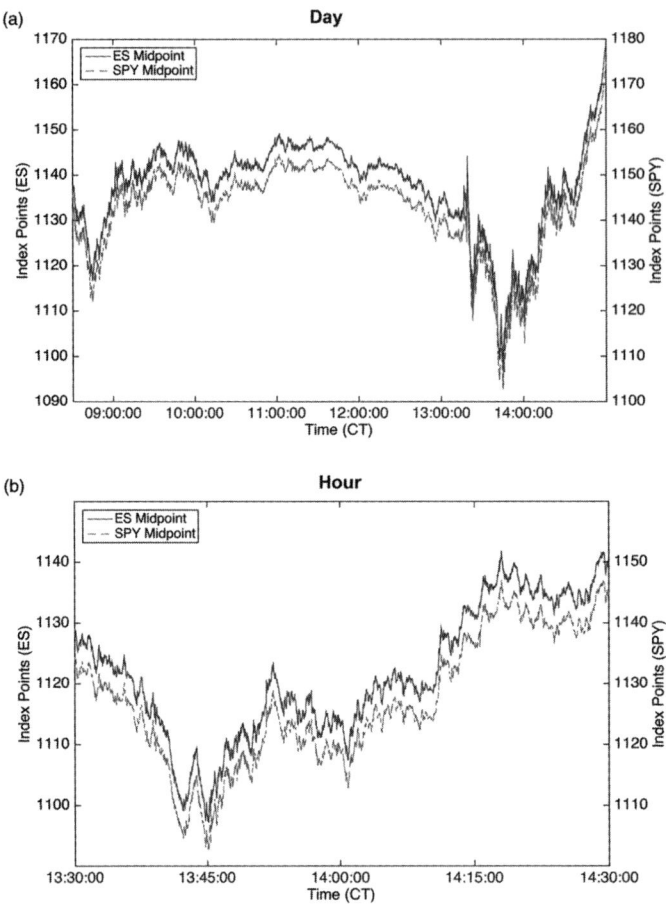

Figure 8.1.
ES and SPY indices—day and hour horizon. Source: Budish, Cramton, and Shim (2015).

expect them to be highly correlated. Figure 8.1 shows the evolution of the prices of these two securities during a typical trading day in 2011. We can see that the two graphs are very close. For sufficiently long time intervals, the correlation between these two securities is close to 1.[7]

However, if we go to a smaller time frame, the correlation between the two securities starts to drop. We can see that in figure 8.2. For a 10 millisecond interval, Budish and his coauthors calculate that the correlation is only 0.1016.

The fact that at very short intervals of time the correlation between these two instruments disappears is inevitable because SPY is traded in New York and ES is traded in Chicago, one of the main exchanges for futures. Since communications

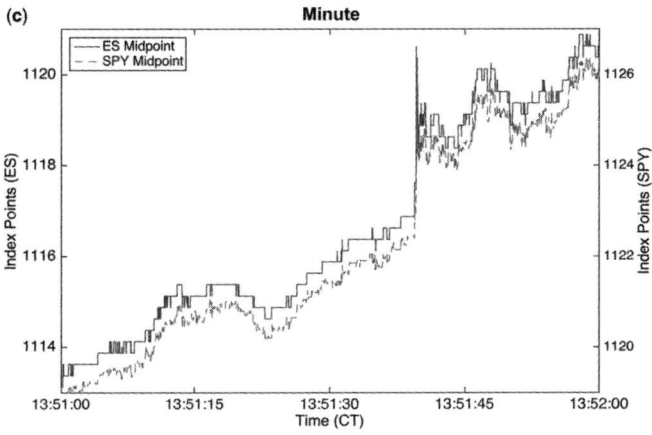

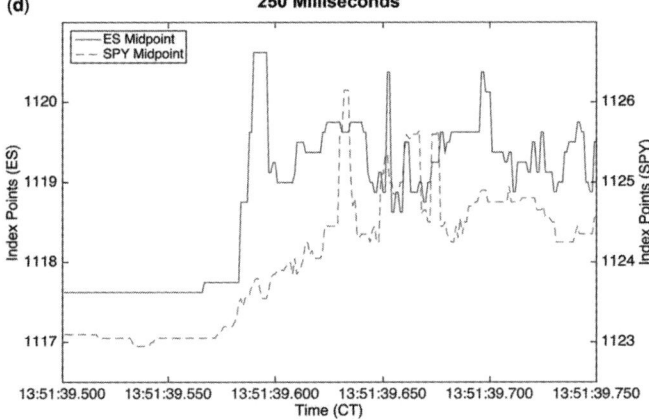

Figure 8.2.
ES and SPY indices—minute and 250 milliseconds horizon. Source: Budish, Cramton, and Shim (2015).

between these two cities *cannot* be instantaneous, anything that happens in one of those cities will necessarily take time before it gets noticed in the other city.

8.3.4 What Is the Matter with High Frequencies?

The race for speed is a direct consequence of how the market works, processing orders continuously. In section 8.1, we saw that orders are processed when they arrive; the exchange does not wait for more orders to arrive. So if traders "know" in advance that the price of a security will move up or down, they may benefit from that knowledge by sending appropriate orders before the others. High-frequency trading is just about this: processing orders as fast as possible.

It is often argued that the rise of high-frequency trading has allowed traders to make substantial profits (at the expense of investors). In section 8.3.3, we saw that the correlation between two instruments (ES and SPY) disappears at high enough frequencies. It is this lack of correlation that creates what the finance literature calls arbitrage opportunities, the same item being priced differently in two different markets (or in two different ways). Arbitrageurs exploit the price difference by buying the good at the low price and selling it at the high price. Clearly, arbitrage opportunities do not last, because the increase in demand (where the price is low) will increase the price, and similarly the increase in supply (where the price is high) will decrease the price. Eventually the two prices will converge, and the arbitrage opportunity will disappear.

Trading at higher and higher frequencies should then increase the number of arbitrage opportunities that traders can spot. It also gives traders "more time" to exploit these arbitrage opportunities. If my reaction time is 5 milliseconds, I can do more things in, say, 1 second than someone whose reaction time is only 20 milliseconds. Therefore, if the increase in the trading frequency is synonymous with higher rent for traders, the volume (in value) of arbitrage should be higher today than it was when trading was slower.

Budish and his coauthors used their data first to identify the arbitrage opportunities caused by the correlation breakdown. Their findings are reported in figure 8.3. We can see that over the years the median duration of an arbitrage fell sharply, from 97 milliseconds in 2005 to 7 milliseconds in 2011. This is not surprising given that during this period traders' speeds have increased. Budish, Cramton, and Shin report, for instance, that in 2005 most arbitrage opportunities they identified in their data lasted at least 50 milliseconds. Six years later, almost all arbitrage opportunities lasted less than 10 milliseconds. The fact that traders manage to trade at higher and higher frequencies clearly has an impact.

The shorter life span of arbitrage opportunities does not necessarily mean higher profits for high-frequency traders. That is, we have not yet shown that higher trading frequency is harmful for investors. The second step of the exercise performed by Budish and his coauthors consists precisely of measuring the profits generated by the arbitrage opportunities. A comparison of those profits is given in figure 8.4. Surprisingly, the authors find that profits are relatively stable; that is, they do not depend on the increase in trading frequency. Panel (a) in figure 8.4 shows that the median profit generated by an arbitrage opportunity does not change over time (with the exception of "unusual" events such as the 2008 financial crisis). Not all arbitrage opportunities yield the same profit. That depends on their duration but also on the price differences (or lack of correlation) between ES in Chicago and SPY in New York. Panel (b) in figure 8.4 shows for each year the distribution of profits from arbitrage. Again, we can see that the distribution is relatively constant over the years.

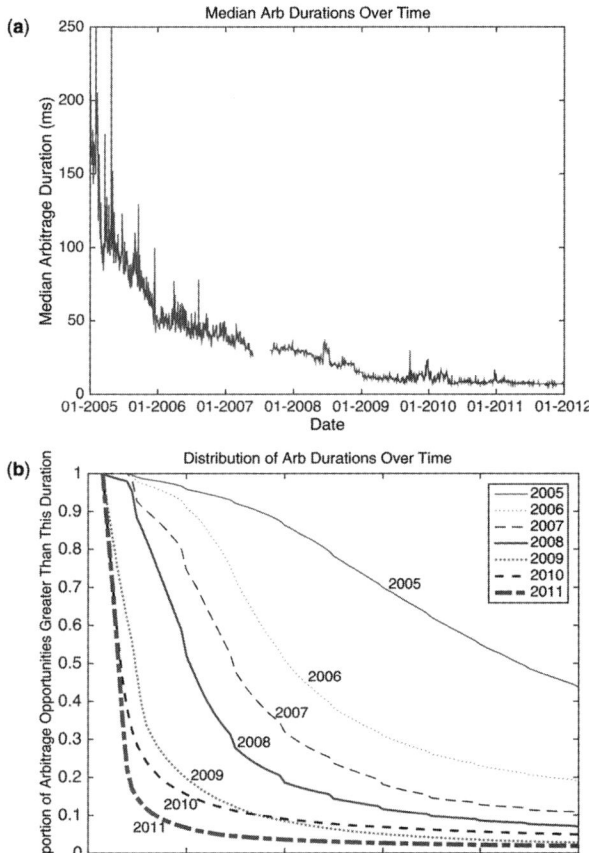

Figure 8.3.
Arbitrage duration. Source: Budish, Cramton, and Shim (2015).

The analysis of arbitrage opportunities performed by Budish, Cramton, and Shim suggests that the increase in trading frequency, however impressive, has not changed the size of the profits made by high-frequency traders. Put differently, a speed increase does not create more revenue for traders; the profit is just here to be grabbed, and it will be for whoever is the fastest.

8.3.5 A Flawed Market Design?

The fact that the same security can be bought or sold at different places only exacerbates the arms race. It gives an edge to the trader that is the fastest, because the orders necessarily take time to reach an exchange. But even if communication could be done simultaneously, we would still have a problem. To see this, we follow here the argument developed by Budish, Cramton, and Shim. Suppose

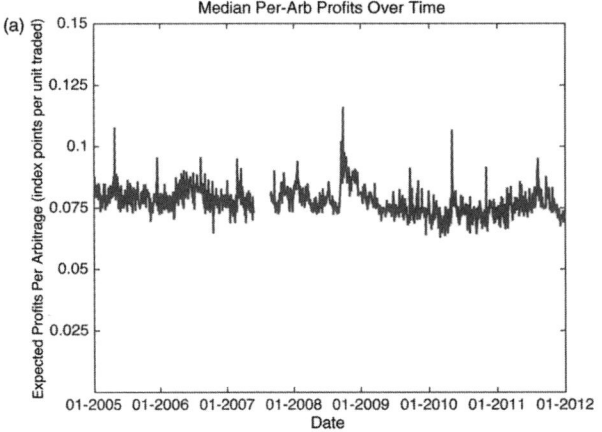

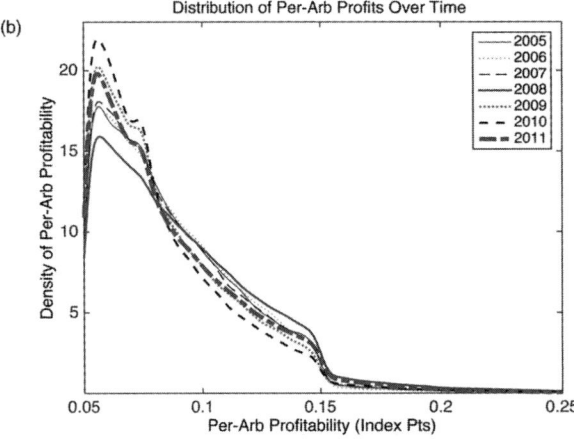

Figure 8.4.
Arbitrage profits. Source: Budish, Cramton, and Shim (2015).

there are four investors, *A*, *B*, *C*, and *D*, who want to sell, say, SPY, and suppose that the current price is $200.00. So they send orders to sell at $200.00. Suppose now that the price of ES jumps. Since ES and SPY are supposedly perfectly correlated, our investors will deduce that the price of SPY will also move up, say to $201.00. They will therefore cancel their sell orders and replace them with new orders to sell at $201.00. But a high-frequency trader can snipe and buy SPY from these investors before they have time to cancel their orders.

The point made by Budish and his coauthors is that *even* if there was no lag, and all investors and traders had the same simultaneous access to markets, we would still have the possibility of sniping. Since the continuous limit order book processes orders one at a time, the processing between our investors and the high-frequency trader when all of them send their orders at the same time will be the random order. With four investors and one trader, there are $5! = 120$ different orders of arrival. For each investor, 50% of the time, the canceled order + new order will be ranked before the high-frequency trader's, and 50% of the time it will be ranked after. So there is a probability of 1/2 of being sniped by the high-frequency trader. If there are more high-frequency traders, say n, an investor will be sniped with a probability of $\frac{n-1}{n}$.

The conclusion from this observation is that the arbitrage opportunities come not only from the delay in communications between traders and exchanges but are also "built in." The very idea of having a limit order book with orders being processed serially as they arrive creates sniping opportunities. These arbitrage opportunities are of course a source of profit for high-frequency traders, but they represent a cost for investors; that is, they constitute friction added to the market.

8.4 Alternative Market Designs

8.4.1 Slowing Down Markets?

Until June 17, 2016, there were only 12 official national security exchanges in the United States. Then that number increased to 13, and in February 2017 it went down to 12 again. There was one new exchange, and one exchange, the National Stock Exchange (NSX), ceased trading operations. More interesting is the new exchange that entered the list, Investor Exchange (IEX), and the way it operates differs substantially from that of the 11 other exchanges. The birth of IEX was largely documented in Michael Lewis's 2014 best seller *Flash Boys: A Wall Street Revolt*.

IEX is an exchange that has been specially designed to protect regular investors against high-frequency traders. To this end, IEX simply slows downs all the orders it receives before feeding them to the *matching engine*, the program that runs the double auction between the bids and the asks.

Traditional exchanges like Bats or NASDAQ do not slow down the orders like IEX does but simply execute the orders as soon as they arrive. Since not all exchanges are located at the same place, there are inevitable delays that result from observing different prices at the same time across the exchanges. Also, investors buying or selling a large number of securities typically split their orders into several smaller orders that they send to different exchanges. For instance, instead of buying 1000 shares of company XYZ on NASDAQ, an investor would buy 200 on NASDAQ, 200 on the NYSE, 200 on Bats, and so on. Since some traders may have slower access to some exchanges than the so-called high-frequency traders (i.e., traders who have very fast reaction times) who invested in fast connections, the price at which a "regular" investor buys or sells a security may differ between the various exchanges to which she sent the orders. This is how, for instance, we can have high-frequency traders sniping regular (slower) investors (see the discussion in section 8.3.5). Those price differences are part of the profit made by high-frequency traders. IEX's solution is aimed precisely at killing the possibility of sniping other traders. To this end, any order arriving at IEX must first pass through a 38 mile optical fiber before being processed, and any signal needs 350 microseconds to complete those 38 miles. Suppose that my connection on my trading terminal (or the computer running my trading algorithm) is such that an order I send to IEX needs 7 microseconds to arrive at IEX and also 7 microseconds to arrive at, say, Edge. If I send an order at 10:00:00.000 a.m. (i.e., 10:00 a.m. and 0 milliseconds), then we have the following.

Order sent to	Order arrives	Order executed
Edge	10:00:00.000007	10:00:00.000007
IEX	10:00:00.000007	10:00:00.000357

Between 10:00:00.000007 a.m. and 10:00:00.000357 a.m., the execution times of my orders sent to Edge and IEX, we have the 350 additional microseconds added by IEX. That is almost nothing for a human, but for computerized trading that is a lot of time, and many things can happen.[8] In particular, compared to the orders and prices at other exchanges, any order arriving at IEX is necessarily late. So strategies that try to play with the arbitrage opportunities between SPY in New York and ES in Chicago that we saw in section 8.3.3 *cannot* work with IEX. IEX believes that with their unique design they will attract "real" investors, those investors who sell and buy securities because they are more interested in the value of those securities and not because they are simply exploiting the details of the market's procedures.

When IEX receives an order, it is fed to its matching engine only 350 microseconds later. What does IEX do in the meantime? Because of its location in New Jersey, not so far from most other exchanges, IEX can obtain information from those other exchanges in about 200 microseconds. So, when an order arrives, IEX has the time to update the information it has about prices and give the investor a better deal.

To see this, consider an investor who wants to sell SPY because that investor anticipates the price will increase. But because our investor is not sure about what the new price would be, she just sends a market order to sell. If our investor sends that market order to an exchange other than IEX, she can be outpaced by a faster trader, who will buy from our investor and resell shortly thereafter once the price of SPY has gone up. So the benefits of the price increase accrue to the fast trader and not our investor. If that investor instead sends her order to IEX, that order will be executed after some delay, and it will typically be executed *once* the price of SPY has gone up. So now the benefit of the price increase accrues to our investor.

If instead the price goes down, the same happens. If the price goes down, the fast trader can send an order that arrives before our investor's order (sent to an exchange other than IEX), but that fast trader will sell SPY at the current price (i.e., the price before SPY's price drops) and then buy from our investor once the price has gone down. For instance, say the current price is $200 and it is anticipated to drop. So the fast trader sells, say, 10 shares of SPY at $200, just before the sell order of the investor arrives. Then the price drops, the investor's order arrives, and the fast trader buys 10 shares from the investor at a price of, say, $199.90. When orders are delayed, IEX claims that such sniping strategies are more difficult (if not impossible) to implement, thus protecting the investor (if the investor is selling, she will be able to sell at a higher price, and if she is buying, she will be buying at a lower price).

8.4.2 Frequent Batch Auctions

Following their analysis, Budish, Cramton, and Shim proposed a radical change in the way markets operate: instead of using the so-called continuous limit order book, exchanges should run double auctions at frequent time intervals. The basic argument is the following. The limit order book is a market design that is adapted for when time is discrete, and this is at odds with a market operating in continuous time. As the analysis of arbitrage made in section 8.3.4 has shown, a speed advantage has value. So any trader would like to be faster than the others. If time is discrete, this is more difficult to achieve, though. In a discrete, time environment, markets process the orders at a fixed moment set in advance. In that case, there is no need or advantage for one trader to arrive before another.

More precisely, the frequent batch auction proceeds as follows. The trading day is divided into a finite number of equal-length intervals, which are called **batch intervals**. Traders can submit orders at any time, but they are processed only at the end of the batch interval. At the end of each batch interval, the exchange runs a double auction similar to the one we saw in chapter 7. This auction is a uniform-price auction; the price is the same for all agents. Orders that could not be executed in the double auction are carried over to the next batch interval (unless they are canceled by the traders who submitted them).

The frequent batch auction is assumed to be a sealed-bid auction. That is, during a batch interval, traders cannot observe the orders that are submitted by the other traders (or whether outstanding orders are canceled). It is only once the auction is over that traders are informed of all the orders that were submitted during the batch interval.

When running the double auction, the bid priority is similar to the one used for the limit order book: orders are first ranked according to the price and then according to the time at which they were submitted (with the earliest orders having the higher priority). One key difference with the continuous limit order book is that two orders arriving *within* the same batch interval are treated as having an identical time priority.

It is not difficult to see that frequent batch auctions can greatly reduce the advantage that fast traders have over slower traders. When the market uses the continuous limit order book, speed has a value at *any time*. To see this, consider again the case of ES and SPY we saw in section 8.3.3. With the current market design, *whenever* there is a lag in the pricing of these two instruments, a fast trader can snipe stale quotes. Frequent batch auctions do not completely eliminate the possibility that fast traders can snipe other traders, but for *most of the time* sniping is virtually impossible. Indeed, sniping is feasible only when the change of the price of, say, ES occurs just before the end of the batch interval. If the price change occurs at, say, the beginning of the batch interval, then all traders have enough time to readjust their orders and sniping cannot occur.

There is another, more subtle channel through which frequent batch auctions protect slow traders against predatory snipers. If investors are significantly slower than high-frequency traders, it is then of course relatively easy for the latter to snipe stale quotes. However, since all sniping attempts would arrive within the same batch interval, there cannot be any speed advantage between snipers. In this case, the order priority is determined by the price, which is in favor of the slow trader. To see this, suppose that there is a jump in the price of ES, and Alice has a sell limit order on SPY at a price p. Alice would like to cancel that order and replace it with the correct ask, p' (with $p' > p$). However, Alice's reaction time is too slow, and any order to cancel that stale order will not arrive before the end of the current batch interval. In the continuous limit order book design, a high-frequency trader who wants to snipe Alice only needs to send a buy order at p that arrives first. But with frequent batch auctions, all the snipers' orders arrive in the same batch interval. Now, since a double auction is nothing but a supply-demand equilibrium, the trader that will be able to trade with Alice is the one offering the highest price. That is, traders who want to snipe Alice in a frequent batch auction have an incentive to increase their bid. This competition pushes the price up to its new (correct) level, p', thereby guaranteeing Alice a trade at the new price p' even though she is not fast enough.

Sniping in a frequent batch auction is thus a little bit similar to sniping on eBay (see section 3.4). For both designs, speed is only valuable (i.e., sniping is only feasible) at the end (of the auction for eBay or the batch interval for trading), whereas for the continuous limit order book design, speed has some value at any time.

9 The Basic Matching Model

So far, we have seen markets where buyers need to pay a price to get an item (or several items), and our main preoccupation was designing and studying methods to elicit the "right" price. We now consider a more challenging environment: markets where there are no prices, where there are no monetary transactions. As we outlined in chapter 1, there are numerous situations where the price is not the only variable one may want to consider when engaging in a transaction. College admission, the assignment of doctors to hospitals, or assigning students to courses or dorms are situations where the price, if there is one (e.g., tuition or salary), may play a role but is not enough to determine transactions. In order to make the presentation of such markets more intuitive, we will consider the more radical situation where there are no prices at all. So the problems we will consider are "pure" matching problems, where the only question is who is matched to whom (or what).

9.1 The Basic Matching Model

In its simplest form, the basic matching problem consists of matching two sets of agents. For this chapter, let us call these two sets *musicians* and *singers*. For the sake of clarity, we will use feminine pronouns for musicians and masculine pronouns for singers.[1] This problem makes two important assumptions:

(i) Each individual (musician or singer) can be matched to at most one person.

(ii) A musician can be either single or matched to a singer. Similarly, a singer can be either single or matched to a musician.

This model is an instance of a *two-sided, one-to-one* matching problem. It is *two-sided* because individuals from one side are to be matched to individuals from the other side of the market—assumption (i)—and it is one-to-one because each individual can be matched to at most one individual.[2]

Departures from this model are, for instance:

- *One-sided matching problems.* Here we only have one set of individuals and we have to form groups of two or more individuals. A classic example is the so-called *roommate problem,* where we have to form pairs of students who would share the same dormitory.
- *Many-to-One matching problems.* Here several individuals from one side can be matched to the same individual from the other side. This is the case, for instance, of firms hiring workers or students attending schools. A firm hires several workers, but a worker can work for only one firm. Similarly, several students can attend the same college, but a student is assigned to at most one college.
- *Many-to-many matching problems.* In such problems, any individual from any side can be matched to several individuals from the other side. A classic situation describing many-to-many matching problems is that of students and professors, where each professor is assigned to many students, and each student is assigned to many professors (via the courses those professors give).

9.1.1 Preferences

In the basic matching problem, it is assumed that there are no monetary transactions when a singer and a musician are matched. This means that for a singer there are not many ways to be matched to the same musician: either he is matched to her or he is not. Hence, our preferences over potential partners do not change.[3] Consider, for instance, a singer having the choice between two musicians, say Alice and Barbara. If he chooses Alice over Barbara, then he will always prefer Alice over Barbara.

This contrasts with the situations we considered in the previous chapters, where a match (i.e., being assigned a good, a position in search results, etc.) also specifies the price. In those situations, buying an item at $10 is not the same as buying it at $8. When there are prices, my demand for a good depends on the price that I have to pay for it. The same happens for the sellers: the decision to sell a good depends on the price the buyer will pay. In the matching model, this is not the case. It is as if there was a monetary transaction but the amount is fixed and cannot be changed. With variable monetary transactions, the choice between two singers (or, for a singer, the choice between two musicians) will depend on the terms of the contract between the two parties. In this case, it may be that under some contract specifications our musician prefers Alice over Barbara but for another type of contract he prefers Barbara over Alice.

If we assume that there are no monetary transactions (or that the terms of the contract between a singer and a musician are fixed and not modifiable), then the individuals' preferences over their potential partners are sufficient information.[4]

The Basic Matching Model

The matching model thus assumes that each singer has preferences over musicians and each musician has preferences over singers. The preferences have to be understood as follows: if a singer prefers musician m_1 to musician m_2, we understand that he prefers being matched to m_1 over being matched to m_2.

There are two extensions to these preferences that we will adopt. One is purely technical, and the other will help us to have a slightly more general model. We start with the latter extension.

- Given a set of singers and musicians, we can assume that some singers deem some musicians **unacceptable**. This means that if a musician m is unacceptable for a singer s, the latter prefers to remain single rather than be matched to m. Similarly, some musicians may find some singers unacceptable: they prefer to remain single rather than be matched to those singers. If a musician prefers being matched to a singer to remaining single, then that singer is **acceptable** for that musician. Similarly, if a singer prefers to be matched to a musician rather than remain single, then that musician is acceptable for that singer.

- The other (technical) extension we will do is to assume that for each musician her preferences are over singers *and* herself. We assume the same for singers: the preferences of a singer also include himself. This extension will allow us to easily represent a singer's preferences with a unique ranking, which will contain all musicians and that singer. The notation is the following. Given a singer, say s_1, his preferences will be denoted by P_{s_1}. If there are, for instance, four musicians, m_1, m_2, m_3, and m_4, then s_1's preference could be something like this:

$$P_{s_1} = m_1, m_3, m_4, s_1, m_2.$$

Singer s_1's preferences read as follows. Musician m_1 is singer s_1's most preferred musician, m_3 is his second most preferred musician, and m_4 his third most preferred musician. Then comes s_1 himself. This means that any musician who is ranked above s_1 is acceptable. So s_1 finds musicians m_1, m_3, and m_4 acceptable. Conversely, any musician who is ranked below s_1 is considered unacceptable for s_1. In this case, m_2 is unacceptable, so s_1 prefers to be single (we will also say that s_2 prefers to be *matched with himself*).

When comparing two alternatives, say m_1 and m_4, we often write

$$m_1 P_{s_1} m_4$$

to denote that s_1 prefers m_1 to m_4 (i.e., according to the preferences of s_1, denoted P_{s_1}, musician m_1 comes *before* musician m_4).

Formally, the preferences of a singer $s \in S$ can be represented by a *strict ordering* over the set $M \cup \{s\}$, which is the set of all musicians + singer s himself. Musicians' preferences are defined similarly: the preferences of a musician $m \in M$ consist of a strict ordering over the set $S \cup \{m\}$, the set of all singers + himself. Note that we do not need to make any further assumptions. For instance, we do not need to introduce a utility function that would tell us *how much* a singer prefers one musician to another one. All we need is an ordering. We say that the ordering is *strict*, meaning that a musician is never indifferent between two singers (or a singer and herself).

9.1.2 Matching

We are now ready to define matchings. A matching is simply a mapping (or a function) from the set of all individuals (singers and musicians) to the set of all individuals. That function is traditionally denoted by the Greek letter μ, which reads as "mu" (the "m" in the Greek alphabet).

A matching is thus a function that says, for each individual, who his or her partner is under the matching μ. For instance, $\mu(\text{Alice})$ is the match of Alice. It could be herself or it could be a musician. If it is herself, then we have $\mu(\text{Alice}) = $ Alice. If she is matched under the matching μ to a singer, say Bob, we have $\mu(\text{Alice}) = \text{Bob}$.

Definition 9.1   A **matching** is a function $\mu : M \cup S \to M \cup S$ such that:

(a) For each singer $s \in S$, $\mu(s) \in M \cup \{s\}$, and for each musician $m \in M$, $\mu(m) \in S \cup \{m\}$.

(b) If $\mu(s) = m$, then $\mu(m) = s$.

Condition (a) says that singer s can be matched to at most one agent, which is an element of the set $M \cup \{s\}$. Similarly, each musician can be matched to a singer or to herself (for each $m \in M$, $\mu(m) \in S \cup \{m\}$). Condition (b) says that if a singer s is matched to a musician m, $\mu(s) = m$, then that musician m is matched to that singer s, $\mu(m) = s$.

We may sometimes want to compare matchings. To this end, we will need to add a symbol or a superscript or subscript to the letter μ to distinguish matchings. We can have, for instance, the matching μ and the matching μ'. So, for instance, we can have that under the matching μ Alice is matched to Bob, but under another matching, say μ', Alice is matched to Derek. So we would write in this case $\mu(\text{Alice}) = \text{Bob}$ and $\mu'(\text{Alice}) = \text{Derek}$.

The preferences individuals have over potential partners implicitly induce preferences over matchings. To see this, consider the preferences of John (a singer),

$P_{\text{John}} = $ Alice, Barbara, Carol, John, Dana.

So John prefers to be matched first to Alice, then Barbara, and then Carol. He finds Dana unacceptable. Now consider two matchings, μ and μ'. In the matching μ, we have $\mu(\text{John}) = \text{Carol}$, and in the matching μ', we have $\mu'(\text{John}) = \text{Alice}$. John prefers the musician he is matched with under μ' to the musician he is matched with under μ. So John prefers the matching μ' to the matching μ.

We can have a third matching, say μ'', such that $\mu''(\text{John}) = \text{Alice}$. In this matching, John's partner is the same as under the matching μ' (but the matches of other singers and musicians could be different between μ' and μ''). So John is **indifferent** between μ' and μ''. It is important to notice the following:

- Singers and musicians have **strict** preferences over each other, meaning no singer is indifferent between two musicians (or a musician and himself) and no musician is indifferent between two singers (or a singer and herself). But once we consider preferences over matchings, indifferences appear because under two different matchings an individual can be matched to the same partner.
- An individual only cares about his or her match. In other words, individuals do not have preferences regarding the matches of other individuals.

The description of the model (singers' and musicians' preferences and the matching function) is complete. The question that we have to address now (and perhaps the most interesting one) is who will be matched to whom.

9.1.3 Stability

So far, we have considered a set of singers and a set of musicians and have defined preferences individuals have over each other and which matchings are feasible. It is obvious that for a given problem there are many matchings that we can think of, but it is also easy to see that some matchings are more interesting or relevant than others. For instance, a matching where everybody remains single may not be very interesting (that matching is called the *empty matching*). Some matchings although less trivial than the empty matching are also unlikely to be interesting. Those are matchings where some singers or some musicians are matched to unacceptable partners.

But there is another type of matching we may want to discard: matchings where there are a singer and a musician who are not matched together but would prefer to be matched together. The following example illustrates this situation.

Example 9.1   We consider two singers (Arthur and Bob) and two musicians (Alice and Barbara). Their preferences are given in the following table.

P_{Arthur}	P_{Bob}
Alice	Alice
Barbara	Barbara
Arthur	Bob

P_Alice	P_Barbara
Arthur	Bob
Bob	Arthur
Alice	Barbara

The preferences read as follows. Arthur's most preferred musician is Alice, and then Barbara. Bob has the same preferences over the musicians as Arthur. However, Alice and Barbara disagree about who is the best singer: it is Arthur for Alice and Bob for Barbara. Also, we can see that each individual finds all the individuals from the other side acceptable: each musician ranks the two singers above themselves, and each singer ranks the two musicians above themselves.

Consider the matching μ defined by

$$\mu(\text{Arthur}) = \text{Barbara} \quad \text{and} \quad \mu(\text{Bob}) = \text{Alice}.$$

Note that from our definition of a matching, the information we just gave about the matching is sufficient: adding $\mu(\text{Alice}) = \text{Bob}$ and $\mu(\text{Barbara}) = \text{Arthur}$ would be redundant.

Now consider Arthur and Alice. Under the matching μ, they are not matched together, yet both would prefer to be matched together. In other words, for both Arthur and Alice, any matching μ' such that $\mu'(\text{Arthur}) = \text{Alice}$ is preferred to the matching μ.[5]

Matchings that suffer from the situation described in example 9.1 or matchings where an individual is matched to an unacceptable partner are called "unstable matchings."

Definition 9.2 A matching μ is **stable** if:

(i) it is **individually rational**, meaning that for each individual $v \in M \cup S$, v weakly prefers $\mu(v)$ to v; and

(ii) there are **no blocking pairs**, meaning that there is no singer-musician pair (s, m) such that $\mu(s) \neq m$ and

$$m \, P_s \, \mu(s) \quad \text{and} \quad s \, P_m \, \mu(m). \tag{9.1}$$

Stability in matching can be understood as the equivalent of the equilibrium in a "classic" supply-demand problem. To see this, consider the supply and demand of, say, gasoline, and imagine the situation where there are many gas stations and consumers. Gas stations are not all the same: some are on the roads I usually take, and others are in locations that are less attractive to me. But other consumers may have other preferences: the gas stations that are not convenient for me can be nicely located for them. An equilibrium situation in this market for gasoline

The Basic Matching Model

would be such that, given the prices proposed by each gas station, consumers do not change their habits. That is, consumers always go to the same gas stations. We do not have an equilibrium when the prices are such that I prefer to go to a gas station other than the usual one, even if it is not on my usual route. That gas station is better off, as it has an additional client (me), and I am also better off, because I can get cheaper gas there (compared to my usual gas station). In the case of gasoline, it is likely that either my usual gas station will lower its prices or that the remote but cheaper gas station will slightly increase its prices. The fact that we did not have any equilibrium induced some people to change their decisions (my going to another gas station and/or gas stations changing their prices).

The same happens in a "pure" matching problem like the one we are studying with musicians and singers. The notion of *equilibrium* directly translates into the notion of *stable matching*. If two individuals who are not interacting with each other (e.g., me and the remote gas station) can be better off by interacting with each other, then the initial situation is not an equilibrium or, in the language of matching theory, a stable matching.

Another situation is when I find the price of gasoline so expensive that I am better off not using my car (and thus buying gasoline) and instead use alternative modes of transportation. A matching can also suffer from this type of situation: a musician (or a singer) can be better off by severing his or her relation with his or her partner. To sum up, the notion of stable matching does not really introduce a brand new concept; it simply adapts the classic notion of market equilibrium to the case of the matching problem.

The expression "weakly prefers" in condition (i) means that individual v is either indifferent between v (himself or herself) and $\mu(v)$, or v (strictly) prefers $\mu(v)$ to himself or herself. We need to say that v weakly prefers because the way the definition of individual rationality is written does not allow us to deduce whether v is matched to himself or herself or someone from the other side of the market.

A matching μ such that v *strictly* prefers himself or herself to his or her partner under μ, $\mu(v)$ is *not* individually rational.

Condition (ii) says the following. A singer s and a musician m **block** a matching μ if they are not matched together under μ (i.e., $\mu(m) \neq s$) and if they *both* prefer each other to their partner under μ. Equation (9.1) describes this situation. The left part says that singer s prefers musician m to his partner under the matching μ, $\mu(s)$. The right part says that musician m prefers singer s to his partner under the matching μ, $\mu(m)$.

In example 9.1, the matching μ is not stable, because it is blocked by the pair Arthur-Alice:

Arthur P_{Alice} $\mu(\text{Alice})$ and   Alice P_{Arthur} $\mu(\text{Arthur})$.

Stable matchings thus capture the intuitive notion that agents cannot improve their situations. A stable matching is, in a certain sense, optimal.

For example 9.1, a stable matching would be the following:

$$\mu'(\text{Arthur}) = \text{Alice} \quad \text{and} \quad \mu'(\text{Bob}) = \text{Barbara}. \tag{9.2}$$

It is easy to see that this matching is individually rational. Each singer and each musician prefers his or her partner under μ to remaining single (for each individual, remaining single is the worst option given the preferences we depicted). Now consider Arthur and Alice. For both of them, their partner under μ' is the best possible partner. So none of them would be part of a blocking pair. If we consider Bob, his partner under μ is Barbara. For him, Barbara is only his second most preferred musician; he would like to block with a more preferred musician, Alice. So we have

Alice P_{Bob} $\mu'(\text{Bob})$.

We have here the first half of equation (9.1). But do we have the other half? That is, do we have that Alice also prefers Bob to her partner under μ'? Formally, do we have Bob P_{Alice} $\mu'(\text{Alice})$? The answer is no, because $\mu'(\text{Alice}) = $ Arthur and Alice prefers Arthur to Bob. So the pair (Bob, Alice) is *not* a blocking pair. As for Barbara, she is also matched to her most preferred singer; she does not want to block the matching μ'. So the matching described in equation (9.2) is stable.

Remark 9.1 It is important to note that when a singer and musician block a matching (they would be better off together), or when a singer or a musician blocks the matching (they are matched to an unacceptable partner), we do not consider what would happen next. In other words, we are not considering the matchings that would arise once the pair blocks (or individuals become single) beyond their new match.

9.2 Algorithms and Mechanisms

From a formal point of view, the analysis of matching markets differs substantially from that of other areas such as auctions, consumer demand, or equilibrium theory in competitive markets. In this section, we outline how we proceed when economists analyze matching problems and how such markets can be organized.

9.2.1 Algorithms?

The description of a matching model is fairly simple. We just need to specify the individuals involved (in this chapter, musicians and singers) and their preferences over potential partners. We have seen that the preferences simply consist of *orderings*.

With such a formalism, it is extremely difficult to work with formulas and do calculus. This contrasts with traditional models such as the consumer model in microeconomics. In that model, we have a utility *function*, a demand *function*, and a budget constraint that can be represented by equations that can be studied. For instance, we can compute the slope of a demand function at a particular point and interpret it. To determine the equilibrium, we can solve a system of two equations, one representing the demand and the other representing the supply

Clearly, such calculations are hard (if not impossible) to do with the matching model we have. If we want to compute the demand of a consumer given some prices, we just have to (in the standard, simple case) solve an equation like this one:

$$\frac{\text{marginal utility for good 1}}{\text{marginal utility for good 2}} = \frac{\text{price of good 1}}{\text{price of good 2}}.$$

But how do we proceed when we do not have functions (such as a utility function) but instead have preference lists like this one: $P_{\text{Albert}} =$ Alice, Barbara, Carol?

For matching markets, economists often use *algorithms*. An algorithm is simply a *recipe* that tells us how we have to use individual preference lists. A recipe for, say, a cake describes all the steps we need to follow to make a cake. A recipe looks like the following:[6]

1. Melt 1 cup of chocolate for pastry (125 grams) and 1 quarter of a cup of butter (60 grams).

2. Mix 1/2 cup of sugar (125 grams) with 3 egg yolks (keep the whites).

3. Add to the mix of sugar and egg yolks 1/2 cup of cornstarch (60 grams) and the melted chocolate-butter. Mix well.

4. Whisk the whites.

5. Delicately mix the whisked whites with the mix of sugar, egg yolks, chocolate, and cornstarch.

6. Pour the mix in a mold, and put it in the oven for 20–30 minutes (180°C, 350°F).

In a matching problem, a recipe/algorithm could be, for instance, this:

1. Take any musician, and take her most preferred singer. If they are mutually acceptable, match them together. If not, take her second most preferred singer. If they are mutually acceptable, match them together. If not, take her third most preferred singer. Continue like this until the musician is either matched to a singer or you have tried all the acceptable singers for that musician.

2. Take another musician at random and do as in step 1.

3. Continue like this until you have considered all the musicians.

The advantage of working with algorithms is that we do not need to know a lot of mathematics. We just need to read the algorithm carefully and do what it says. There are two difficulties we may encounter, though:

- It could be that the algorithm is not very well written, so it is difficult to understand what we have to do. This does not mean that algorithms are a bad thing! To avoid this type of problem, algorithms are sometimes long. You do not have to be scared by the length of an algorithm.
- Some algorithms are not well designed. For instance, the algorithm described has a flaw: it may not produce a matching as we have defined it. To see this, consider the following small example.

We have two musicians, Alice and Barbara, and two singers, Albert and Bob. For both Albert and Bob, the two musicians are acceptable. The preferences of the musicians are:

$P_{\text{Alice}} =$ Albert, Bob, Alice,

$P_{\text{Barbara}} =$ Albert, Bob, Barbara.

We can see that Alice and Barbara find both Albert and Bob acceptable. Let us execute the algorithm.
1. We take a musician at random, say Alice. Her most preferred singer is Albert. They are mutually acceptable (Albert is acceptable for Alice, and Alice is acceptable for Albert). So the algorithm says that we have to match them. Step 1 is done.
2. We take another musician at random. Observe first that the algorithm does not say that it has to be a musician different from the one we picked in step 1. This means that there is some probability (small) that we always take the same musician, and thus our algorithm will never stop! For simplicity, assume that at random we pick Barbara.

Her most preferred singer is Albert. They are mutually acceptable (Albert is acceptable for Barbara, and Barbara is acceptable for Albert). So the algorithm says that we have to match them.
3. We have considered all the musicians, so our algorithm stops.

We can see that we do not have a matching as defined in section 9.1.2. Indeed, we see that Albert is matched to both Alice and Barbara.

The algorithms we will study in this chapter and the following ones will not suffer from this kind of flaw. They are well designed and, if we are a little bit careful, are relatively easy to follow. What will be interesting is to analyze the types of matchings they produce.

9.2.2 Matching Mechanism

The way to proceed is to construct a **matching mechanism**. Such a mechanism is a procedure that works as follows:

• **Step 1** Each individual submits a preference list over potential partners. The mechanism specifies how the submitted preferences should look.

• **Step 2** An algorithm is used to compute a matching using the preference lists submitted in step 1.

• **Step 3** The outcome of the algorithm (a matching) is implemented.

This description of a mechanism makes it clear that the key element is the algorithm we use. Using the terminology of appendix B, a matching mechanism is simply a direct mechanism, and the algorithm is the outcome function. When analyzing a matching mechanism, the two main directions we consider are:

• Given some preference lists, what matching do we obtain? What are the properties of the matching calculated by the algorithm?

• Given the algorithm used in step 2, how will participants behave? Will they submit a preference list that is identical to their preferences, or will they misrepresent their preferences?

When there are no restrictions on how the submitted preference should look, then the question of whether individuals reveal their true preferences makes sense. However, it could be that the mechanism restricts individuals. For instance, a mechanism could ask individuals to submit a preference list over only two potential partners. In this case, it is obvious that individuals cannot reveal all their preferences.

Even if individuals are not constrained in the type of preference lists they can submit to the mechanism, we will see that the choice of the algorithm can affect what is "best" for the individuals. That is, some algorithms may be such that it is best to submit one's true preferences, while other algorithms may give incentives to submit a preference list that is not the same as the true preferences.

A key property for mechanisms we will consider is the following.

Definition 9.3 A (matching) mechanism is **strategyproof** if for each individual it is a dominant strategy to reveal her true preferences.

9.3 Finding Stable Matchings

The question we face now is how to find stable matchings. Another related question is whether there is always a stable matching for any problem (i.e., preferences) we may consider. The answer to that second question is yes, and to prove it we will present a simple method that identifies a stable matching (and thus address the first question). That method is known as the **Deferred Acceptance algorithm**, and it was introduced by David Gale and Lloyd Shapley in 1962, in the very same research paper where they defined the matching model we have just presented.

9.3.1 The Deferred Acceptance Algorithm

One can present the Deferred Acceptance algorithm as a story of (modern) courtship. We first have to choose a side, namely singers or musicians, that will court the other side. To explain the procedure, we will assume that singers court musicians. In this courting game, each singer will try to be matched to a musician, starting with the most preferred musician. Some musicians may receive proposals from multiple singers and will thus reject all proposals but the most preferred one. Some musicians may receive only one proposal and thus only have to see whether that proposal comes from an acceptable partner. Musicians not receiving any proposal will have (for the moment) nothing to do.

Singers whose proposal has been rejected will then propose to their next most preferred musician. At each stage, a musician compares the singer whose proposal she currently holds with the new offer she received (if any), and rejects all but the most preferred one. The procedure stops when no singer is rejected. At the end of our courtship game, the proposals that are currently held are accepted, and the matching is realized. It is in this sense that acceptance (of a proposal) is *deferred*: we have to wait until the end of the algorithm to see who is matched to whom. A formal description of the algorithm (with singers proposing) is the following.

Algorithm 9.1 Deferred Acceptance

Step 1 Each singer proposes to the musician who is ranked first among his preferences (if a singer finds all musicians unacceptable, then he remains single).

Each musician who received at least one offer temporarily holds the offer from the most preferred singer among the singers who made a proposal to her and who are acceptable, and rejects the other proposals (those that are not from the most preferred singer and the offers from unacceptable singers).

Step k, $k \geq 2$ Each singer whose proposal was rejected in the previous step proposes to his most preferred musician among the musicians to whom he has not yet proposed. If there are no such musicians, the singer remains single.

Each musician receiving new proposals temporarily holds the proposal of the singer she prefers the most among

The Basic Matching Model

- the singers who just proposed to her and are acceptable to her and
- the singer whose proposal she held from the previous step.

The other proposals (those that are not from the most preferred singer and the offers from unacceptable singers) are rejected.

End The algorithm stops when no singer's offer is rejected. Each musician is matched to the singer who made the offer she was holding when the algorithm stopped.

The following example illustrates how the algorithm works.

Example 9.2 There are three singers, Bob, John, and David, and three musicians, Dinah, Melanie, and Janis. Their preferences are in the following table.

P_{Bob}	P_{John}	P_{David}
Dinah	Dinah	Dinah
Melanie	Melanie	Janis
Janis	Janis	Melanie
Bob	John	David

P_{Dinah}	$P_{Melanie}$	P_{Janis}
Bob	Bob	Bob
John	David	John
David	John	David
Dinah	Melanie	Janis

We now run the Deferred Acceptance algorithm with these preferences and with singers proposing.

- **Step 1** Each singer proposes to his most preferred musician. So Bob, John, and David all propose to Dinah (they all prefer Dinah to any other musician).

  Dinah has three proposals. The one she prefers is from Bob, so she temporarily accepts him and rejects the offers from John and David.

  The matching at the end of this step is $\mu(Bob) = $ Dinah, and Melanie, Janis, John, and David are all single ($\mu(David) = $ David, $\mu(John) = $ John, $\mu(Melanie) = $ Melanie, and $\mu(Janis) = $ Janis).

- **Step 2** John and David are the only singers who were rejected in the previous step. Each proposes to their most preferred musician among those who have not rejected him. For John it is Melanie, and for David it is Janis.

Dinah still holds the offer from Bob and does not receive any new offer, so she keeps that offer. Melanie now has her first offer, from John. It is the only offer, and since John is acceptable to her, she holds that offer. As for Janis, she has one offer, from David. That singer is acceptable to her, so she keeps that offer.

The matching at the end of this step is $\mu(\text{Bob}) = \text{Dinah}$, $\mu(\text{John}) = \text{Melanie}$, and $\mu(\text{David}) = \text{Janis}$. No singer is rejected in this step, so the algorithm stops.

The matching at the end of this step is the final matching:

$$\mu(\text{Bob}) = \text{Dinah}, \ \mu(\text{John}) = \text{Melanie}, \text{ and } \mu(\text{David}) = \text{Janis}. \tag{9.3}$$

Of course, there is no reason why it should be singers proposing. We can switch the roles in our courtship game and run the Deferred Acceptance algorithm with musicians proposing. In this case, we will obtain (with the preferences of example 9.2) the following matching:

$$\mu(\text{Bob}) = \text{Dinah}, \ \mu(\text{John}) = \text{Janis}, \text{ and } \mu(\text{David}) = \text{Melanie}. \tag{9.4}$$

9.3.2 Deferred Acceptance and Stable Matchings

The matching obtained when running the Deferred Acceptance algorithm has a very interesting property: it is a stable matching. Before explaining why, let us check with example 9.2.

Bob will never be part of a blocking pair, for he is matched to his most preferred musician. The cases of John and David can be treated simultaneously. The only musician they would consider for blocking is the musician they strictly prefer to their current partner, Dinah. But Dinah prefers Bob, her current match, to both John and David. So musician Dinah will refuse to block with any of those singers. Finally, since all singers and musicians prefer their match to being single, the matching μ described in equation (9.3) is individually rational. So μ is a stable matching.

We now show that the matching obtained through the Deferred Acceptance algorithm is necessarily stable. This will be relatively straightforward. To this end, consider any matching problem. That is, there are an arbitrary number of musicians and singers, and individuals' preferences are arbitrary.

Let μ be the matching we obtain with these preferences when running the Deferred Acceptance algorithm with singers proposing (the analysis is similar when musicians propose), and suppose that μ is not stable. Definition 9.2 says that a matching is not stable if it is either not individually rational or there is a blocking pair. So we have two cases we must consider.

Case 1 μ is not individually rational

Suppose first that there is a singer, say Leonard, who is matched to an unacceptable musician, say Nina. So in Leonard's preferences, Leonard (himself) is ranked *above* Nina. This means that during the algorithm Leonard has been rejected by all the musicians he prefers to himself (if any), and thus there is a step at which there was no acceptable musician he has not yet proposed to. In this case, the algorithm says that the singer remains single; he cannot propose to another musician. But for Leonard and Nina to be matched, it must be that at some step he proposed to her. We just explained that this is impossible, so a singer cannot end up being matched to an unacceptable musician.

Similarly, a musician cannot be matched to an unacceptable singer. The description of the algorithm makes it clear that a musician always rejects offers from unacceptable singers.

Therefore, the matching constructed by the Deferred Acceptance algorithm is necessarily individually rational.

Case 2   There is a blocking pair

Let Barbara and Paul be a blocking pair for the matching μ, so they both prefer each other to their match under μ. For Paul to be matched to μ(Paul), it must be that he proposed μ(Paul) (if this is a musician) at some step of the algorithm or has been rejected by all his acceptable musicians during the algorithm if μ(Paul) = Paul. Clearly, if Paul wants to block with Barbara, it must be that Barbara is acceptable and is ranked above μ(Paul). The algorithm is clear: Paul *cannot* propose (at some step) to μ(Paul) *before* having proposed, at an earlier step, to Barbara. So there is a step at which Paul proposed to Barbara. Since Paul ends up *not* being matched to Barbara, it must be that at some step she rejected him.

The key argument is the following. A musician only rejects an offer if she has a better one (according to her preferences). So if Barbara rejected an offer from Paul, it must be that she had, at some step, an offer from a singer she prefers to Paul. Notice that during the course of the algorithm with singers proposing, musicians can only improve their situation: they are the ones deciding who to reject, and they reject a singer only if they have a better alternative. So if Barbara rejected Paul, it must be that *when* she rejected him she had an offer from a preferred musician, say Alex.

Alex may not be Barbara's final match. She may receive later in the algorithm an offer from a singer she prefers to Alex. Whether she ends up being matched to Alex or a singer she prefers to Alex, Barbara necessarily ends up being matched to a singer she preferred to Paul. In other words, Barbara prefers μ(Barbara), her final match, to Paul. So the pair (Barbara, Paul) cannot block, and thus *there cannot be any blocking pair in the matching constructed by the Deferred Acceptance algorithm*.

The analysis of the algorithm we just conducted gives the following result.

Result 9.1 (*Gale and Shapley*)   For any matching problem:

- There always exists a stable matching.
- The Deferred Acceptance algorithm (whether musicians or singers propose) always produces a matching that is stable with respect to the preference lists used by the algorithm.

9.4 Preferences Over Stable Matchings

From section 9.3, we know that there always exists a stable matching for any problem. This result raises two additional questions:

- If there are several stable matchings, can we select one of them?
- In order to use the Deferred Acceptance algorithm, we first need to know individuals' preferences. How do we proceed?

We will see, that the answers to these questions are closely related. In this section, we address the question of multiplicity of stable matchings. The question of incentives will be treated in section 9.5.

9.4.1 Musician-Optimal and Singer-Optimal Matchings

The preferences given in example 9.2 yield two stable matchings, which we will call μ_S and μ_M (this notation will quickly become self-explanatory),

$\mu_S(\text{Bob}) = \text{Dinah}, \quad \mu_M(\text{Bob}) = \text{Dinah},$

$\mu_S(\text{John}) = \text{Melanie}, \quad \mu_M(\text{John}) = \text{Janis},$

$\mu_S(\text{David}) = \text{Janis}, \quad \mu_M(\text{David}) = \text{Melanie}.$

In these two matchings, Bob is matched to the same musician, so he is indifferent between μ_S and μ_M. However, John and David are not matched to the same musician. If we look closely, we can see that both John and David strictly prefer μ_S to μ_M. If we consider musicians, the converse happens: Melanie and Janis (the partners of John and David under μ_S and μ_M) prefer μ_M to μ_S (Dinah is indifferent between these two matchings, as she is matched to the same singer, Bob).

Notice also that for singers there is (weak) unanimity to say which matching, among the stable matchings, is the best. The unanimity is only weak because Bob

The Basic Matching Model

is indifferent between the two stable matchings. The same occurs for musicians: there is (weak) unanimity to say which stable matching is the best. It turns out that the unanimity we just observed holds not only for that particular problem but for *any* matching problem.

But we can say more: the best stable matching for the singers is also the worst stable matching for the musicians, and the best stable matching for the musicians is also the worst stable matching for the singers.

Result 9.2 For any matching problem, there is a stable matching that is the most preferred stable matching for the singers and the worst stable matching for the musicians.

Conversely, there is a stable matching that is the most preferred stable matching for the musicians and the worst stable matching for the singers.

Before going further, we need to insist that the preferences regarding stable matchings presented in result 9.2 hold only for *stable matchings*. The result does not say that all singers prefer the same matching to any other stable matchings. The "unanimity" (among singers or among musicians) that result 9.2 refers to is only valid when we confine ourselves to the set of stable matchings (and not the set of all matchings).

Notice that the matching μ_S was obtained when running the Deferred Acceptance algorithm with singers proposing, and the matching μ_M was obtained running the Deferred Acceptance algorithm with musicians proposing. Result 9.2 can be extended in the following way.

Result 9.3 The matching obtained with the Deferred Acceptance algorithm with singers proposing is the singers' most preferred stable matching (and musicians' least preferred stable matching).

The matching obtained with the Deferred Acceptance algorithm with musicians proposing is the musicians' most preferred stable matching (and singers' least preferred stable matching).

The matching obtained with singers proposing is called the **singer-optimal matching** and is denoted by μ_S. Similarly, the matching obtained with musicians proposing is called the **musician-optimal matching** and is denoted by μ_M.

9.4.2 Proofs

The proofs of results 9.2 and 9.3 are not difficult. We start with result 9.3. To proceed, we need the following concept.

Definition 9.4 A singer s and a musician m are **achievable** (to each other) if there exists a stable matching where s and m are matched together.

To prove result 9.3, it will be sufficient to show that when running the Deferred Acceptance algorithm with *singers* proposing, no singer is rejected by an achievable musician. If this holds, then we will deduce that a singer cannot be matched, at a stable matching, to a musician who is more preferred than the one he is matched to at the end of the algorithm (or himself if he is single).

Consider then *any* preferences for singers and musicians, and execute the Deferred Acceptance algorithm with singers proposing. If the algorithm stops at the end of step 1, then it means that all singers are matched to their most preferred musicians and we are done. So the problem becomes interesting if there is more than one step.

We will look at what happens the first time a singer, say Paul, is rejected by an achievable musician, say Alice. That is, there exists a stable matching where Alice and Paul are matched together, but in the execution of the algorithm there is a step at which Alice rejects Paul, and it is at the earliest step that such a thing occurs. If Paul is unacceptable for Alice, then the pair cannot be achievable (because stable matchings are, by definition, individually rational). So it must be that Alice rejected Paul *because* she received the offer of a singer she prefers, say John. Therefore,

$$\text{John } P_{\text{Alice}} \text{ Paul}. \tag{9.5}$$

Let k_P be the step at which Paul is rejected by Alice, and let k_J the step at which John made an offer to Alice. Note that we can have $k_J \leq k_P$. We have either $k_J = k_P$ (Paul's offer is immediately rejected because he makes his offer at the same step as John) or $k_J < k_P$ (Alice is already holding John's offer when Paul makes his). But we cannot have $k_P < k_J$, because John is *the* singer that makes Alice reject Paul's offer.

We will show now that Alice cannot be achievable for Paul. Since John made an offer to Alice at step k_J, she is John's most preferred musician among all the musicians to whom he has not yet proposed. All the other musicians rejected John at an earlier step. Since Paul is the *first* singer to be rejected by an achievable musician, the musicians who rejected John before step k_J are unachievable for John. So John's set of achievable musicians is the set made by Alice and all the musicians less preferred by Alice, or a subset thereof.

We supposed that Alice is achievable for Paul. So there exists a stable matching (call it $\hat{\mu}$) such that $\hat{\mu}(\text{Alice}) = \text{Paul}$. Since $\hat{\mu}$ is a stable matching, all the other musicians and singers are matched to an achievable mate. We have seen that all John's achievable partners cannot be more preferred by Alice. Since Alice is matched to Paul under $\hat{\mu}$, John is matched under $\hat{\mu}$ to someone less preferred than Alice. So we have

$$\text{Alice } P_{\text{John}} \hat{\mu}(\text{John}). \tag{9.6}$$

The Basic Matching Model

But now we have equations (9.5) and (9.6) telling us that John and Alice can block $\hat{\mu}$ (because $\hat{\mu}(\text{Alice}) = \text{Paul}$). That is, $\hat{\mu}$ cannot be stable! This contradicts our hypothesis that $\hat{\mu}$ was a stable matching, so the initial hypothesis (that a singer is rejected by an achievable musician in the singer-proposing version of the algorithm) cannot be true.

We are now ready to conclude. We just showed that no singer can be rejected by an achievable musician when we run the algorithm with singers proposing. Since the outcome of the algorithm is a stable matching, all singers are matched with their most preferred achievable musician. That is, the matching we obtain when singers propose gives each singer his most preferred achievable musician, his most preferred stable partner. So the singer-optimal matching is singers' most preferred stable matching.

Result 9.2 is a simple consequence of a more general result: if all singers prefer some stable matching μ to another stable matching μ' (these need not be the singer-optimal or musician-optimal matchings), then all musicians have opposite preferences. That is, all musicians prefer the matching μ' to the matching μ. To see this, let μ and μ' be two stable matchings such that all singers prefer μ to μ'. We need to show that *all* musicians prefer μ' to μ. To this end, suppose that this is not true; that is, there is a musician, say m, who prefers μ to μ'. This means that m is not matched to the same singer under μ and μ'. Let $s = \mu(m)$ and $s' = \mu(m)$. So we have

$$\mu(m) = s\, P_m\, s' = \mu'(m). \tag{9.7}$$

Since s and s' are two different singers, s is matched to a different musician under μ'. And since all singers prefer the matching μ to the matching μ', we have

$$\mu(s) = m P_s \mu'(s). \tag{9.8}$$

But then equations (9.7) and (9.8) imply that the pair (m, s) blocks the matching μ'; that is, μ' cannot be stable. This contradicts our hypothesis that μ and μ' are stable matchings, so if all singers prefer a matching μ to another matching μ', no musician can also prefer μ to μ'.

9.5 Incentives with the Deferred Acceptance Algorithm

The Deferred Acceptance algorithm, although presenting singers' and musicians' decisions as being their own decisions (making and accepting or rejecting offers), is better understood as being what it is: an algorithm. That algorithm is fed with singers' and musicians' preferences, and it produces a matching. So we have to understand the use of the Deferred Acceptance algorithm as a situation in which participants report their preferences to a clearinghouse that uses these preferences to run the Deferred Acceptance algorithm. Economists call such situations **centralized markets**, as opposed to **decentralized markets**, which are situations where individuals decide and realize their transactions by themselves.

If we proceed that way, an immediate problem that arises is whether singers and musicians have incentives to reveal their *true* preferences. One could well imagine a situation where a singer or a musician prefers (with respect to their true preferences) the matching that we obtain when running the algorithm with untruthful preferences to the matching we would have obtained had we used the individuals' true preferences. The next example illustrates this point.

Example 9.3 Consider two singers (s_1 and s_2) and two musicians (m_1 and m_2). Their preferences are given in the tables that follow.

P_{s_1}	P_{s_2}
m_1	m_2
m_2	m_1
s_1	s_2

P_{m_1}	P_{m_2}
s_2	s_1
s_1	s_2
m_1	m_2

Suppose we run the Deferred Acceptance algorithm with singers proposing. In the first step, s_1 and s_2 propose to m_1 and m_2, respectively. Each musician has only one offer, and the offer comes from an acceptable singer. So both m_1 and m_2 accept their offers and thus no singer is rejected. Hence, the algorithm stops, and the singer-optimal matching is

$$\mu_S(s_1) = m_1 \quad \text{and} \quad \mu_S(s_2) = m_2.$$

Suppose now that musician m_1, instead of submitting her true preferences to the algorithm, submits the following preference ordering:

$$P'_{m_1} = s_2, m_1, s_1.$$

If we run the Deferred Acceptance algorithm with the preferences ($P_{s_1}, P_{s_2}, P'_{m_1}, P_{m_2}$), the following happens. In the first step, the offers made by the singers are as before: s_1 proposes to m_1 and s_2 to m_2. But now, according to P'_{m_1}, singer s_1 is unacceptable, so he is rejected. In the meantime, m_2 temporarily accepted s_2's offer. So now there is a second step.

The Basic Matching Model

In step 2, singer s_1 (the only rejected singer) makes an offer to the next acceptable musician, m_2. Now m_2 has two offers: one from s_2 (from the previous step) and one from s_1. She prefers s_1, so she accepts s_1 and rejects s_2's offer.

In step 3, singer s_2 makes an offer to the next acceptable musician, m_1. This singer is acceptable according to P'_{m_1}, so she accepts him. No singer is rejected, so the algorithm stops and the final matching (which we call μ'_S) is

$$\mu'_S(s_1) = m_2 \quad \text{and} \quad \mu'_S(s_2) = m_1.$$

One can see that, according to her true preferences, m_1 prefers her match when she is not truthful (s_2) to the match she obtains when she is truthful (s_1).

The previous example showed that some individuals may be better off not revealing their true preferences. It turns out that if we use the Deferred Acceptance algorithm, individuals making offers never have an incentive to misrepresent their preferences.[7]

Result 9.4 (Dubins and Freedman; Roth) A matching mechanism that uses the Deferred Acceptance algorithm to match individuals is strategyproof for the proposing side.

We skip the proof of that result, as it is a little bit involved. We know from example 9.3 that it is not possible to induce truthful preference representations from both sides when we use the Deferred Acceptance algorithm. But could there be other matching algorithms that do the job? The answer is no.

Result 9.5 (Roth) There is no matching mechanism that satisfies, for any matching problem, the following two properties at the same time:

(i) The matching is stable with respect to the submitted preference lists.

(ii) The mechanism is strategyproof for *all* individuals.

The proof of that result is relatively straightforward. To see this, consider the (true) preferences from example 9.3. With these preferences, there are two stable matchings, μ_M and μ_S.

Suppose there exists some "magic" algorithm such that with that algorithm a matching mechanism satisfies properties (i) and (ii) of result 9.5. If the submitted preferences are the (true) preferences $(P_{s_1}, P_{s_2}, P_{m_1}, P_{m_2})$, then the magic algorithm should produce either μ_M or μ_S.

Suppose that our magic algorithm selects μ_S. We know from example 9.3 that either musician can be better off by misrepresenting her preferences. So the only way to avoid musician m_1 or m_2 misrepresenting her preferences is to use an algorithm that selects the matching μ_M.

But then we have a problem. Careful scrutiny of the preferences of example 9.3 shows that the problem is symmetric for musicians and singers. That is, if the algorithm selects μ_M, then the situation is reversed: both singers now have an incentive to misrepresent their preferences. For instance, singer s_1 can be better off by submitting the preference ordering $P'_{s_1} = m_1, s_1, m_2$. With the preferences $(P'_{s_1}, P_{s_2}, P_{m_1}, P_{m_2})$, there is a unique stable matching where s_1 is matched to m_1.

To sum up, we have the following situation for the preferences described in example 9.3:

- To satisfy property (i) of result 9.5, the algorithm must select either μ_M or μ_S.
- To satisfy property (ii) of result 9.5, we cannot select μ_M (otherwise the mechanism is not strategyproof for the singers), nor can we select μ_S (otherwise the mechanism is not strategyproof for the musicians).

In a recent study, Fuhito Kojima and Parag Pathak showed that, as the market becomes large (i.e., as the number of singers and musicians increases), on average there are fewer and fewer situations where manipulations can be profitable.[8] This result is closely related to the fact that, as the number of individuals increases, the average number of stable matchings tends to decrease. Since there is always at least one stable matching (see result 9.1), for sufficiently large markets there is, in general, only one stable matching.

Problems

1. Take any matching problem in which agents on one side, say musicians, have the same preferences over the singers. Is there a unique stable matching? If yes, provide a sketch of the proof. If not, provide an example.

2. We consider the following matching problem with four musicians and five singers. The preferences are as follows.

P_{m_1}	P_{m_2}	P_{m_3}	P_{m_4}
s_1	s_1	s_4	s_5
s_5	s_3	s_2	s_3
s_4	s_5	s_3	s_2
s_3	s_2	s_5	s_1
s_2	s_4	s_1	s_4

The Basic Matching Model

P_{s_1}	P_{s_2}	P_{s_3}	P_{s_4}	P_{s_4}
m_4	m_1	m_1	m_4	m_4
m_2	m_2	m_3	m_2	m_2
m_1	m_3	m_4	m_1	m_3
m_3	m_4	m_2	m_3	m_1

(a) Find the musician-optimal matching.

(b) Find the singer-optimal matching.

(c) Consider the Deferred Acceptance algorithm with singer proposing. Does any singer or musician have an incentive to misrepresent his or her true preferences?

10 The Medical Match

We started with an abstract model to study matchings between individuals from two different sets. One question that arises is whether such a model can be applied to real-life settings and/or whether it can help us to understand better how some matching markets work. If we consider a matching market between firms and workers, dating websites, college admission, or any other situation where individuals seek to be matched to other individuals or institutions (firms, colleges, etc.), real-life situations are obviously more complex than the simple matching model introduced in chapter 9. There are many relevant aspects of such situations that are not captured by the simple matching model we have seen. Yet, we will show that the simplicity of the matching model is not an impediment to the analysis of such markets. Even better, economists can help policy makers to better design such real-life markets.

This chapter is built around one of the most successful applications of the matching model: the matching of medical residents to hospitals. This case will lead us to introduce additional results that can be relevant for policy makers but also shed light on why so far we have insisted so much on stability when analyzing matchings.

10.1 History

Upon completing their degrees at a medical school, students in most countries must spend some time at a hospital as *residents* to complete their training. In the United States, there are each year a bit more than 20,000 candidates seeking to make their residency in one of the (roughly) 3800 residency programs (most programs have several vacancies, so there are almost as many vacancies as there are candidates). This market is rather complex. To begin, it would be more accurate to talk of markets, for there is a market for each medical specialty. While most candidates only look for first-year positions (and do all their residency in the same program), some candidates also seek second-year positions (which add additional constraints, as they have to be matched to prerequisite first-year positions). The fact that some residency programs also want to fill only a specific number of positions adds to the complexity of the problem. However, in spite of these complications (and many others), the study of the medical match constitutes one of the earliest success stories of the matching literature.

In the early 1980s, Alvin Roth started to look at the U.S. assignment of medical students to residency programs. For the first half of the twentieth century, Roth observed that the market for residents was **decentralized**; that is, programs were *deciding themselves* whether to make a job offer to a particular candidate. The competition was so intense that hospitals would hire students several years before graduation. Such a phenomenon, called **unraveling** in the matching literature, generated many inefficiencies. For instance, a student hired by a hospital several years before graduation would have less incentive to study hard, and thus students who would be considered as very good when being hired by a hospital could end up not being as good as expected when they start their residency. Unraveling could also be prejudicial for students in the sense that early decisions can have long-term negative consequences. Some students could be hired as, say, pediatricians but discover upon finishing medical school that they would have preferred other specialties.

To fix these problems, American medical schools agreed in 1945 not to disclose information about their students before a certain date, with the hope that it would limit unraveling effects. While well intentioned, this reform created bottlenecks, forcing residency programs and candidates to play a complicated timing game. On the one side, hospitals are obviously reluctant to make offers to candidates who would eventually reject them, and even more so if such candidates take too much time notifying hospitals of their decisions. Time is precious in decentralized markets, because a rejection that takes time to arrive is equivalent to losing the opportunity to make offers to other good candidates. The more a hospital waits, the more likely that those candidates will have already accepted an offer at some other place. On the other side, candidates usually prefer to wait before rejecting an offer. This is so first because decisions in such markets are often irreversible: if a student rejects an offer from a hospital, it is unlikely that this hospital would make a job offer again to that same student at a later date. Similarly, accepting an offer often consists of exiting the market. So the problem for a student is to accept or reject an offer without having the time to consider all possible offers. Accepting an offer often means forgoing the possibility to receive better offers, while rejecting an offer means taking the risk of ending up with a less favored outcome. Frequently a candidate receives an offer from a hospital that is not the most preferred one and is also notified that she is on the waiting list at a more preferred hospital. That candidate would then wait until the last minute to make her decision. If it happens that this candidate finally receives an offer from her preferred hospital, the rejected hospital would not be happy: it lost time waiting for the answer, losing the opportunity to make offers to other candidates (who are no longer available because they accepted other offers). Some candidates would even reject an offer they initially accepted. Because of this, many hospitals would pressure candidates, forcing them to make hasty decisions.

Between 1945 and 1951, several adjustments were made, most of them shortening the delays given to hospitals and candidates to make their decisions. The delay given to candidates decreased rapidly, from 10 days in 1945 to less than 12 hours in 1950, but none of those adjustments proved satisfactory. In 1952, after much deliberation, the various American medical associations agreed to change the rules of the game by switching to a **centralized market mechanism**. In such a market, hospitals do not make offers directly to candidates, and candidates do not directly accept or reject offers. Instead, hospitals send to a clearinghouse an ordered list of the candidates they want to hire, and candidates submit a preference list of the hospitals they would like to work for. An algorithm is then used to match students and hospitals. This procedure still exists today and is called the **National Resident Matching Program** (NRMP).[1] A crucial fact of the NRMP procedure is that participation is voluntary: hospitals and candidates are not compelled to find a match through the NRMP. In spite of this, the NRMP quickly reached a participation rate of over 95%. So, what happened exactly?

In a famous paper published in 1984, Roth studied the algorithm used in the NRMP.[2] The algorithm used then by the NRMP was not the Deferred Acceptance algorithm we saw in chapter 9, but Roth managed to prove that the algorithm used by the NRMP is equivalent to the Deferred Acceptance algorithm (given the preference lists submitted by candidates and hospitals). Economists believe that the use of an algorithm that produces stable matchings is one of the key elements of the NRMP's success. The use of an algorithm and a central clearinghouse to match students and hospitals has great benefits for all participants, as it can easily overcome most of the problems we mentioned earlier.

10.2 The Many-to-One Matching Model

Studying the medical match creates the opportunity to introduce a more general model than the matching model we studied in chapter 9, the so-called **many-to-one matching model**. This model is similar to the previous one in that there are two sets of agents, except that now agents from one side will be able to be matched with *several agents* from the other side. Since we are considering the medical match, the two sides of our matching market in this chapter will be hospitals and doctors. In the model we will develop, each doctor can be matched to at most one hospital, but each hospital can hire more than one doctor. So we have a many-to-one matching model: *many* doctors matched to *one* hospital.

Formally, the medical match model can thus be described as follows:

- There is a set of *doctors* $D = \{d_1, \ldots, d_n\}$.
- There is a set of *hospitals* $H = \{h_1, \ldots, h_m\}$.
- There is a vector of *capacities*, $q = (q_{h_1}, \ldots, q_{h_m})$, that specifies, for each hospital, the maximum number of doctors the hospital can hire.

10.2.1 Preferences in the Many-to-One Matching Model

So far, our model is not complete, for we also need to specify doctors' and hospitals' preferences over potential matches. The preferences of doctors are as in chapter 9. That is, each doctor $d \in D$ has a preference ordering P_d over the hospitals and herself, $H \cup \{d\}$. If doctor d prefers remaining by herself to being matched to hospital h, then hospital h is said to be *unacceptable* for doctor d, and we write $dP_d h$.

For the hospitals, their preferences are more complex because now hospitals contemplate hiring not only one doctor but several doctors. This implies that for hospitals we need to consider preferences over *sets* of doctors. The problem of having preferences over sets of individuals (instead of over individuals) is very frequent, and an obvious example is the hiring of athletes for a team: the preferences of a coach regarding athletes usually depend on the composition of the team. We have to be careful when defining preferences over sets of doctors, however.

There are many ways to model preferences over sets of individuals, but to facilitate the exposition, we will consider the simplest way, which in the matching literature is called **responsive preferences**. Such preferences are relatively straightforward to describe and, fortunately for us, are almost entirely described by preferences over individuals (and not sets of individuals). As we will see, responsive preferences amount roughly to saying that how attractive a doctor is does not depend on who the other doctors hired by the hospital are.

We first start with a preference relation over doctors for each hospital, as in chapter 9. That is, each hospital $h \in H$ has a preference relation P_h over doctors and itself, $D \cup \{h\}$. So if for hospital h we have $dP_h d'$, that means it prefers doctor d to doctor d'. Also, a hospital may see some doctors as unacceptable. If doctor d is unacceptable for hospital h, we then have $hP_h d$.

With such preferences over doctors, it is possible to describe a preference relation over *sets* of doctors. To stress that preferences over doctors is not the same as preferences over *sets* of doctors, we will denote by $P_h^\#$ the preferences of hospital h over *sets of doctors*.

Responsive preferences are defined by imposing a condition on how we compare two sets of doctors. But it will not be necessary to compare any set of doctors. It will be sufficient to compare two sets of doctors that differ by only one doctor. We will do the following. Suppose that a hospital has hired a group of doctors that includes Dr. Alice. Now comes Dr. Bob, and the hospital is wondering whether to replace Dr. Alice with Dr. Bob. Note that now the hospital has to compare two groups of doctors: the group with Dr. Alice and the group where Dr. Alice is replaced by Dr. Bob. Which group is preferred by the hospital? Responsive preferences simply require that in this case it suffices to compare Dr. Alice and Dr. Bob with the preference relation P_h (and not $P_h^\sharp$), and the group with Dr. Alice will be preferred to the other group if, and only if, Dr. Alice is preferred to Dr. Bob.

Formally, the preference relation of hospital h, $P_h^\sharp$ is responsive if for any group of doctors (let us call that group S) and any two doctors d and d' such that

$d \in S$, meaning doctor d (Dr. Alice in our example), is in the group, and
$d' \notin S$, meaning doctor d' (Dr. Bob in our example), is not in the group,

then we have

$$S\, P_h^\sharp\, S \cup \{d'\} \backslash \{d\} \quad \Leftrightarrow \quad d' P_h d. \tag{10.1}$$

Here the notation $S \cup \{d'\} \backslash \{d\}$ means that we take the group of doctors S where we add doctor d' (the notation $\cup \{d'\}$) and we withdraw doctor d (the notation $\backslash \{d\}$).

We illustrate how responsive preferences work in the following example.

Example 10.1 We consider a hospital, called hospital h, and four doctors, called d_1, d_2, d_3, and d_4. What is given to construct responsive preferences (and we have seen that this is enough) are the preferences over doctors, P_h. Let us assume that hospital h's preferences over these four doctors are

$P_h = d_1, d_2, d_3, d_4.$

So hospital h prefers doctor d_1 to doctor d_2, d_2 to d_3, and d_3 to d_4. Recall that those are preferences over doctors and not over sets of doctors (which would be denoted $P_h^\sharp$).

We are now ready to study the preference relation $P_h^\sharp$. Consider a first set of doctors, say $\{d_1, d_3, d_4\}$, and a second set, say $\{d_1, d_2, d_4\}$. These two sets differ by one doctor only: the first set contains d_3 but not d_2, while the second set contains d_2 but not d_3. So doctors d_2 and d_3 play the roles of Dr. Alice and Dr. Bob in our previous example.

Responsive preferences say that to compare $\{d_1, d_3, d_4\}$ and $\{d_1, d_2, d_4\}$ it suffices to compare d_3 and d_2. The preference relation P_h we have specified says that d_2 is preferred to d_3, so the definition of responsive preferences implies that hospital

h prefers the group of doctors $\{d_1, d_2, d_4\}$ to the group of doctors $\{d_1, d_3, d_4\}$. So we have

$\{d_1, d_2, d_4\} \, P_h^\sharp \{d_1, d_3, d_4\}$.

Let us consider two other sets, say $\{d_1, d_3\}$ and $\{d_2, d_4\}$. These two sets differ by more than one doctor, so in principle the responsiveness condition does not tell us how to compare those two sets. But we can use a third set, $\{d_2, d_3\}$, that will allow us to make a comparison. We first compare $\{d_2, d_4\}$ and $\{d_2, d_3\}$. These two sets differ by only one doctor, so we can compare them. We have $d_3 P_h d_4$, so we obtain

$$\{d_2, d_3\} P_h^\sharp \{d_2, d_4\}. \tag{10.2}$$

Now we compare $\{d_2, d_3\}$ and $\{d_1, d_3\}$. Since $d_1 P_h d_2$, we have

$$\{d_1, d_3\} \, P_h^\sharp \{d_2, d_3\}. \tag{10.3}$$

Combining equations (10.2) and (10.3), we obtain

$$\{d_1, d_3\} P_h^\sharp \{d_2, d_3\} P_h^\sharp \{d_2, d_4\} \quad \Rightarrow \quad \{d_1, d_3\} \, P_h^\sharp \{d_2, d_4\}. \tag{10.4}$$

With responsive preferences, we can also compare sets that are not of the same size. To see this, consider the sets of doctors $\{d_2\}$ and $\{d_1, d_3\}$. In this case, we now have to consider the capacity of hospital h. If $q_h = 1$, then the set $\{d_1, d_3\}$ is not acceptable for the hospital and thus we obtain

$\{d_2\} \, P_h^\sharp \{d_1, d_3\}$.

If, however, $q_h \geq 2$, we can use the responsiveness condition in the following way. The set $\{d_2\}$ has to be understood as being equivalent to the set $\{d_2, \emptyset\}$, where $\emptyset$ denotes an unfilled vacancy. Since doctors d_1 and d_3 are acceptable for hospital h, we have $d_1 P_h \emptyset$ and $d_3 P_h \emptyset$. So we can "replace" $\emptyset$ by d_3 to obtain a more preferred set for hospital h. So we have

$\{d_2, d_3\} \, P_h^\sharp \{d_2, \emptyset\}$.

Then we are in a situation similar to the earlier one, where we can replace d_2 by d_1, and we obtain

$\{d_1, d_3\} \, P_h^\sharp \{d_2, d_3\}$,

which yields

$\{d_1, d_3\} \, P_h^\sharp \{d_2\}$.

There are some cases where we cannot compare sets, though. For instance, the groups $\{d_1, d_4\}$ and $\{d_2, d_3\}$ (from example 10.1) are not comparable with the simple criterion of responsiveness. Surprisingly, the fact that we cannot compare any two pairs of groups of doctors is not an impediment to analyzing many-to-one matching models. As we will see in section 10.2.4, many results similar to those obtained for the one-to-one matching model can be obtained in a many-to-one model with responsive preferences. Another issue we have not discussed is whether hospitals can be *indifferent* between two groups of doctors. Surprisingly, we only need to assume that hospitals have strict preferences over individuals (i.e., hospitals are never indifferent between two doctors).

10.2.2 Matchings and Stability in a Many-to-One Matching Model

A matching in a many-to-one matching model is defined similarly as in the one-to-one matching model we saw in chapter 9, except that now we need to take into account that

- hospitals can be matched with more than one doctor, and
- hospitals have capacities, a maximum number of doctors they can hire.

We use the same notation to describe a matching: a mapping $\mu: H \cup D \to H \cup D$. In a many-to-one matching model, a matching μ must satisfy the following restrictions:

(i) For each doctor $d \in D$, $\mu(d) \in H \cup \{d\}$.

This means that doctor d can be matched to at most one "agent," which can be either a hospital ($\mu(d) \in H$) or herself ($\mu(d) \in \{d\}$).

(ii) For each hospital $h \in H$, $|\mu(h)| \leq q_h$, and if $|\mu(h)| \geq 1$, then $\mu(h) \subseteq D$.

The first requirement, $|\mu(h)| \leq q_h$, says that the number of individuals hospital h is matched with (the notation $|\mu(h)|$) cannot exceed the capacity of hospital h, q_h.

The second requirement applies only if the hospital is matched to at least one individual. In this case, all the individuals that are matched with hospital h must constitute a subset of the set of doctors (the notation $\mu(h) \subseteq D$).

Notice the difference in notation. For a doctor d, we write $\mu(d) \in H \cup \{d\}$, which means that $\mu(d)$, the match of doctor d, is an *element* of a set (all the hospitals *and* doctor d herself). If we write $\mu(h) \subseteq D$, then we mean that all the individuals that are matched to hospital h constitute a *subset* of the set of doctors (i.e., the subset can contain more than one element).

(iii) $\mu(d) = h$ if, and only if, $d \in \mu(h)$.

This requirement states that doctor d is matched to hospital h, $\mu(d) = h$, which is equivalent to saying that doctor d is one of hospital h's hires, $d \in \mu(h)$.

We are now ready to define stability for the many-to-one matching model. As we will see, it is very similar to the concept introduced in chapter 9 for the one-to-one matching model, except that now we will also have to take into account the capacity constraints of the hospitals. Three properties are used to define stability in a many-to-many matching problem.

- A matching μ is individually rational if each doctor $d \in D$ weakly prefers $\mu(d)$ (her match) to d (herself), and for each hospital $h \in H$, there is no doctor $d \in \mu(h)$ that is unacceptable

The condition for doctors is the same as the condition we saw in chapter 9. For hospitals we also need to check that no doctor hired by the hospital is unacceptable (with the preference relation over doctors P_h).

- A doctor d and hospital h are a **blocking pair** for a matching μ if they are not matched together under μ (i.e., $\mu(d) \neq h$) and if they both prefer each other to their partner under μ. For the doctor, comparing hospital h and her current match is easy. For the hospital, it is more tricky because the hospital may be matched to more than one doctor. To this end, it suffices to find one doctor matched to the hospital that is less preferred by doctor d. That is, we need a doctor, say d', such that hospital h prefers d to d' (i.e., $dP_h d'$). The responsiveness condition will then ensure that the hospital prefers the set of doctors it has under μ without doctor d' and adding doctor d to the set of doctors it has under μ.³ Formally, a doctor d and hospital h **block** a matching μ if $\mu(d) \neq h$ and there is a doctor $d' \in \mu(h)$ such that

$$hP_d\mu(d) \quad \text{and} \quad dP_h d'. \tag{10.5}$$

- A matching μ is **wasteful** if there exists a doctor d and a hospital h such that $\mu(d) \neq h$, h is acceptable for d and d is acceptable for h, and

$$hP_d\mu(d) \quad \text{and} \quad |\mu(h)| < q_h.$$

In words, a matching μ is wasteful if the following occurs. There is a doctor d who prefers to be matched to a hospital h rather than her partner under the matching μ (herself or another hospital), and hospital h has an unfilled vacancy (the notation $|\mu(h)| < q_h$). This concept is different from the concept of a blocking pair because here doctor d being matched to hospital h is not at the expense of another doctor.

Example 10.2   There are three doctors, d_1, d_2, and d_3, and two hospitals, h_1 and h_2. Their preferences are in the table that follows. These preferences are such that for each doctor both hospitals are acceptable. Similarly, for each hospital, all doctors are acceptable. Note that the hospitals' preferences depicted here are the preferences over doctors and *not* over sets of doctors.

Hospital h_1 has a capacity of 2, $q_{h_1} = 2$, and hospital h_2 has a capacity of 1, $q_{h_2} = 1$.

P_{d_1}	P_{d_2}	P_{d_3}
h_1	h_1	h_1
h_2	h_2	h_2

P_{h_1}	P_{h_2}
d_1	d_1
d_2	d_3
d_3	d_2

A wasteful matching
Consider the matching

$$\mu(d_1) = h_1, \ \mu(d_2) = h_2, \ \mu(d_3) = d_3.$$

In the matching μ, doctors d_1 and d_2 are matched to hospitals h_1 and h_2, respectively. Doctor d_3 is matched to herself; that is, she is not matched to any hospital. Note that, from the definition of a matching, we deduce that $\mu(h_1) = d_1$ and $\mu(h_2) = d_2$. This matching is wasteful because there is a doctor, d_3, who is unmatched, a hospital, h_1, that does not fill its capacity, and d_3 and h_1 are mutually acceptable.

A matching blocked by a pair
Now consider the matching

$$\mu'(d_1) = h_1, \ \mu'(d_2) = h_2, \ \mu'(d_3) = h_1.$$

In the matching μ', both hospitals fill all their vacancies, so this matching is not wasteful. Doctor d_1 is matched to her most preferred hospital, h_1, and she is the best doctor for hospital d_1. So doctor d_1 does not want to block this matching, and hospital h_1 does not want to replace d_1 with any other doctor. Doctor d_3 is also matched to her most preferred hospital, but she is the least preferred doctor for hospital h_1. The responsiveness condition implies that hospital h_1 would like to replace doctor d_3 with a more preferred doctor, d_2. What about doctor d_2? She is matched to hospital d_2, which is her least preferred hospital. So the pair (d_2, h_1) is a blocking pair. This implies that the matching μ' is not stable.

A stable matching

Finally, consider the following matching:

$\mu''(d_1) = h_1$, $\mu''(d_2) = h_1$, $\mu''(d_3) = h_2$.

This matching is not wasteful, and it is not difficult to check that it is not blocked by any doctor-hospital pair. So μ'' is a stable matching.

10.2.3 Finding Stable Matchings

It is not too difficult to construct an algorithm that can help find stable matchings. Some minor modifications of the Deferred Acceptance algorithm we saw in chapter 9 are needed to find stable matchings, though. The reason is simple: we need to take into account that now hospitals can be matched to more than one doctor. As for the one-to-one case, we can consider two versions of a Deferred Acceptance algorithm, depending on which side makes the proposals.

10.2.3.1 Doctors proposing For the doctors, the algorithm is identical to the Deferred Acceptance algorithm of chapter 9, the only difference being for the hospitals. In the first step of the algorithm, when a hospital receives offers from doctors, the decision to accept or reject offers works as follows. Each hospital h accepts the doctors making an offer to it, up to its capacity, one at a time, following the preference order over doctors. That is, hospital h first admits the most preferred doctor among the doctors making an offer to it, then the second most preferred doctor (among the doctors making an offer to it), and so on until either hospital h has accepted q_h doctors (where q_h is the capacity of hospital h) or, if less than q_h acceptable doctors have applied to hospital h, it has accepted all the acceptable doctors. The remaining doctors, if any, are rejected.

For the steps of the algorithm, the decision to accept or reject is made with the following rule. Each hospital receiving applications considers *the set of doctors it accepted at the previous step together with the set of new applicants*. From this larger set, the hospital accepts doctors up to its capacity, one at a time, starting with the most preferred doctors.

Example 10.3 There are three doctors, d_1, d_2, and d_3, and two hospitals, h_1 and h_2. Doctors' preferences over hospitals and hospitals' preferences over doctors (i.e., not over sets of doctors) are displayed in the table that follows. Hospital h_1 has a capacity of 2, $q_{h_1} = 2$, and hospital h_1 has a capacity of 1, $q_{h_2} = 1$.

P_{d_1}	P_{d_2}	P_{d_3}
h_1	h_2	h_2
h_2	h_1	h_1

P_{h_1}	P_{h_2}
d_1	d_2
d_2	d_3
d_3	d_1

We now run the Deferred Acceptance algorithm with doctors proposing.

Step 1 Doctor d_1 proposes to hospital h_1, and doctors d_2 and d_3 both propose to hospital h_2.

Hospital h_1 has only one offer, which is acceptable, so h_1 accepts d_1. Hospital h_2 has two offers, both of them acceptable. But h_2 can only hire one doctor. The most preferred doctor is d_2, so doctor d_2 is accepted and doctor d_3 is rejected.

Step 2 Doctor d_3 is the only doctor rejected at the previous step. She proposes to the next hospital in her preference order, hospital h_1.

Hospital h_1 has one new proposal, from d_3, and holds one offer from the previous step, from d_1. Both offers are acceptable, and since h_1 can accommodate two doctors, both offers are accepted. No doctor is rejected, so the algorithm stops. The final matching is

$$\mu = \{(d_1, h_1), (d_2, h_2), (d_3, h_3)\}.$$

10.2.3.2 Hospitals proposing The version of the algorithm with hospitals proposing follows the same principle as the other algorithms: hospitals make offers to doctors, who either accept or reject them. The main difficulty here will be to carefully specify to whom hospitals make their offers, as now they have preferences over groups of doctors.

Algorithm 10.1 Deferred Acceptance algorithm with Hospital Proposing

Step 1 Each hospital proposes to its most preferred set of doctors, and each doctor rejects all but the most preferred acceptable hospital that proposed to her.

Step k, $k \geq 2$ Each hospital that had one or more rejections at the previous step proposes to its most preferred set of doctors that satisfies the following conditions:

- The set must contain all doctors the hospital proposed to at an earlier step (i.e., at steps $1, 2, \ldots, k-1$) and who have not rejected it.
- Any additional doctor in the set must be a doctor to whom the hospital did not propose earlier.

Once the hospitals have made their proposals, each doctor rejects all but the most preferred acceptable hospital that proposed to her so far.

Example 10.4   To illustrate the Deferred Acceptance algorithm with doctors proposing, we use the situation in example 10.3.

Step 1 Hospital h_1 wants to hire two doctors. The most preferred set is made with doctors d_1 and d_2, so hospital h_1 proposes to these two doctors. For hospital h_2, the choice is easier, as it wants to hire only one doctor. So h_2 proposes to d_2.

Doctor d_1 receives one offer, from h_1. This offer is acceptable, so d_1 accepts it. Doctor d_2 has two offers, from d_1 and d_2. Hospital h_2 is preferred, so d_2 accepts h_2 and rejects h_1's offer.

Step 2 Hospital h_1 is the only hospital with offers being rejected at step 1. So now h_1 has to make an offer to two doctors. So far, hospital h_1 has proposed to d_1 and d_2. Doctor d_1 is the only doctor who has not rejected h_1, so doctor d_1 must again receive an offer from h_1. Doctor d_3 is the only doctor who has not yet received an offer from h_1, so hospital h_1 must make an offer to the most preferred set of doctors among the following two sets: $\{d_1, d_3\}$ and $\{d_1\}$. The responsiveness condition implies that $\{d_1, d_3\}$ is the most preferred set, so hospital h_1 makes an offer to these two doctors.

Both doctors d_1 and d_3 accept h_1's offer. No offer is rejected, so the algorithm stops. The final matching is

$$\mu = \{(d_1, h_1), (d_2, h_2), (d_3, h_1)\}.$$

10.2.4   One-to-One v. Many-to-One Matchings: Similarities and Differences

The many-to-one matching model we have described and the Deferred Acceptance algorithm enjoy most of the properties we have seen for the one-to-one matching model, but not all.

First, in a many-to-one matching model with responsive preferences, stable matchings always exist, for any preferences and for any collection of hospitals' capacities. The existence is relatively simple to show as it follows the same lines as the proof of existence for the one-to-one matching model.

In our many-to-one matching model, there is also a stable matching that is the most preferred stable matching for the doctors (the *doctor-optimal matching*), and a stable matching that is the most preferred stable matching for the hospitals (the *hospital-optimal matching*). How do we find these matchings? The answer is simple: with the Deferred Acceptance algorithm! To find the doctor-optimal matching, we simply have to run the algorithm with doctors proposing, and for the hospital-optimal matching, it has to be the doctors proposing.

These findings are in fact expected. To see this, observe that if each hospital can accommodate only one doctor, then the problem becomes a simple one-to-one matching model. In other words, the one-to-one model of chapter 9 is a special case of the many-to-one model we study here.

Some of the incentive properties of the Deferred Acceptance algorithm also carry over in our many-to-one model. If we run the version of the algorithm with doctors proposing, it is still a dominant strategy for the doctors to reveal their true preferences over hospitals. However, if we run the version of the algorithm with hospitals proposing, hospitals may have an incentive to misrepresent their preferences. To see this, consider the following example.

There are three hospitals, h_1, h_2, and h_3, and four doctors d_1, d_2, d_3, and d_4. The capacities are $q_{h_1} = 2$, $q_{h_2} = 1$ and $q_{h_3} = 1$. The preferences are given by the table below (we assume that each doctor find each hospital acceptable and each hospital finds each doctor acceptable).

P_{d_1}	P_{d_2}	P_{d_3}	P_{d_4}
h_3	h_2	h_1	h_1
h_1	h_1	h_3	h_2
h_2	h_3	h_2	h_3

P_{h_1}	P_{h_2}	P_{h_3}
d_1	d_1	d_3
d_2	d_2	d_1
d_3	d_3	d_2
d_4	d_4	d_4

If we run the Deferred Acceptance algorithm with hospitals proposing we obtain the matching μ_D,

$$\mu_H(h_1) = \{d_3, d_4\}, \quad \mu_H(h_2) = d_2 \quad \text{and} \quad \mu_H(h_3) = d_1.$$

Consider now the matching μ',

$$\mu'(h_1) = \{d_2, d_4\}, \quad \mu'(h_2) = d_1, \quad \text{and} \quad \mu'(h_3) = d_3.$$

Notice first that all hospitals strictly prefer μ' to μ_D. It is easy to see that this is the case for hospitals h_2 and h_3. For hospital h_1, observe that μ' is obtained by replacing d_3 by d_2. Since hospital h_1 prefers d_2 to d_3, the definition of responsive preferences implies that hospital h_1 prefers μ' to μ_D.

The matching μ' can in fact be obtained with the Deferred Acceptance algorithm with hospital proposing when hospital h_1 submit the preferences $P'_{h_1} = d_2, d_4, d_3, d_1$. With the submitted preference profile $(P'_{h_1}, P_{h_2}, P_{h_3})$ the Deferred Acceptance with hospital proposing stops at the first step: no doctor receives an offer from more than one hospital (and thus no hospital has a proposal that is rejected).

The above example also illustrates an additional difference between the one-to-one and the many-to-one matching models. In the one-to-one matching model there is no matching that is preferred by *all* singers to the singer-optimal matching (and similarly there is no matching that is preferred by *all* musicians to the musician-optimal matching). In the many-to-one case that result does no longer hold: the matching μ' is preferred by all hospitals to the hospital-optimal matching.

10.3 Why Stability Matters

10.3.1 A Natural Experiment

The development and success of the NRMP suggests that having a centralized procedure that computes stable matchings is the key to a well-functioning matching market. But it was not until the early 1990s that Roth came up with a convincing argument.[4] The idea was to consider the market for physicians and surgeons in the United Kingdom. The problem there was similar to that of the U.S. market for residents, except that the market was split into several regional markets and those several regions were using a centralized procedure where market participants submit their preferences over potential matches. Roth observed that the different regions did not use the same algorithm, and some regions had stopped using some procedures that they initially devised. Most of the time, a procedure would stop being used simply because participants (students and hospital) were able to find better matches than the ones proposed by the procedure. But observe that this is precisely what the concept of stability aims at avoiding! Indeed, when the matching is stable, no doctor or hospital can find a better partner than the one that is given by the matching. A comparison of the various procedures analyzed by Roth is given in table 10.1.

Table 10.1
Stable and unstable algorithms

Market	Use stable algorithm?	Still in use? (in 1990)
Edinburgh (1969)	Yes	Yes
Cardiff	Yes	Yes
London Hospital	No	Yes
Cambridge	No	Yes
Birmingham	No	No
Edinburgh (1967)	No	No
Newcastle	No	No
Sheffield	No	No

Remarkably, procedures that produce stable matchings tend to perform relatively well and are maintained over time, while markets that do not produce stable matchings are eventually abandoned.[5]

10.3.2 Unraveling in the Lab

The analysis of the medical markets in the United Kingdom done by Roth strongly suggests that the use of a stable matching mechanism is the right approach. But this analysis left several questions unanswered. First, it is possible that the evolution of the medical markets across the United Kingdom we documented in section 10.3.1 is governed not only by the choice of the algorithm but also by other factors that are not directly related to the way the markets are organized. Second, the analysis of the British medical match does not fully address the unraveling observed in the American medical match we discussed in section 10.1. Before the NRMP was implemented, unraveling was prevalent across U.S. hospitals, and this phenomenon nearly disappeared once the centralized matching mechanism was implemented. There are reasons to believe that the demise of unraveling was caused by the NRMP's use of an algorithm that produced stable matchings, but it could simply be that unraveling is most prevalent in decentralized markets.

The American and British experiences nevertheless suggest that the use of a centralized mechanism that utilizes a stable matching algorithm is the right design for markets like the medical match. To confirm this intuition, John Kagel and Alvin Roth conducted a lab experiment that would mimic both the American and the British markets and their transition to a centralized stable matching mechanism.[6]

In their experiment, Kagel and Roth assigned roles to participants: half of the subjects were assigned the role of "firms" and the other half the role of "workers." In this experiment, each worker had to match to a firm (and conversely, each firm had to match to a worker). Half of the firms and half of the workers were identified as "high productivity." The other subjects were identified as "low productivity." The situation at hand was a one-to-one matching model, and the payoffs were designed as follows:

- A subject (worker at a firm) matched to a high-productivity partner would earn *about* $15.
- A subject (worker at a firm) matched to a low-productivity partner would earn *about* $5.
- A subject (worker at a firm) not matched earns $0.

The payoffs of $15 and $5 when matched to a high- or low-productivity partner are only averages. Small differences were introduced so that the preferences among potential partners were strict. These small differences were such that:

• All workers or firms agree that being matched to a high-productivity partner yields a higher payoff than being matched to a low-productivity partner.

• Workers disagree about which high-productivity firm is the best or worst, and also which low-productivity firm is the best or worst. Similarly, firms disagree about the ranking of high-productivity workers and the ranking of low-productivity workers.

To mimic the American and British situations, Kagel and Roth used two designs:

Design 1. A decentralized market that runs over three periods.

At each period, the subjects playing the roles of firms would have to make offers to the workers. A worker receiving one or several offers only has to decide whether to accept the highest offer she received or reject all offers.

The payoff of a worker or a firm that has been matched is reduced by
• $2 if the match is made in the first period;
• $1 if the match is made in the second period.

The payoff reduction was intended to mimic the early match observed in the American medical market (the unraveling phenomenon). As we discussed in section 10.1, accepting an early match proposal has a cost.

Design 2. A centralized market where both firms and workers have to submit a preference ordering among potential partners, and an algorithm is used to match workers and firms.

For this design, two algorithms were considered:
• the Deferred Acceptance mechanism;
• the algorithm that was used in Newcastle, which works as follows. First, for each worker-firm pair, we calculate the *priority product* as being equal to the rank of the firm in the workers' preferences × the rank of the worker in the firm's preferences.

(a) **Step 1** Match a worker and a firm if their priority product is 1.

(b) **Step 2** Among the remaining workers and firms, match a worker and a firm if their priority product is 2.[7]

(c) **Step** $k = 3, 4, \ldots$ Among the remaining workers and firms, match a worker and a firm if their priority product is k.

This algorithm does *not* guarantee producing stable matchings.

The instruction given to subjects about the matching algorithm was the same for both treatments, noting that they were actually used in real-life settings and even

The Medical Match

suggesting to subjects that ranking one's choices according to their payoffs was the best strategy (i.e., the top choice being the partner bringing the highest payoff, the second choice being the partner bringing the second-highest payoff, etc.).

In their experiment, Kagel and Roth first ran Design 1 by itself ten times, and then ran fifteen times a combination of Design 1 and Design 2 that consisted first of running Design 1 but only with the first two periods and then running Design 2 for the subjects that were still not matched at the end of the second period. The experiment thus mimics the American situation (over ten repetitions) with the introduction of a centralized mechanism.

There are several ways to assess the performance of the market design in this experiment, with the most immediate metrics being:

- the cost of an early match ($2 for a period 1 match, $1 for a period 2 match); and
- the number of pairs being matched at each period.

In this experiment, there are two sources of inefficiency. We have already mentioned the first source, the fact that some matches can be inefficient because they are done too early (through the discount applied to the payoffs if matched in period 1 or period 2). The second source of inefficiency occurs when a high-productivity subject is matched to a low-productivity one. In the experiment, the payoffs (and thus the preferences that are derived from them) are such that, at the (unique) stable matching, high (low)-productivity firms are matched to high (low)-productivity workers. Economists call such matchings **assortative matchings**. Since there are throughout the experiment some periods where the matching market is decentralized (i.e., subjects make their matching decisions themselves), some individuals can make suboptimal decisions (e.g., accepting an early offer from a low-productivity firm) to protect against the risk of ending up with a worse match at a later period.

Kagel and Roth observed unraveling in their experiment: there are always subjects that manage to get matched in the first or second period of phase 1. When analyzing the first ten repetitions (i.e., when there is only Design 1), Kagel and Roth observe that the cost of an early match increases. That is, when subjects start to become more familiar with the setup, the number of early matches increases. Unraveling is thus pervasive. However, the introduction of a centralized market does reduce the cost of an early match. When the Deferred Acceptance algorithm is used, unraveling is severely minimized: after about ten repetitions, no matches are done in period 1. In contrast, when the unstable matching mechanism is used, the number of early matches (in periods 1 and 2) increases (albeit modestly).

Further scrutiny of the experimental results also shows different behavior between low- and high-productivity subjects. Most of the early matches are made by high-productivity individuals. This is in line with the stakes faced by the individuals: high-productivity subjects have more to lose than low-productivity ones and thus face a higher cost of not being matched.

When the centralized market is introduced after ten repetitions, Kagel and Roth observe a slight initial decrease of early matches for both matching algorithms. That is, the use of a centralized market to match individuals is initially attractive for individuals. However, as time passes by, subjects act differently whether the algorithm used is the Deferred Acceptance or the unstable one. When the Deferred Acceptance algorithm is used, early matches for high-productivity subjects sharply decrease, while they increase when the unstable algorithm is used. In other words, the use of the Deferred Acceptance algorithm makes the market **safe** for participants. There is no risk in delaying their matching decisions to the third period. In contrast, an unstable matching algorithm maintains unraveling (i.e., early match) even if it is costly for participants to do so.

10.4 The Rural Hospital Theorem

A remarkable aspect of the matching procedure set by the NRMP is that it produces stable matchings. As the analysis of the medical match in the United Kingdom suggests, the use of a stable matching procedure is a key element that guarantees the viability of a matching procedure. But for policy makers this may not be enough. As we saw in chapter 9, for a given situation (i.e., preferences of agents) there might be more than one stable matching. The multiplicity of stable matchings is not confined to the one-to-one matching model. Many-to-one matching problems can also have multiple stable matchings. In this case, it could be that the policy makers (or the people running the matching procedure) would prefer some stable matchings over others.

One of the issues that quickly arose with the development of the American medical match concerns the distribution of doctors across hospitals. As it turns out, candidates usually strongly favor hospitals in large urban areas, neglecting hospitals in rural areas. As a consequence, hospitals in urban areas had no problems filling their vacancies, but hospitals in rural areas often end the hiring season hiring fewer doctors than they were planning. Clearly, such situations can have a negative impact for health policies. The question that arose was then the following: could the NRMP modify its algorithm such that it would still produce stable matchings but maximize the number of hires in rural hospitals? It turns out that the answer to that question is negative, and it is summarized in a famous theorem in the matching literature.[8]

The Medical Match

Result 10.1 (Rural Hospital Theorem)   For any preferences of doctors and hospitals, if at a stable matching a hospital does not fill all its vacancies, then it does not fill all its vacancies at *any* stable matching.

Furthermore, if a hospital does not fill its vacancies at some stable matching, it is matched to the same set of doctors at *any* stable matching.

The proof of that surprising result is not too difficult when hospitals' preferences are responsive (which is the case we consider in this chapter). For simplicity, we will outline the proof when each hospital has only one vacancy.[9]

To prove the rural hospital theorem, we first need another (famous) result: the *Decomposition Lemma*[10]

Decomposition Lemma.   *Let μ and μ' be two stable matchings for the same problem. Let A be the set of doctors who prefer μ' to μ and B the set of hospitals that prefer μ to μ'. Each doctor in A is matched, under both μ and μ', to a hospital in B (but not the same hospital!). Similarly, each hospital in B is matched, under both μ and μ', to a doctor in A.*

The intuition behind the Decomposition Lemma is relatively straightforward to show. Consider a doctor, say d, in A, so d prefers μ' to μ. Note that it could be that she is matched to herself in μ. But since she prefers μ' to μ, it must be that she is matched to a hospital under μ' (because $\mu'(d) \neq \mu(d)$). Let h be the hospital doctor d is matched to under μ'; that is $\mu'(d) = h$. Now we must have that hospital h prefers μ to μ', for otherwise the pair (d,h) would block the matching μ' (and thus would contradict the fact that μ' is stable). So hospital h is in the set B.

We now know that each doctor in A is matched, under μ, to a hospital in B. This implies that if there are, say, ten doctors in A, we then have *at least* ten hospitals in B. It now suffices to repeat the same argument but with the hospitals, and we deduce that each hospital in B is matched to a doctor in A under μ'. Note that this implies that there cannot be more than ten hospitals in B, so there are as many doctors in A as there are hospitals in B, and the statement of the Decomposition Lemma follows.

Now that we have the Decomposition Lemma, we can outline the proof of the rural hospital theorem. Take a stable matching, call it μ, and suppose that there is a doctor, say d, such that $\mu(d) = d$, meaning the doctor is not matched to any hospital under μ. Let μ' be another matching. We need to show that for that other *stable* matching we have $\mu'(d) = d$ as well. Suppose this is not the case, so, under μ', doctor d is matched to a hospital. Call that hospital h. Observe that we cannot have $dP_d h$, for otherwise the matching μ' would not be individually rational (doctor d would prefer being alone to being matched to h). We then have $hP_d d$, meaning doctor d prefers μ' (she is matched to h) to μ (she is not matched to any hospital). So doctor d belongs to the set A (the same set we had for the Decomposition Lemma). But observe that the Decomposition Lemma says that, under the matching μ, all doctors in A are matched to a hospital in B. That is, doctor d must be matched to a hospital (in B) under the matching μ, so we should not have $\mu(d) = d$. Hence, it is not possible to obtain two stable matchings where a doctor is matched to a hospital under one of the stable matchings but not under the other stable matching.

10.5 The Case of Couples and the Engineering Method

The NRMP, while successful, is not exempt from some issues. One of the major problems was that in the early 1970s an increasing number of married couples abstained from participating in the NRMP. The main reason was that the initial algorithm set by the NRMP was not handling the case of couples who were seeking to be matched to either the same hospital or nearby hospitals. It was easier for the couples to circumvent the NRMP and find a match by negotiating directly with hospitals.

10.5.1 A Very Complex Problem

The people running the NRMP were aware of this problem and introduced several modifications to their algorithm. The solution they came up with was the following. Couples could either negotiate directly with hospitals (i.e., not participate in the NRMP) or register as a couple and enter a special "couples algorithm" offered by the NRMP. This algorithm consisted of asking couples to designate a "leading member," who would be matched to a hospital in the usual way (i.e., like the candidates that are singles). The preference list among hospitals of the other member of the couple would then be edited to remove distant positions and then be matched to a possible hospital near the leading member.

Unfortunately, this fix to the matching procedure did not solve the participation problem. One of the main reasons for that is that the couples algorithm did not really allow couples to submit preferences over *pairs of positions*. In the mid-1980s, the NRMP modified its procedure again so that couples would be able to express such preferences. But solving the problem turns out to be harder than one might expect. The issues brought on by the presence of couples are deeper than they might appear, as one flirts with what it is possible to do when designing a matching market. In a "simple" two-sided matching problem like the one we saw in chapter 9, stable matchings always exist, and we know how to find them. When there are couples, stable matchings may not exist, and many other attractive properties no longer hold.[11]

The difficulties encountered by the societal evolution of the medical match led the NRMP to ask Alvin Roth and his colleagues to overhaul the matching algorithm used to that point and design a new matching procedure.[12] As they explain, most of the existing results in the matching literature only have bite in simple

The Medical Match

matching problems. The medical match is a more complex problem for which the theory has little to say and, when it does, the counterexamples (i.e., negative results) constitute most of what the theory can say about such markets. In the medical match, not only do some doctors search for jobs in pairs (and have preferences among pairs of positions), but some hospitals can also have complex requirements. For instance, for some hospitals, the number of doctors it can hire in one medical specialty may depend on the number of doctors hired in another specialty.

10.5.2 When Theory Fails

To see how the theory can fail with complex situations like the one brought on by the presence of couples, consider the following situation. In a one-to-one matching problem, we have seen that during the course of the deferred acceptance, the accepting side never regrets rejecting someone. That is, if the medical match is a one-to-one matching problem and we use the Deferred Acceptance algorithm with doctors proposing, a hospital rejecting a doctor at some step of the algorithm always does that because it can be matched to a more preferred doctor. When there are couples, this may no longer be the case. To see this, consider Alice and Albert, a couple participating in the medical match. Suppose that their more preferred pair of positions are hospitals h_1 and h_2 in New York. If they cannot both have a job at these two hospitals, then their second most preferred option is to have jobs at hospital h_3 in Boston. Hospitals h_1 and h_2 have only one position, but hospital h_3 can hire two doctors. We also have two other (single) doctors: Bill and Carol. Bill's top choice is hospital h_1, and Carol's top choice is hospital h_2. The preferences of the hospitals (and those of the doctors) are given in the following table.

Alice and Albert	Bill	Carol
(h_1, h_2)	h_1	h_2
(h_3, h_3)		
(not hired, h_2)		

h_1	h_2	h_3
Bill	Albert	Alice
Alice	Carol	Albert

If we run the Deferred Acceptance algorithm with doctors proposing, we have in the first step hospital h_1 receiving a proposal from Bill and Alice, and hospital h_2 receiving a proposal from Albert and Carol. Hospital h_1 prefers Bill, so it rejects Alice, and hospital h_2 rejects Carol in favor of Albert. At the end of this step, Albert is tentatively matched but not Alice. If we stop here, this matching is not stable:

Alice and Albert prefer to be hired by hospital h_3, a better outcome for that hospital. Modifying the algorithm to allow Alice and Albert to propose to hospital h_3 in the next step then breaks with the spirit of the Deferred Acceptance algorithm: hospital h_2 is now matched with no one; it regrets having rejected Carol's proposal in the first step of the algorithm.

10.5.3 Fixing the NRMP

Roth and his colleagues introduced several profound modifications in the algorithm in order to cope with the presence of couples and other complexities. Two of the most important changes were:

- **Switch to a doctor-proposing algorithm.** The original algorithm developed for the NRMP and the subsequent modifications were based on the Deferred Acceptance algorithm with hospital proposing. It was thought to be fairer for the candidates and also, as Roth and Peranson (1999) explain, increase the odds of finding optimal, stable matchings.
- **Process some doctors' proposals sequentially.** In the original Deferred Acceptance algorithm, all proposing agents make their offers at the same time. In the new NRMP algorithm, some offers are made one at a time. As Roth and Peranson explain, processing that way allows the (new) algorithm to better correct the potential sources of instability and correct them along the way.

The new NRMP algorithm is rather complex, but a rough description can be offered.

Algorithm 10.2   NRMP Algorithm with Couples

Step 1 Run the Deferred Acceptance algorithm with doctors proposing considering all hospitals but only the single doctors.

Step 2 One by one, match couples to pairs of hospitals in order of their preferences.

For instance, if a couple has the preferences (h_1, h_2), (h_3, h_4), (h_5, h_6), then we first try to match the couple to hospitals h_1 and h_2. If there are vacancies at these hospitals or if the couple members are preferred to (single) doctors already matched to these hospitals, then the couple is matched (and single doctors rejected if necessary).

In this process of matching couples, some single doctors may become unmatched because a doctor who is part of a couple took her position.

Step 3 For all single doctors displaced in step 2, try to match them, one by one, to a hospital in order of their preferences.

The new algorithm was first used in 1998 and turned out to be a success. Most of the problems encountered earlier vanished, and the participation rate went up again. But the story of the improvement of the NRMP procedure does not end here. As Roth and Peranson explain, the design of the new algorithm was made possible thanks to a new approach in the design of matching markets. As we have seen, the many complexities of the medical match mean that the theory of simple matching models does not apply directly. To overcome that difficulty, Roth and Peranson used computer simulations using data from previous years to test various possible designs. Computer experiments were first designed following the intuitions coming from the standard matching theory, and the results indicated which change in the algorithm was better. The theory was used in the sense that it helps direct the practical design and gives hints about which problems can be encountered (e.g., the source of instability, as in our preceding example with Alice and Albert). In other words, Roth and Peranson worked like engineers: the theory is here to provide guidance, but because the real problem is too complex to be fully addressed by the theory, experiments are run in order to fine-tune the details.

Surprisingly, the analysis of the real-life data showed that the structure of candidates' and hospitals' preferences avoided the nonexistence results predicted by the theory. In other words, the types of preferences in a matching with couples problem where stable matchings do not exist do not seem to resemble the types of preferences we can observe in real life. Also, the algorithm the authors designed does produce matchings that are stable. This was not obvious at the outset, because for some preferences the algorithm does not output a well-defined matching (let alone a stable one).

11 Assignment Problems

This chapter is devoted to a model similar to the matching model we saw in chapter 9. That is, we will consider two sets and study how to match elements from one set to elements from the other set. The main difference with the matching model we have studied so far is that one set consists of individuals while the other set contains "objects" (e.g., house, occupation). To distinguish it from the matching problem, the problem of matching individuals to objects is referred to as an *assignment problem*. As we will see, the assignment problem is very similar to the matching problem, but it will trigger different questions and give the opportunity to study new algorithms.

11.1 The Basic Model

The basic model is relatively simple. We have a set of individuals and a set of objects. As for the matching model, we will not consider monetary transactions. Therefore we can assume that individuals have fixed preferences over objects. An **assignment problem** is given by

- a set of individuals $I = \{i_1, i_2, \ldots, i_n\}$;
- a set of objects $K = \{k_1, k_2, \ldots, k_m\}$; and
- for each individual a preference relation over the objects.

For now, the model looks like the matching model of chapter 9, except that we replaced one side (the musicians or the singers) with objects. However, such a change has an important impact: objects have no preferences over their potential owners/partners!

The main problem we have in such models is to find an assignment that allocates objects to individuals. As for matching problems, there are several cases we can consider: an agent can be assigned at most one object, an agent can be assigned several objects, and other variations. For the sake of simplicity, we will study the simple case where an individual can be assigned to at most one object and each object is assigned to at most one individual. That is, we will consider here **one-to-one assignment problems**.

Definition 11.1 An **assignment** is a function $\mu : I \cup K \to I \cup K$ such that:

(a) For each individual $i \in I$, $\mu(i) \in K \cup \{i\}$, and for each object $k \in K$, $\mu(k) \in I \cup \{\emptyset\}$.
(b) If $\mu(i) = k$, then $\mu(k) = i$.

The interpretation of these two conditions is similar to the one we had for matching. Condition (a) says that an individual i is either assigned an object ($\mu(i) \in K$) or not assigned any object ($\mu(i) \in \{i\}$, which is equivalent to $\mu(i) = i$). An object k can be assigned to an individual $\mu(k) \in I$ or not assigned, $\mu(k) = \emptyset$. Condition (b) says that if individual i is assigned object k ($\mu(i) = k$), then object k is assigned to individual i ($\mu(k) = i$).

11.1.1 Public versus Private Endowments

There are two families of assignment problems, depending on who the objects initially belong to when we consider a specific problem, *before* we start addressing the issue of assigning objects to individuals.

- **None of the objects belong to anyone**. This case is known as the **public endowment** problem. This means that endowments (who own the objects) belong to the whole society (i.e., the group of individuals). A typical example of such a situation is the case of assigning students to schools.
- **The individuals own the objects**. For such problems, we say that individuals are *endowed* with an object (or several objects). We call these problems **private endowment** problems. They are similar to barter, where individuals exchange goods without monetary transactions.

These two variations can also be merged into a problem where some objects are privately owned by some individuals and other objects constitute a public endowment (for which we will see an example in chapter 16). This is the case, for instance, on many campuses when we want to allocate dorms to students: some students are already assigned a room (sophomores and older students), which they may exchange for another one but can keep if they want (so it is as if they own their rooms). The rooms occupied by the students who just graduated (and thus left the campus) are assigned to nobody and are thus part of the public endowment.

11.1.2 Evaluating Assignments

Whatever the problem (public or private endowments), we need some criterion to evaluate assignments. There are two cases we should consider for such problems:

Assignment Problems

- Each object has a *priority ordering* over agents. Such orderings specify who has more "rights" to possess the object or, put differently, which individuals should be considered first when allocating this object. For instance, if two individuals claim the same object, then it would be assigned to the one with the higher priority.

  Formally, those priority orderings work like preference orderings. One may interpret them as objects having some sort of "preference" over the individuals. But since objects do not have "preferences," these orderings will not be taken into account whenever we analyze the overall welfare of an assignment. In chapter 13, we will study in detail a particular case of the assignment model where objects have priority orderings over individuals.

- The objects are "completely free"; that is, they do not come with an ordering over individuals.

The main property that economists usually consider when analyzing assignments in this context is **efficiency**, which refers to *Pareto efficiency*.

Definition 11.2 An assignment μ is efficient if there is no other assignment μ' such that:

(a) each individual either prefers μ' to μ or is indifferent between the two assignments and

(b) there is at least one individual who strictly prefers μ' to μ.

Example 11.1 There are three individuals (Alice, Bob, and Carol) and three objects (A, B, and C). Their preferences over these objects are given in the following table.

P_{Alice}	P_{Bob}	P_{Carol}
A	C	B
C	A	C
B	B	A

The assignment $\mu(\text{Alice}) = C$, $\mu(\text{Bob}) = A$, and $\mu(\text{Carol}) = B$ is *not* efficient, because the assignment $\mu'(\text{Alice}) = A$, $\mu'(\text{Bob}) = C$, and $\mu'(\text{Carol}) = B$ is better for both Alice and Bob (while making Carol indifferent). So μ' satisfies both conditions (a) and (b) of definition 11.2.

11.2 Finding Efficient Assignments

11.2.1 Serial Dictators

If we want to find an efficient assignment, the most basic solution is very simple and is often used in practice: take an order of individuals and let them pick, one by one, their most favorite object among the objects that are (still) available. The algorithm we just described is known as the **Serial Dictatorship algorithm**. The word *dictator* comes from the fact that at each step an agent chooses what is best for her, without considering the other individuals.

Algorithm 11.1   Serial Dictatorship

Step 0 Pick an order of the individuals.

Step 1 The first individual in the order is assigned her most preferred object.

Step k, $k \geq 2$ The individual ranked k in the order is assigned her most preferred object among all objects except the ones taken by the first $k-1$ individuals in the order.

End The algorithm stops when all individuals have chosen an object or when there is no object left.

Serial Dictatorship is more useful when endowments are public, but there is nothing to prevent us from using it when endowments are private (in this case, the initial ownership is not taken into account).

Result 11.1   For any order of individuals (dictators), the assignment obtained with the *Serial Dictatorship* algorithm is efficient.

The proof of result 11.1 is relatively easy. To see this, suppose that the assignment (which we call μ) is not efficient. This means that there is another assignment, say μ', that makes some individuals better off (and leaves others indifferent). Consider the set of individuals who strictly prefer μ' to μ, and let Alice be the first person in the sequence who prefers μ' to μ. So all the individuals (dictators) who chose *before* Alice obtain the same object with μ and μ'. This means that when we construct μ and it is Alice's turn to choose, the object she gets with μ' (the assignment she prefers) is available: it is not taken by any individual who chose before her (otherwise Alice would not be the *first* individual in the sequence who gets a different object between μ and μ'). But that contradicts the fact that Alice picks $\mu(i)$ with the serial dictatorship rule when we construct μ. Indeed, if Alice prefers $\mu'(i)$ to $\mu(i)$, she must choose $\mu'(i)$ when it is her turn—or an object she prefers to $\mu'(i)$—which cannot be $\mu(i)$. So we have a contradiction with the assumption that Alice prefers μ' to μ, and thus μ is necessarily efficient.

11.2.2   Trading Cycles

When there are private endowments, we can of course use the Serial Dictatorship algorithm to find an efficient assignment, but this solution may create a problem: an individual can end up with an object that is worse than the object she initially owned. Avoiding an agent being worse off (when comparing an assignment and her endowment) is crucial for market participation. If an individual faces the risk of being worse off, she may not want to participate in a market mechanism, fearing to swap her endowment for a less preferred good.

One of the most famous solutions for this problem is the so-called **Top Trading Cycle** algorithm. The general idea consists of finding a sequence of individuals such that:

- Each individual in the sequence is assigned the object owned by the next individual in the sequence.

For instance, if in the sequence we have Alice, Bob, Carol, Denis, ... then Alice takes Bob's object, Bob takes Carol's object, Carol takes Denis's object, and so on. The last individual in the sequence would take Alice's object.

- The sequence is such that each individual prefers the object of the next person in the sequence.

In the preceding example, we would then require that Alice prefer Bob's object to hers, Bob prefer Carol's object to his, and so on.

Clearly, if each agent in this sequence obtains a more preferred object, then they are all better off. But to obtain an efficient assignment, we need to take the "best" sequence. To see this, consider the following example with three individuals (Denis, Erin, and Francine) and three objects (A, B, and C). At the outset, Denis owns A, Erin owns B, and Francine owns C. Their preferences over these objects are given in the following table.

P_{Denis}	P_{Erin}	$P_{Francine}$
B	C	A
C	A	B
A	B	C

Suppose the sequence is (Denis, Francine, Erin). Denis is the first to decide, so he takes Francine's object, C. The second person to decide is Francine. She takes Erin's object, B, and then Erin, the last person in the sequence, is assigned Denis's object, A. We thus obtain the assignment $\mu(\text{Denis}) = C$, $\mu(\text{Erin}) = A$, and $\mu(\text{Francine}) = B$.

Clearly, the assignment μ is not efficient, because all individuals prefer the assignment $\mu'(\text{Denis}) = B$, $\mu'(\text{Erin}) = C$, and $\mu'(\text{Francine}) = A$ to the assignment μ. That assignment could have been obtained with the sequence (Denis, Erin, Francine), so the question is to choose the right sequence.

The solution is to construct the sequence using the individuals' preferences: we take an individual, say Denis, and the next individual will be the agent possessing Denis's most preferred object. In our example, it is Erin. So we put Erin next in the sequence, just after Denis. Then next to Erin we put the agent who possesses Erin's most preferred object, and we continue that way, adding individuals to our sequence.

One issue we may face when constructing the sequence is whether the owner of the most preferred object of the last person in the sequence is the first individual in the sequence. The answer is yes and no. It is *no* because individuals may have any type of preference, so there is no way to ensure that the *first* person in the sequence is the owner of the *last* person's most preferred object.

But the answer is also *yes*, because even though we may not go back to the first person in the sequence, we necessarily go back to *someone* in the sequence. All that matters is to have a *cycle* of individuals. To see this, suppose that we start with individual i_1 to construct our sequence. Then i_1's most preferred object is owned by i_2, and i_2's most preferred object is owned by i_3. Continuing this way, suppose that so far we have constructed the sequence

$(i_1, i_2, i_3, \ldots, i_k)$.

Suppose that the object that i_k prefers the most is not the one owned by i_1 but the object owned by, say, i_3. In this case, we could have started with i_3 to construct our sequence! That way we would obtain a sequence $(i_3, i_4, \ldots, i_k)$.

Also, since we always consider for this type of problem a finite number of individuals and objects, if all objects are initially owned by agents, for any individual we may start with we *must* eventually have an agent added to the sequence such that her most preferred object is owned by an individual who is already in the sequence.

Once we have a cycle, we allocate to each individual in the sequence the object owned by the person next to it, and we withdraw these agents (and the objects they have exchanged) from our problem. Note that these agents have managed to be assigned their most preferred objects, so, *for these agents*, there is no way to improve their situation (in a Pareto sense).

If there are some agents left, individuals who were not part of this first cycle/sequence, we just start searching for another sequence, but considering only the remaining objects. We keep doing this until all agents have been allocated an object.

The Top Trading Cycle algorithm is usually described using graphs, where individuals "point" to the owner of the object they prefer the most. We now give a more formal presentation of this procedure and then illustrate it with an example (where we will draw those graphs).

Assignment Problems

Algorithm 11.2   Top Trading Cycle

Step 1 For each individual, we draw an arrow starting from that individual to the individual who owns her most preferred object. If an agent's most preferred object is already hers, then the arrow goes to herself (we call that a *self-cycle*).

We have seen that there must be at least one cycle: if we start from an agent in the cycle and we follow the arrows, we eventually visit that agent again.

Each individual in a cycle is assigned the object owned by the individual she is pointing to. Agents in a cycle and the objects they are assigned to are removed from the problem.

Step k, $k \geq 2$ For each individual, we draw an arrow starting from that individual to the individual who owns her most preferred object *among the objects that have not been removed*. If an agent's most preferred object is already hers, then the arrow goes to herself (we have a self-cycle).

There must be at least one cycle. Each individual in a cycle is assigned the object owned by the individual she is pointing to. Agents in a cycle and the objects they are assigned to are removed from the problem.

End The algorithm stops when all individuals have been removed or there are no acceptable objects left for any individual who has not yet been removed.

Example 11.2   Consider five individuals, Alice, Bob, John, Lisa, and Suzanne, and five objects, A, B, C, D, and E. The following table gives the preferences over those objects, with the first row indicating the endowments (who owns the objects). So, for instance, Alice initially owns object A.

Endowment →	A	B	C	D	E
	P_{Alice}	P_{Bob}	P_{John}	P_{Lisa}	$P_{Suzanne}$
	B	C	A	C	A
	E	A	E	A	C
	D	D	D	E	B
	C	E	B	B	D
	A	B	C	D	E

Step 1 Alice's most preferred object is B, and the owner of B is Bob. So we draw an arrow from Alice to Bob. Repeating the procedure for the other individuals, we obtain the graph of figure 11.1, where next to the arrow, we indicate the object the

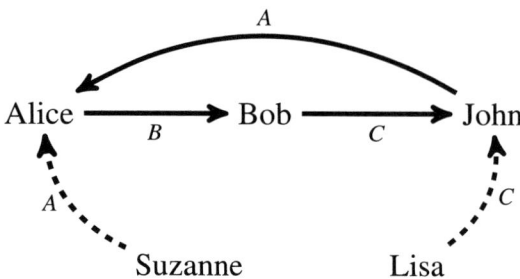

Figure 11.1.
Step 1

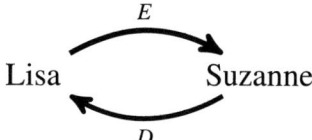

Figure 11.2.
Step 2

pointing individual wants (e.g., Alice wants object B, so the arrow from Alice to Bob is labeled B).

We have a cycle with Alice, Bob, and John (in bold). The dashed arrows are not part of any cycle. So each individual in the cycle gets the object owned by the person they are pointing to. That is, Alice gets object B, Bob gets C, and John gets A. We withdraw Alice, Bob, and John from the problem, and we also withdraw objects A, B, and C. So only Lisa and Suzanne remain, and so do the objects D and E.

Step 2 Now we ask Lisa and Suzanne to point to the person owning the object that they prefer the most between D and E. So we obtain the (very) simple graph of figure 11.2.

This is clearly a cycle, so Lisa obtains the object owned by Suzanne, and Suzanne gets object D.

So the assignment under the Top Trading Cycle algorithm is

$\mu(\text{Alice}) = B$, $\quad \mu(\text{John}) = A$,

$\mu(\text{Bob}) = C$, $\quad \mu(\text{Lisa}) = E$,

$\mu(\text{Suzanne}) = D$.

Result 11.2 The assignment obtained with the Top Trading Cycle algorithm is efficient.

The intuition behind result 11.2 is very similar to the argument used to show that the Serial Dictatorship algorithm yields an efficient assignment (result 11.1). It is as if at each step we have as many dictators as there are individuals in the cycle we consider: all individuals in the cycle get their most preferred object among the objects that are still available. In the preceding example, the first cycle contains Alice, Bob, and John, and the allocation they get is, from the definition of the algorithm, their top choice. So all the individuals who are part of the first cycle obtain their best possible outcome.

What about the other individuals? We start with the individuals who are in a cycle identified in step 2. Those agents are assigned either to their most preferred object (among those not assigned in step 1) or to their most preferred object. The latter case occurs when the object they were pointing at in the first step was not part of a cycle in the first step. So the object is not removed and is still available, and thus such individuals obtain their best possible outcome. If an individual in this second step is assigned her second most preferred object, it means that his most preferred object is no longer available in step 2. That is, it was removed in the first step. If we want to find another assignment such that this individual is better off, we must give her the most preferred object. But doing so will make the individual who got it in the first step worse off. So it is not possible to find another assignment that makes the individuals who are assigned an object in step 1 better off. Using the same argument for the individuals assigned in steps 3, 4, ... yields the desired result: the outcome of the Top Trading Cycle is efficient.

11.2.3 Implementing Allocation Rules

Now that we have defined rules to allocate objects to individuals, we can start talking about their implementation. The Serial Dictatorship algorithm can be implemented in a decentralized way: we just need the individuals to agree on an order, and then they just proceed to pick their favorite object one after the other. But we can also implement the Serial Dictatorship algorithm in a centralized way, asking individuals to submit a preference ordering over the objects. In this case, it is not difficult to obtain the following result.

Result 11.3 An assignment mechanism that uses the Serial Dictatorship algorithm to assign objects to individuals is strategyproof.

Proof Take the first individual in the ordering. She can obtain her most preferred object, so she has no incentive to lie. The same occurs for the second individual in the ordering: she can get her most preferred object among the remaining objects, so she has no incentive to lie. It suffices to repeat the argument for the other individuals. ∎

In fact, we can obtain a result stronger than result 11.3. The Serial Dictatorship mechanism is also **group strategyproof**. By this we mean that no group of agents can jointly misrepresent their preferences in such a way that all agents in the group are weakly better off and some agents in that group are strictly better off. The intuition behind this result is relatively simple. When assigning an object to an individual, the Serial Dictatorship mechanism only needs the preferences of that individual. That is, a group of individuals can be better off by misrepresenting their preferences, and then the first individual (with respect to the order of dictators) can also be better off by being the only one misrepresenting her preferences. In other words, if the Serial Dictatorship mechanism is not group strategyproof, then it is not strategyproof. That would contradict result 11.3, so the Serial Dictatorship mechanism is group strategyproof.

The case of the Top Trading Cycle algorithm is a bit more complicated. A priori, there is no reason why this procedure cannot be implemented in a decentralized way. But it is reasonable to doubt that it will succeed, as soon as there are "many" individuals and objects. The reason is that it needs a lot of coordination between individuals. In a decentralized market, each individual would first try to get her most preferred object. Each person would check whether the person who has the object she wants also wants her object (e.g., Alice would try to see if Bob, who has her favorite object, also wants her object). Having two individuals who own something and would like the object the other individual has is what economists call "double coincidence of wants." A major problem with this strategy is that there might not be situations where this occurs (e.g., example 11.2), and this is increasingly likely when the number of different objects is large.

If I want my most preferred object, first I would have to know who has it and ask that person if she wants my object. If I own her most preferred object, then we have the so-called double coincidence of wants, and we can proceed to a trade. But what if that person's most preferred object is not the one I own? We would need to find a third person, which may lead us to search for a fourth person. The problem can therefore become quite complicated, and a nonexpert eye may not be able to keep track of all individuals' desires, let alone identify a cycle. The fact that long chains of trades like the one identified in step 1 of example 11.2 are difficult to identify is one of the reasons why money exists. Having a medium like money that is easily exchangeable with any other object eliminates the need to worry about the double coincidence of wants (let alone considering long exchange chains).

The complexity of the problem then calls for a centralized solution. The question is then similar to the one we had with the marriage model (or with the auctions): do individuals have incentives to reveal their true preferences? For the Top Trading Cycle algorithm, the answer turns out to be positive.

Result 11.4 An assignment mechanism that uses the Top Trading Cycle algorithm to assign objects to individuals is strategyproof.

The Top Trading Cycle algorithm clearly solves the problem we outlined at the beginning of this section: avoiding the outcome that an individual obtains an object less preferred than her endowment. The intuition is the following. As for the Deferred Acceptance algorithm, the Top Trading Cycle algorithm will first try to allocate to the individuals their most preferred object. If this is not possible, the algorithm will consider the second most preferred object, if that is not possible, then the third most preferred, and so on. In the earlier example, Suzanne and Lisa first try to get their most preferred object. Once it is not possible to get these objects, they try (and manage) to get their less preferred object. For an individual, an object that is less preferred than the endowment is necessarily ranked *below* the endowment. This implies that in the Top Trading Cycle algorithm an agent will be in a self-cycle *before* she points to the owner of the object she does not like. But as soon as an agent points to herself, the algorithm assigns the agent her own object, so nobody can get an object that is less preferred than her endowment.

The Top Trading Cycle algorithm is in fact an extremely compelling way to allocate objects when individuals have endowments.[1]

Result 11.5 (*Ma*) An assignment mechanism is strategyproof, efficient, and individually rational if and only if it uses the Top Trading Cycle algorithm.

The Top Trading Cycle algorithm can also be used when individuals do not have initial endowments. But in this case we need to revise the way we construct the arrows. With private endowments (what we have seen so far), in the Top Trading Cycle algorithm, an individual first looks at her most preferred object (among the objects that are left available if we are not in step 1) and *then* looks at who the owner is. Once the owner is identified, our individual points to that person. But in the case of public endowments, there are no owners. So how do we proceed? The solution consists of using priorities that objects have over the individuals and having the objects also pointing to the agent with the highest priority.

So the graph we construct this way will be a bit bigger, for it contains individuals *and* objects. But the intuition to draw the arrows is fairly unchanged:

- Each individual points to her most preferred object.
- Each object points to the individual with the highest priority.

Once we have drawn all the arrows, we simply search for a cycle (which necessarily exists, for the same reasons as earlier). When a cycle is identified, we simply assign to each agent in the cycle the object she is pointing to. Example 11.3 illustrates how this works.

Example 11.3   Consider five individuals, Alice, Bob, John, Lisa, and Suzanne, and five objects, A, B, C, D, and E. The following table gives the preferences of the individuals over those objects.

P_{Alice}	P_{Bob}	P_{John}	P_{Lisa}	$P_{Suzanne}$
B	C	A	C	A
E	A	E	A	C
D	D	D	E	B
C	E	B	B	D
A	B	C	D	E

and the next table gives, for each object, the priority ordering over the individuals.

P_A	P_B	P_C	P_D	P_E
Alice	Bob	John	Lisa	Bob
John	Lisa	Suzanne	Suzanne	Suzanne
Bob	Suzanne	John	Alice	Alice
Suzanne	Alice	Lisa	John	John
Lisa	John	Alice	Bob	Lisa

Step 1 For each individual, we draw an arrow toward the individual's most preferred object (e.g., an arrow from Alice to object B, from Bob to object C, etc.), and for each object we draw an arrow from that object to the individual with the highest priority for that object (e.g., for object D, Lisa has the highest priority, so we have an arrow from D to Lisa). We obtain the graph in figure 11.3.

We can see that there is one cycle identified with the solid arrows: (Alice, B; Bob, C; John, A). The arrows that are not part of any cycle are dashed. All the individuals and objects are assigned and removed from the problem, so for now the assignment is

$\mu(\text{Alice}) = B, \quad \mu(\text{Bob}) = C, \quad \text{and} \quad \mu(\text{John}) = A.$

Step 2 We have two individuals left (Lisa and Suzanne) and two objects (D and E). We draw an arrow from Lisa to her most preferred object among D and E (and do the same for Suzanne). For the object, we draw an arrow from the object to the individual with the highest priority among Lisa and Suzanne (so there is an arrow from D to Lisa and an arrow from E to Suzanne). We obtain the graph in figure 11.4.

Assignment Problems

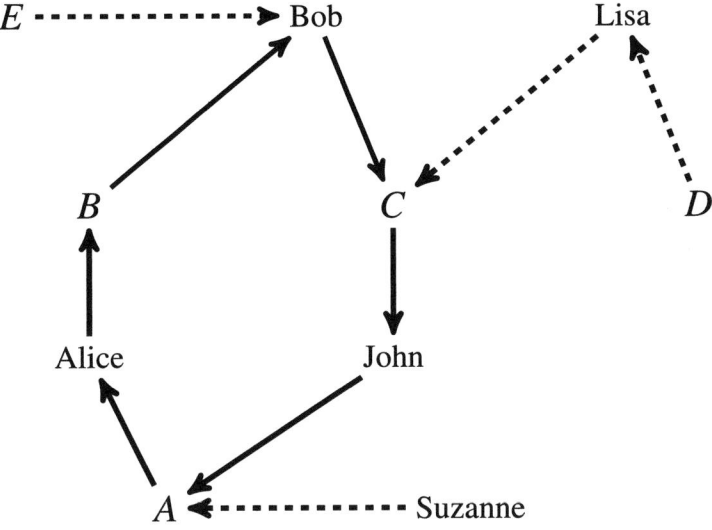

Figure 11.3.
Step 1

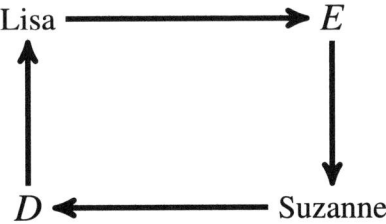

Figure 11.4.
Step 2

There is a new cycle, so Lisa and Suzanne are assigned to the object they are pointing to, and since there is no individual who is left in the problem, the algorithm stops. The final assignment is

$\mu(\text{Alice}) = B, \quad \mu(\text{Bob}) = C, \quad \mu(\text{John}) = A,$

$\mu(\text{Lisa}) = E, \quad \mu(\text{Suzanne}) = D.$

For the types of situations we have just considered (public endowments and objects have priority orderings over the individuals), a result similar to result 11.4 holds: *An assignment mechanism that uses the Top Trading Cycle algorithm to assign objects to individuals is strategyproof.*

11.2.4 Individual Rationality and the Core

An important question when designing markets is whether the agents will participate in the market. We have seen, for instance, that in the medical match (chapter 10) nonparticipation by couples was putting in jeopardy the matching mechanism used by the NRMP. Similarly, the lack of bidders for the spectrum auction in the Netherlands (chapter 6) was one of the main reasons why the auction did not perform as expected.

In assignment problems, participation in the assignment mechanism is also an important concern. In the case of private endowments, if one or several individuals anticipate that they cannot be assigned to an object that is acceptable and at least as good as the object they currently own, they will be obviously better off by opting out of the mechanism. Requiring that individuals cannot end up worse off compared to their initial situation is called **individual rationality**. Is it always possible to find an assignment that is individually rational? The answer is yes, and it turns out that we already know how to find such an assignment (beyond the trivial assignment where each individual keeps her endowment).

Result 11.6 (From Result 11.5)   The assignment obtained with the Top Trading Cycles algorithm is individually rational.

Note that result 11.6 is not directly related to the efficiency of the assignment obtained with the Top Trading Cycles algorithm (result 11.2). We can indeed have an efficient assignment where one individual obtains an object that is less preferred than her endowment. Suppose there are two objects, A and B, and two individuals, Alice and Bob, who have the same preferences: object A is the most preferred. If Alice's endowment is object A (and B for Bob), then the assignment where Alice gets object B and Bob gets object A is efficient (we cannot make Alice better off without making Bob worse off) but is not individually rational.

We can extend the notion of individual rationality to *groups of individuals*. Doing so is a natural step when we consider *exchanges*: individuals gather in a marketplace (whether it uses a specific mechanism or not) because they can benefit from trade. We can thus also consider the situation where a group of agents finds that, as a group, they do not obtain a satisfactory assignment. In such a case, we would have several individuals who find that on their own they can achieve a better assignment. The following example illustrates this situation.

Example 11.4   We have three individuals, Alice, Bob, and Carol, and three objects, A, B, and C. The preferences and endowments are given in the following table.

Assignment Problems

Endowment →	A	B	C
	P_{Alice}	P_{Bob}	P_{Carol}
	B	A	A
	C	B	B
	A	C	C

The assignment

$$\mu(\text{Alice}) = C, \quad \mu(\text{Bob}) = B, \quad \mu(\text{Carol}) = A$$

is efficient: there is no other assignment that makes at least one individual strictly better off (and the others as well off).

However, we can see that Alice and Bob would be better off without Carol. They can trade their endowments (Alice gets object B and Bob object A), obtaining an assignment they prefer to μ (at the expense of Carol).

An assignment such that no group of individuals may have an interest in splitting from the others is called a **core assignment**.

Definition 11.3 The **core** of an assignment problem is the set of all assignments μ such that there is no coalition S of individuals and an assignment μ' for which:

1. for an individual in $i \in S$, the object $\mu'(i)$ is the endowment of another individual in S; and

2. each individual in S prefers μ' to μ or is indifferent between μ and μ', and there is at least one individual who strictly prefers μ' to μ.

Point 1 in definition 11.3 captures the fact that the coalition is splitting from the group composed of all the individuals and trades only between themselves. Point 2 is similar to the notion of efficiency but only applies to the "dissident group." In example 11.4, the assignment μ is not in the core.

The concept of the core seems very appealing because it provides a strong justification for creating marketplaces that involve as many individuals as possible. However, it seems a bit demanding. Is there always at least one assignment that is in the core for any assignment problem? The answer is yes, and it turns out that we already know how to find such an assignment.[2]

Result 11.7 (Roth and Postlewaite) For any assignment problem, there is always a unique assignment that is in the core, and this assignment can be obtained with the Top Trading Cycles algorithm.

The intuition behind result 11.7 is rather simple. It suffices to look at the execution of the Top Trading Cycles algorithm. Consider any cycle that is obtained in the first step. All the individuals involved in this cycle will obtain their most preferred object. Therefore, any assignment that does not give these individuals their most preferred object cannot be in the core. All of these individuals will prefer the assignment they obtain through the cycle (it satisfies condition 2 of definition 11.3). Observe now that, in the cycle, any individual obtains the endowment of another individual who is also in the cycle. So the assignment they obtain through the cycle also satisfies condition 1 of definition 11.3.

To summarize, if an assignment is in the core, it must be that all the individuals assigned in the first step (i.e., in the first cycle) are assigned the same object as the one they get with the Top Trading Cycles algorithm. It now suffices to repeat this reasoning with the agents who are assigned in a cycle found in step 2 of the Top Trading Cycles algorithm. Those individuals can only be better off if we assign them an object that was assigned in step 1. This is not possible, so the best they can get is the assignment they obtain in step 2. It is not difficult to see that if the step 2 individuals get something different from their assignment under the Top Trading Cycles algorithm (but individuals from step 1 are assigned their Top Trading Cycles assignment), then the assignment they obtain in step 2 satisfies conditions 1 and 2 of definition 11.3.

Repeating the procedure with the assignment found in steps 3, 4,... we end up with a well-defined assignment that is in the core and such that any other assignment cannot be in the core.

11.3 Mixed Public-Private Endowments

So far, we have seen two polar cases of assignment problems: one situation where no individual initially owns any object (public endowments) and the opposite situation, where individuals initially own one object (private endowment).

There are many situations, however, where we have a mix of these two cases. A common situation is that of student dorms on campuses. Many colleges and universities offer housing to their students, and since each year some students graduate and leave their dorms and new students arrive, we have a situation with a mix of private and public endowments:

- Private endowments: the students who were already on campus the previous academic year. The room they occupied the previous year is their private endowment.

- Public endowments: the rooms that are left vacant by the students who just graduated. The newly arrived students do not have any endowments.

How should we proceed in this case? This problem is delicate because of the individuals who already have an endowment. If we set up a procedure to assign the objects, those individuals may end up being assigned an object that is worse (according to their preferences) than the object they initially owned. For this reason, it is customary to give such individuals an additional choice: participate in the mechanism that assigns the objects or just stay apart and continue to enjoy the object they are currently endowed with. That way those individuals can avoid taking risks. But such a policy creates an additional problem in that the assignment we obtain may not be efficient.

The problem of mixed public-private endowments was first analyzed by Atila Abdulkadiroğlu and Tayfun Sönmez.[3] Their study was motivated by the allocation of houses, with special attention to the case of on-campus housing. Here we follow their approach by calling the individuals that already own an object **existing tenants** (and the objects are **houses**), and the individuals who do not have any endowment **new applicants**. Interestingly for us, they analyzed various mechanisms in real-life applications, which we now review.

11.3.1 Inefficient Mechanisms

The first assignment mechanism that is extremely common is the so-called **Random Serial Dictatorship with Squatting Rights**. This mechanism is or was used for undergraduate housing at Carnegie-Mellon, Duke, and Harvard among others. The idea is to run a Serial Dictatorship where the order of the dictators is random (we will study the use of randomness in detail in section 12.2), but before doing that, the existing tenants have to decide whether they want to participate or not. If an existing tenant chooses to opt out, then she keeps her house; otherwise she participates in the Serial Dictatorship mechanism but then loses her endowment. A formal description of the mechanism follows.

Algorithm 11.3   Random Serial Dictatorship with Squatting Rights

Step 1 Existing tenants announce whether they want to keep their house. If they do so, they are assigned their house; otherwise their house is added to the pool of vacant houses.

Step 2 We draw a random ordering of all individuals, consisting of the new applicants and the existing tenants who chose not to keep their house.

Step 3 The Serial Dictatorship algorithm with the order chosen in step 2 is run.

The problem with this mechanism is that it cannot guarantee an existing tenant that she will get a house that is at least as good as the house she was initially endowed with. Consequently, risk-averse existing tenants may prefer to opt out, which can easily result in inefficient assignments.

Example 11.5 Suppose that there are three houses, A, B, and C, and three individuals, Alice, Bob, and Carol. Alice is the only existing tenant, and her endowment is house A. The preferences are as follows.

P_{Alice}	P_{Bob}	P_{Carol}
B	B	A
A	C	B
C	A	C

Suppose that the order of the dictators is Carol, Bob, and Alice and that Alice decides to participate. If we run the Serial Dictatorship algorithm with this order, Carol takes house A, Bob takes house B, and Alice is left with house C. Clearly, she would be better off keeping her house. If Alice does not participate, then Carol takes the first available house, B, and Bob is left with house C. We can see that this assignment is Pareto dominated by the assignment where Alice is assigned house B, Bob is assigned house C, and Carol is assigned house A.

To avoid the risk of having existing tenants be worse off if they participate the mechanism, Abdulkadiroğlu and Sönmez identified two other mechanisms, used for graduate housing at the University of Rochester and MIT.

The Rochester solution is called the **Random Serial Dictatorship with Waiting List**. The idea is to run a Serial Dictatorship algorithm with a random order of the dictators, but at any step an individual can only take a house that is *available*. A house is available if it is either a house in the public endowment or if it is a house left vacant by an existing tenant.

In the algorithm, for a new applicant, all houses are **obtainable**, and for an existing tenant, a house is obtainable if it is her house (i.e., her endowment) or a house she prefers to her house.

Algorithm 11.4 Random Serial Dictatorship with Waiting Lists

Step 1 We draw a random ordering of all individuals, consisting of the new applicants and the existing tenants.

Step 2 The set of available houses is the set of vacant houses (i.e., those not currently owned by an existing tenant).

The individual with the highest priority (given by the random order) among those who have at least one acceptable house is assigned her most preferred available house (and is then removed from the procedure). Her new assigned house is removed from the set of available houses. If that individual is an existing tenant, her endowment is added to the pool of available houses (if that house is not her most preferred).

Assignment Problems

Step k, $k \geq 3$ The set of available houses is constructed at the end of step $k-1$.

The individual with the highest priority (given by the random order) among the remaining individuals that has at least one acceptable house is assigned her most preferred available house (and is removed from the procedure). Her new assigned house is removed from the set of available houses. If that individual is an existing tenant, her endowment is added to the pool of available houses (if that house is not her most preferred).

End The algorithm ends when there is either no remaining individual or when there is no available house left for any individual.

While this algorithm clearly ensures existing tenants that they cannot end up with a house less preferred than their endowment, inefficient assignments may be obtained.

Example 11.6 There are existing tenants Alice, Bob, and Carol, and four houses, A, B, C, and D. (There are no new applicants.) The endowments and preferences are given in the following table.

Endowment $\to$	A	B	C
	P_{Alice}	P_{Bob}	P_{Carol}
	B	C	A
	C	A	D
	A	B	C
	D	D	B

We now run the algorithm. Suppose that the order is Alice, Bob, and then Carol. At the beginning, only house D is available, as it is the only vacant house. Only Carol has this house as obtainable (for Alice and Bob, house D is less preferred than their endowments). So Carol takes this house, and now house C is vacant.

In the next step, we consider the remaining individuals, Alice and Bob. The only available house is house C, which is obtainable for both of them. Alice has the highest priority, so she is the one taking that house, now making her house, A, available.

In the final step, the available house is house A, which is achievable for Bob. So he takes that house. The final assignment is then

$$\mu(\text{Alice}) = C, \quad \mu(\text{Bob}) = A, \quad \mu(\text{Carol}) = D.$$

However, this assignment is Pareto dominated by the assignment

$$\mu'(\text{Alice}) = B, \quad \mu'(\text{Bob}) = C, \quad \mu'(\text{Carol}) = A.$$

The mechanism used at NH4, one of the houses at MIT, is another attempt to give existing tenants the possibility to get a room at least good as their existing endowment.[4]

Algorithm 11.5 MIT–NH4

Step 1 We draw a random ordering of all individuals, consisting of the new applicants and the existing tenants.

Step 2 The first individual is *tentatively* assigned to her top choice among all houses, the second individual is assigned to her top choice among the remaining houses, and so on, until a *squatting conflict* occurs.

Step 3 A *squatting conflict* occurs if the requested house has an existing tenant but for that tenant all the remaining houses are worse than her current house. This means that there is an individual, the *conflicting individual*, who chose (earlier) the tenant's house. If this happens, then:

- The existing tenant (the one with the conflict) is assigned her house.
- The tentative assignment of the conflicting individual is canceled as well as the assignments of all the individuals who decided after the conflicting individual.
- The process starts again with the conflicting individual (i.e., we do not change the tentative assignments of the individuals who made their choices before her).

End The algorithm stops when there is no house or individual left. At this point, all tentative assignments are finalized.

Abdulkadiroğlu and Sönmez showed that this algorithm is not guaranteed to produce efficient assignments.

Example 11.7 There are four tenants, Alice, Bob, Carol, and Denis, and one new applicant, Erin. There is one vacant house, house E. The endowments and preferences over the other houses are given in the following table.

Endowment →	A	B	C	D	
	P_{Alice}	P_{Bob}	P_{Carol}	P_{Denis}	P_{Erin}
	C	D	E	C	D
	D	E	C	E	E
	E	B	D	D	C
	A	C	B	B	A
	B	A	A	A	B

Assignment Problems

Suppose that the order is Alice, Bob, Carol, Denis, and Erin. We now run the algorithm.

Step 1 Alice is tentatively assigned house C, then Bob is tentatively assigned house D, and then Carol is tentatively assigned house E. When it is Denis's turn, we have a conflict: his house, D, and all the houses he prefers, C and E, are taken. The person who took his house is Bob, so Bob is the conflicting individual. So we cancel Bob's assignment as well as Carol's assignment (because she decided after Bob), and house D is assigned to Denis. Both Denis and house D are removed from the process.

Step 2 We start again with Bob. The next available house for him is house E, which he gets (temporarily). Then it is Carol's turn, but we have a conflict, where Alice is the conflicting individual. So we cancel Alice's assignment, and house C is assigned to Carol. Both Carol and house C are removed from the process.

Step 3 We start with the conflicting individual, Alice. She is tentatively assigned to house E (houses C and D have been removed). Then Bob is tentatively assigned to house B, and finally Erin is assigned to house A. All houses and all individuals are assigned, so the algorithm stops.

The final assignment is

$\mu(\text{Alice}) = E, \quad \mu(\text{Bob}) = B, \quad \mu(\text{Carol}) = C,$

$\mu(\text{Denis}) = D, \quad \mu(\text{Erin}) = A.$

However, this assignment is Pareto dominated by the following assignment:

$\mu'(\text{Alice}) = C, \quad \mu'(\text{Bob}) = B, \quad \mu'(\text{Carol}) = E,$

$\mu'(\text{Denis}) = D, \quad \mu'(\text{Erin}) = A.$

11.3.2 Two Efficient Solutions

The problem with a solution like the MIT–NH4 algorithm is that when there is a squatting conflict the existing tenant does not have any chance to get something better than her endowment. We forgo any possibility of getting something more efficient. To solve this problem (and get efficient assignments), Abdulkadiroğlu and Sönmez propose a modification of the MIT–NH4 algorithm, coined the **You Request My House—I Get Your Turn algorithm**. This solution starts like the MIT–NH4 algorithm. The difference lies in how conflicts are resolved. In this algorithm, a conflict occurs when someone takes the house of an existing tenant and this latter has not yet chosen a house.

Algorithm 11.6   You Request My House—I Get Your Turn (YRMY–IGYT)

Step 1 We draw a random ordering of all individuals, consisting of the new applicants and the existing tenants.

Step 2 The first individual is assigned to her top choice among all houses, the second individual is assigned to her top choice among the remaining houses, and so on, until someone requests the house of an existing tenant.

Step 3 If the existing tenant has already chosen a house, then we continue (with the next individual order). Otherwise, we first cancel all the assignments starting from the conflicting individual. Second, we modify the ordering of the individuals, moving the existing tenant just above the conflicting individual. We then resume the serial dictatorship starting with the existing tenant.

Step 4 It may happen that at some point a cycle is formed; that is, we have individuals $i_1, i_2, \ldots, i_k$ such that i_1 is assigned the house of i_2, i_2 is assigned the house of i_3, and so on until i_k is assigned the house of i_1. In this case, we remove all the individuals in the cycle by assigning them the house they demand, and we proceed with the procedure.

End: The algorithm stops when there is no house or individual left.

Example 11.8   We use the preferences and endowments of example 11.7 and the order of individuals is:

Alice, Bob, Carol, Denis, and Erin.

Alice is the first to choose, so she picks her most preferred house, house C. That house has a tenant, Carol, who has not yet chosen any house. This is a conflict, so we cancel Alice's assignment and move Carol above Alice. The order of individuals is now:

Carol, Alice, Bob, Denis, and Erin.

Carol picks her most preferred house, E, then Alice chooses C. Bob chooses D, but the tenant, Denis, has not yet chosen a house. There is a conflict. Denis's house was picked by Bob. So we cancel Bob's assignment and we move Denis above Bob. The order of individuals is now:

Carol, Alice, Denis, Bob, and Erin.

Denis chooses D. This is a trivial cycle, so Denis is assigned D. House D and Denis are removed from the problem.

Bob is next. He chooses B. Again this is a trivial cycle, so Bob is assigned B. House B and Bob are removed from the problem.

Erin is next. Her most preferred house among the available houses is A. The tenant is Alice, but since she has already chosen a house there is no conflict.

The algorithm stops. The final assignment is the efficient assignment μ' we mentioned at the end of example 11.7,

$\mu'(\text{Alice}) = C, \quad \mu'(\text{Bob}) = B, \quad \mu'(\text{Carol}) = E,$

$\mu'(\text{Denis}) = D, \quad \mu'(\text{Erin}) = A.$

Abdulkadiroğlu and Sönmez propose a second mechanism, which is an adaptation of the Top Trading Cycles mechanism for the case of mixed public-private endowments.

Algorithm 11.7 Top Trading Cycles with Mixed Endowments

Step 1 We draw a random ordering f of all individuals, consisting of the new applicants and the existing tenants.

Step 2 For each individual, we draw an arrow starting from that individual to her most preferred house.

For each vacant house (i.e., not owned by any existing tenant), we draw an arrow starting from that house to the first individual in the ordering f.

For each occupied house (i.e., a house owned by a tenant), we draw an arrow starting from that house to its tenant.

There must be at least one cycle. Every individual participating in a cycle is assigned the house she is pointing to. All the houses and individuals participating in the cycle are removed from the problem.

Step k, $k \geq 3$ For each individual remaining in the problem, we draw an arrow starting from that individual to her most preferred house *among the houses that have not been removed*.

For each vacant house that remains in the problem, we draw an arrow starting from that house to the first individual in the ordering f *among the individuals who are still in the problem*.

For each occupied house that remains in the problem, we draw an arrow starting from that house to its tenant.

There must be at least one cycle. Every individual participating in a cycle is assigned the house she is pointing to. All the houses and individuals participating in the cycle are removed from the problem.

End The algorithm stops when all individuals have been removed or all houses have been assigned.

This algorithm turns out to be an optimal solution for the problem of mixed endowments.

Result 11.8 (*Abdulkadiroğlu and Sönmez*)   An assignment mechanism that uses the Top Trading Cycles algorithm with mixed endowments is

- efficient;
- individually rational (each existing tenant cannot obtain a house that is less preferred than her endowment); and
- strategyproof.

Example 11.9   We use the preferences and endowments of example 11.7.

Step 1 Let the order of individuals be Alice, Bob, Carol, Denis, and Erin.

Step 2 Each individual points to his or her most preferred house, so Alice points to house C, Bob to house D, ...

Each occupied house points to its tenant. So house A points to Alice, house B points to Bob, and so on. House E is vacant, so it points to the first agent in the ordering, Alice. So we have the graph in figure 11.5. There is a cycle involving Alice, house C, Carol, and house E (identified with the solid arrows), so Alice is assigned house C and Carol house E.

Step 3 Only houses A, B, and D and Bob, Denis, and Erin remain in the problem. Bob is still pointing to house D. Denis is now pointing to his most preferred house that remains, D, and Erin is still pointing to house D. Houses B and D are still pointing to Bob and Denis, respectively. House A is now vacant, so it points to the highest individual in the ordering among the remaining individuals, Bob.

The graph is depicted in figure 11.6. There is one cycle, involving Denis and house D, so Denis is assigned house D.

Step 4 Only Bob and Erin and houses A and B remain in the problem. Bob and Erin are pointing to their most preferred houses among these two: Bob points to house B and Erin to house A. House B is still pointing to its tenant, Bob, and house A is still pointing to Bob because he ranks higher than Erin in the ordering. The graph is depicted in figure 11.7.

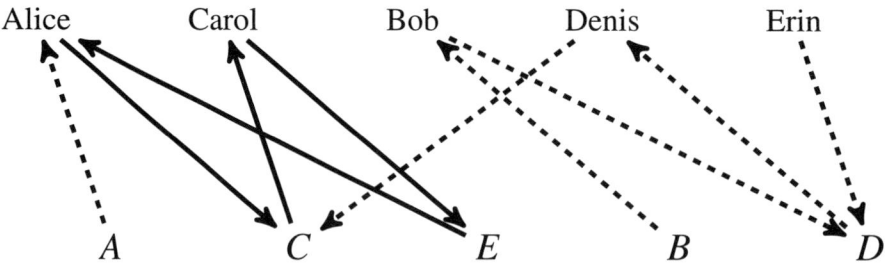

Figure 11.5.
Step 1

Assignment Problems

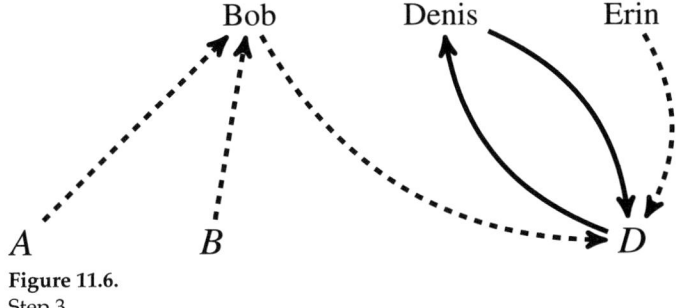

Figure 11.6.
Step 3

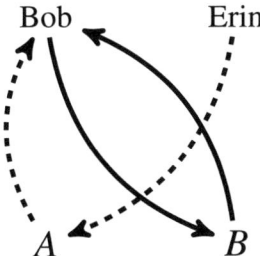

Figure 11.7.
Step 4

We have a cycle involving Bob and house B, so Bob is assigned to that house and they are both removed from the problem.

Step 5 Erin points to house A, and house A points to Erin. That makes a cycle, so Erin gets house A and the algorithm ends.

The final assignment is the assignment μ' we identified in example 11.7:

$\mu'(\text{Alice}) = C, \quad \mu'(\text{Bob}) = B, \quad \mu'(\text{Carol}) = E,$

$\mu'(\text{Denis}) = D, \quad \mu'(\text{Erin}) = A,$

which we know is efficient.

The fact that the YRMH–IGYT algorithm produces an efficient assignment in example 11.8 is not surprising. Abdulkadiroğlu and Sönmez indeed show that this algorithm is equivalent to the Top Trading Cycle algorithm we saw earlier.

Result 11.9 (Abdulkadiroğlu and Sönmez) For any ordering of the individuals, the YRMH–IGYT algorithm yields the same outcome as the Top Trading Cycle algorithm with mixed endowments (using the same ordering of the individuals).

12 Probabilistic Assignments

So far, we have only seen assignment algorithms that are *deterministic*. By this we mean that if we take an assignment problem (i.e., sets of individuals and objects and a preference ordering for each individual) and an algorithm and we run the algorithm with the same problem several times, we always obtain the same assignment.

There is a class of algorithms that does not always yield the same assignment even though we always feed the algorithm with the same data. Such algorithms introduce a probabilistic dimension to the assignment, and we can obtain outcomes of the form *Alice is assigned object A with a probability of 30% and object B with a probability of 70%*. We already (implicitly) saw random assignments in chapter 11: the Serial Dictatorship or any other algorithm that uses a specific ordering of individuals (like the MIT–NH4 algorithm) generates random assignments whenever the ordering of the individuals is random. As we will see, allowing for randomization can solve some problems that deterministic procedures have.

12.1 Random Assignments

12.1.1 Preliminaries

In section 11.1, we saw that an assignment is a function that describes who gets what. Given an assignment problem, a **random assignment** describes for each individual i and each object k the probability that individual i is assigned object k. If there are n individuals and k objects, a random assignment is thus described by $n \times k$ probabilities (for each individual, there are k probabilities, one for each object).

When we defined deterministic assignments, we had that two individuals could not be assigned the same object. For a random assignment, the restrictions we have to impose are a bit more general. To explain them, we do need some additional notation. We will denote probabilities with the letter β (the Greek letter for

b, which is called "beta"), and $\beta_{i,k}$ will denote the probability that individual i obtains the object k. That is, the first index (here i) refers to an individual, and the second object (here k) refers to an object.

For simplicity, we will make several assumptions:

- A random assignment problem consists of a set of individuals $I = \{i_1, i_2, \ldots, i_n\}$ and a set of objects $\{k_1, k_2, \ldots, k_n\}$ *of the same size.* That is, *there are as many objects as there are individuals.*
- Each individual has preferences over the objects, and all objects are acceptable.

These two conditions imply that a minimum optimality requirement for a random assignment is that each individual must end up being assigned to an object.

Definition 12.1 A **random assignment** is a collection of $n \times n$ probabilities β such that:

(a) For each agent-object pair (i, k), there is a probability $\beta_{i,k}$ that indicates the probability that individual i is assigned object k.
(b) For each individual i, $\sum_{k \in K} = \beta_{i,k_1} + \beta_{i,k_2} + \cdots + \beta_{i,k_n} = 1$.
(c) For each object k, $\sum_{i \in I} = \beta_{i_1,k} + \beta_{i_2,k} + \cdots + \beta_{i_n,k} = 1$.

Condition (b) says that with probability 1 each individual will be assigned an object. An example where that condition is not satisfied is the following. There are three objects, k_1, k_2, and k_3, and a random assignment such that individual i obtains object k_1 with a probability of 30%, object k_2 with a probability of 20%, and object k_3 with a probability of 10%. This means that the sum of the probabilities is less than 1: $\beta_{i,k_1} + \beta_{i,k_2} + \beta_{i,k_3} = 0.3 + 0.2 + 0.1 = 0.6$. The difference, 0.4 (= 40%), is the probability that individual i is not assigned any object.

Condition (c) is similar to condition (b), but for objects. This means that each object has a 100% probability of being assigned to an individual.

Observe that our random assignment model encompasses the deterministic assignment model we saw in section 11.1. To see this, note that a deterministic assignment is simply a random assignment such that an individual is assigned an object either with probability 0 or with probability 1. For instance, if individual i_3 has object k_7 with probability 1, then condition (c) implies that for each individual i that is *not* individual i_3, we will have $\beta_{i,k_7} = 0$. Similarly, condition (b) implies that the probability that individual i_3 is assigned an object k that is *not* object k_7 must be zero; that is, $\beta_{i_3,k} = 0$ if $k \neq k_7$.

Observe that a random assignment β can be easily represented by a matrix, where the rows represent the individuals and the columns the objects.

Probabilistic Assignments

Example 12.1 Suppose that there are three individuals (Albert, Bob, and Carol) and three objects (an apple, an orange, and a pear).

An example of a random assignment can be the following:

$$\begin{array}{c} \text{apple} \quad \text{orange} \quad \text{pear} \\ \downarrow \quad \downarrow \quad \downarrow \end{array}$$
$$\begin{array}{c} \text{Alice} \to \\ \text{Bob} \to \\ \text{Carol} \to \end{array} \begin{pmatrix} 0.15 & 0.40 & 0.45 \\ 0.10 & 0.35 & 0.55 \\ 0.75 & 0.25 & 0 \end{pmatrix}$$

In this random assignment, Alice obtains the apple with a probability of 0.15, an orange with a probability of 0.40, and a pear with a probability of 0.45. So we have

$$\beta_{\text{Alice, apple}} = 0.15, \quad \beta_{\text{Alice, orange}} = 0.40, \quad \beta_{\text{Alice, pear}} = 0.45.$$

The orange is assigned to Alice, Bob, and Carol with a probability of 0.40, 0.35, and 0.25, respectively.

As we said, deterministic assignments are particular cases of random assignments. For instance, the assignment that gives the apple to Bob, the orange to Alice, and the pear to Carol is represented by the following random assignment:

$$\begin{pmatrix} 0 & 1 & 0 \\ 1 & 0 & 0 \\ 0 & 0 & 1 \end{pmatrix}$$

12.1.2 The Birkhoff–von Neumann Theorem

A random assignment assigns to each pair of an individual and an object a *probability* that this individual is assigned that object. Eventually, the individuals must end up with an object. How do we ensure that two individuals do not get the same object? If we consider example 12.1, we see that the apple is assigned to Alice, Bob, and Carol with probabilities 0.15, 0.10, and 0.75, respectively. These probabilities are independent, so there is in principle nothing that prevents the apple from being assigned simultaneously to Alice, Bob, and Carol.

Our problem here is that a random assignment is defined as a probability for each pair of an individual and object. It does not guarantee that two individuals obtain the same object. So, did we miss something in the definition of a random assignment? Fortunately for us, the answer to that question is no. This is because of a famous result, the Birkhoff–von Neumann Theorem.

Before we explain the Birkhoff–von Neumann Theorem, we need to introduce an alternate way to describe random assignments. Consider a collection of deterministic assignments, say μ, μ', and μ''. Suppose now that we will pick each of these assignments with some probability. For instance, we pick the assignment μ with a probability of 0.4 and the assignments μ' and μ'' with probabilities 0.5 and 0.1, respectively. A collection of deterministic assignments together with a probability for each assignment is called a **lottery** (over deterministic assignments). Clearly, with a lottery we avoid the problem we mentioned earlier by assuring that no object is allocated to more than one individual.

To sum up, we have two alternatives to represent random assignments:

- For each pair of an individual and object, we have a probability. These probabilities must satisfy the conditions of definition 12.1.
- We can use a lottery over deterministic assignments.

The following result says that there is some sort of equivalence between these two ways of representing random assignments.

Result 12.1 (Birkhoff–von Neumann Theorem) Any random assignment can be decomposed into a lottery over deterministic assignments.

Example 12.2 Consider the following random assignment with three individuals (Alice, Bob, and Carol) and three objects (apple, orange, and pear).

$$
\begin{array}{c}
\phantom{\text{Alice}\to}\text{apple}\quad\text{orange}\quad\text{pear}\\
\phantom{\text{Alice}\to}\downarrow\qquad\downarrow\qquad\downarrow\\
\begin{array}{c}\text{Alice}\to\\\text{Bob}\to\\\text{Carol}\to\end{array}\begin{pmatrix}0.3 & 0.7 & 0\\ 0.5 & 0.3 & 0.2\\ 0.2 & 0 & 0.8\end{pmatrix}
\end{array}
$$

Now consider the following three deterministic assignments:

$$\mu = \begin{pmatrix}1 & 0 & 0\\ 0 & 1 & 0\\ 0 & 0 & 1\end{pmatrix},\quad \mu' = \begin{pmatrix}0 & 1 & 0\\ 1 & 0 & 0\\ 0 & 0 & 1\end{pmatrix},\quad \mu'' = \begin{pmatrix}0 & 1 & 0\\ 0 & 0 & 1\\ 1 & 0 & 0\end{pmatrix},$$

and let the probabilities for μ, μ', and μ'' be 0.3, 0.5, and 0.2, respectively.

If we pick one of those deterministic assignments according to the probabilities that are given, what is, for instance, the probability that Alice is assigned an orange? She gets an apple in assignments μ' and μ'' (the first row, second column). Assignment μ' is realized with a probability of 0.5 and μ'' with a probability of 0.2, so she gets an orange with a probability of $0.5 + 0.2 = 0.7$. That is exactly the probability given in the random assignment!

More precisely, we can observe that the matrix describing random assignment is in fact the sum of the three matrices describing the deterministic assignments,

each weighted by its probability:
$0.3 \times \mu + 0.5 \times \mu' + 0.2 \times \mu''$

$$= 0.3 \times \begin{pmatrix} 1 & 0 & 0 \\ 0 & 1 & 0 \\ 0 & 0 & 1 \end{pmatrix} + 0.5 \times \begin{pmatrix} 0 & 1 & 0 \\ 1 & 0 & 0 \\ 0 & 0 & 1 \end{pmatrix} + 0.2 \times \begin{pmatrix} 0 & 1 & 0 \\ 0 & 0 & 1 \\ 1 & 0 & 0 \end{pmatrix}$$

$$= \begin{pmatrix} 0.3 \times 1 + 0.5 \times 0 + 0.2 \times 0 & 0.3 \times 0 + 0.5 \times 1 + 0.2 \times 1 & 0.3 \times 0 + 0.5 \times 0 + 0.2 \times 0 \\ 0.3 \times 0 + 0.5 \times 1 + 0.2 \times 0 & 0.3 \times 1 + 0.5 \times 0 + 0.2 \times 0 & 0.3 \times 0 + 0.5 \times 0 + 0.2 \times 1 \\ 0.3 \times 1 + 0.5 \times 0 + 0.2 \times 1 & 0.3 \times 0 + 0.5 \times 0 + 0.2 \times 0 & 0.3 \times 1 + 0.5 \times 1 + 0.2 \times 0 \end{pmatrix}$$

$$= \begin{pmatrix} 0.3 & 0.7 & 0 \\ 0.5 & 0.3 & 0.2 \\ 0.2 & 0 & 0.8 \end{pmatrix}.$$

To summarize, the Birkhoff–von Neumann Theorem guarantees us that there are no hidden conflicts in the definition of a random assignment. One caveat is that the decomposition of a random assignment into a lottery over deterministic assignments may not be unique.

12.1.3 Evaluating Random Assignments

For an individual, it is very easy to compare two deterministic assignments: she only has to compare the two objects she gets under each assignment. Using the preferences over objects, it is thus very easy for an individual to see if she prefers one assignment over the other. How do we proceed with random assignments? The question is trickier. To see this, suppose that Alice's preferences are

$P_{\text{Alice}} = $ apple, orange, pear.

Now she wants to compare the following two random assignments β and β':

- $\beta_{\text{Alice, apple}} = 0.5$, $\beta_{\text{Alice, orange}} = 0.3$, $\beta_{\text{Alice, pear}} = 0.2$,
- $\beta'_{\text{Alice, apple}} = 0.3$, $\beta'_{\text{Alice, orange}} = 0.4$, $\beta'_{\text{Alice, pear}} = 0.3$.

If the realization of β is such that she gets the apple and the realization of β' assigns her the orange, then she prefers β to β'. But Alice could receive the orange under β and the apple under β'. In that case, she would prefer β' over β.

The way to proceed is to compare the two probability distributions by using the concept of **stochastic dominance**. The idea is to start with Alice's preferences over objects. As we just said, we assume that she prefers the apple to the orange. We proceed as follows:

Step 1 We start by comparing the probability of getting the most preferred object, the apple. Under β that probability is higher than under β' (0.5 vs. 0.3).

Step 2 Now we consider the *two most preferred* objects, the apple *and* the orange, and we compare the sum of the probabilities for these two objects. That is, we look at the probability of getting either of the two objects, so we have

$$\beta_{\text{Alice, apple}} + \beta_{\text{Alice, orange}} = 0.5 + 0.3 = 0.8,$$

$$\beta'_{\text{Alice, apple}} + \beta'_{\text{Alice, orange}} = 0.3 + 0.4 = 0.7.$$

Step 3 We could continue by comparing the sum of the probabilities of the three most preferred objects. Since in our example there are only three objects, we know that the sum is 1 under both β and β'.

We can see that in step 1 the assignment β has a higher probability. In step 2, the sum of the probabilities of β is still higher than the sum of the probabilities of β'. That is, for both step 1 and step 2 the comparison of β and β' gives a higher sum for β. If, for each step, it is always β that gives a higher sum, then we say that β **stochastically dominates** β'. This means that, for any object, the probability that Alice gets that object or a more preferred object is higher for β than for β'.

Formally, if individual i's preference over the objects is

$$P_i = k_1, k_2, \ldots, k_n,$$

and if $\beta_{i,k_1}, \beta_{i,k_2}, \ldots, \beta_{i,k_n}$ and $\beta'_{i,k_1}, \beta'_{i,k_2}, \ldots, \beta'_{i,k_n}$ are two random assignments (for individual i), then β stochastically dominates β' for individual i if

$$\text{for each } h = 1, \ldots, n-1, \quad \sum_{\ell \leq h} \beta_{i,k_\ell} \geq \sum_{\ell \leq h} \beta'_{i,k_\ell}. \tag{12.1}$$

The intuition is the following. Take the ℓth most preferred object. We ask the following question: for a given random assignment, what is the probability that individual i obtains the ℓth object or an object that she prefers? The answer is simply the sum of the probabilities up to object ℓ:

$$\beta_{i,k_1} + \beta_{i,k_2} + \cdots + \beta_{i,k_{\ell-1}} + \beta_{i,k_\ell} = \sum_{h \leq \ell} \beta_{i,k_h}.$$

We compute the same sum for the assignment β' and compare these two sums. Stochastic dominance says that we have to make the comparison for all levels in the preference ordering: for $\ell = 1$, then for $\ell = 2$, then for $\ell = 3, \ldots$ until $\ell = n-1$. So we have to make $n-1$ comparisons. If for each of these comparisons it is always the same random assignment, say β, that gives the highest sum, then the random assignment β stochastically dominates the other random assignment.

Probabilistic Assignments

Example 12.3 Consider four objects, apple, orange, pear, and cherry, and Alice's preferences are

$P_{\text{Alice}} =$ apple, orange, pear, cherry.

Consider the following three random assignments.

	β	β'	β''
apple	0.2	0.3	0.2
orange	0.3	0.3	0.5
pear	0.1	0.2	0
cherry	0.4	0.2	0.3

We can see that β' stochastically dominates β because we have

$\beta'_{\text{Alice, apple}} \geq \beta_{\text{Alice, apple}}$,

$\beta'_{\text{Alice, apple}} + \beta'_{\text{Alice, orange}} \geq \beta_{\text{Alice, apple}} + \beta_{\text{Alice, orange}}$,

$\beta'_{\text{Alice, apple}} + \beta'_{\text{Alice, orange}} + \beta'_{\text{Alice, pear}} \geq \beta_{\text{Alice, apple}} + \beta_{\text{Alice, orange}} + \beta_{\text{Alice, pear}}$.

Using the numerical probabilities, we have

$0.3 \geq 0.2$,

$0.3 + 0.3 \geq 0.2 + 0.3$,

$0.3 + 0.3 + 0.2 \geq 0.2 + 0.3 + 0.1$.

Likewise, it can be checked that β'' stochastically dominates β.

It is important to note that if we have two random assignments, say β and β', such that β does *not* stochastically dominate β', then it is not necessarily the case that β' stochastically dominates β. To see this, consider the random assignments β' and β'' from example 12.3. We can see that, up to the orange, β'' seems to dominate β': $0.2 + 0.5 \geq 0.3 + 0.3$. However, if we consider the probabilities up to the pear, the inequality is reversed: $0.2 + 0.5 + 0 < 0.3 + 0.3 + 0.2$. So we cannot compare β' and β'' by using stochastic dominance.

12.2 Random Serial Dictatorship

One obvious critique of the Serial Dictatorship method we saw in section 11.2 is that it gives a huge advantage to the individuals who are at the top of the queue. The last individuals do not have much choice; they only have access to the leftovers. To circumvent this problem, the rule is often modified, and we first randomize the order of the individuals. That way, all individuals have the same chance to be the first, the last, the second, and so on in the queue. Randomizing the queue thus yields the **Random Serial Dictatorship** algorithm.

Besides making the assignment mechanism more fair, allowing for randomization also has a (surprising) consequence: we can see how seemingly unrelated assignment mechanisms are actually equivalent. One such equivalence is between the Random Serial Dictatorship mechanism and the Top Trading Cycle mechanism. This equivalence has been proved by Atila Abdulkadiroğlu and Tayfun Sönmez.[1]

Before explaining this equivalence result in detail, we need to describe how we introduce randomization in the Top Trading Cycle algorithm. We have seen that there are two versions of this algorithm, one with public endowments and one with private endowments. The "trick" used by Abdulkadiroğlu and Sönmez consists of taking the problem with public endowments and making the endowments random. Once this operation is realized, we run the Top Trading Cycle algorithm.

Example 12.4 Consider three individuals, Alice, Bob, and Carol, and three objects, k_1, k_2, and k_3. The preferences over these objects are as follows.

P_{Alice}	P_{Bob}	P_{Carol}
k_1	k_1	k_3
k_2	k_2	k_2
k_3	k_3	k_1

Suppose we run the Random Serial Dictatorship algorithm, and suppose that the random order over the individuals is first Alice, then Bob, and finally Carol. So Alice chooses first and takes k_1. Then Bob picks an object. His most preferred object, k_1, is no longer available, so he picks the second most preferred, k_2. Then Carol is left with k_3.

We now consider the Top Trading Cycle mechanism. To this end, we first need to distribute the endowments. Suppose that the (random) distribution yields the following endowments: Alice owns k_1, Bob owns k_3, and Carol owns k_2. We now run the Top Trading Cycle algorithm (algorithm 11.2).

In the first step, Alice points to herself: she owns her most preferred object. Bob points to Alice, and Carol points to k_3's owner, Bob. The self-cycle involving Alice is the only cycle, so Alice is assigned k_1.

In the second step, object k_1 is no longer available, so Bob now points to k_2's owner, Carol. Carol still points to k_3's owner, Bob. We then have a cycle involving Bob and Carol, so Bob is assigned k_2 and Carol k_3.

Probabilistic Assignments

In example 12.4, we picked the random order (for the Serial Dictatorship) and the random endowment (for the Top Trading Cycle algorithm) such that they yield the same final assignment. It is obvious that this may not always be the case. But what happens when we consider all possible random orders and all possible random endowments? Is it the case that for any order of play (for Serial Dictatorship) there exists an endowment allocation such that the Top Trading Cycle algorithm yields the same final assignment? The answer is yes.

Result 12.2 (Abdulkadiroğlu and Sönmez) For any assignment problem with the same number of individuals and objects, the random assignments generated by the Random Serial Dictatorship algorithm are the same (with the same probabilities) as the random assignments generated by the Top Trading Cycle algorithm with random endowments.

In other words, result 12.2 says that if the Random Serial Dictatorship algorithm produces an assignment μ with a probability of, say, 0.25, then when randomizing endowments and running the Top Trading Cycle algorithm there is a probability of 0.25 that we obtain the assignment μ.

12.3 The Probabilistic Serial Mechanism

When we introduce random assignments, there are two ways we can evaluate an assignment:

- **ex-ante**: We look at the assignments that can be obtained (together with their probabilities). The expression "ex-ante" comes from Latin and means "before the event."
- **ex-post**: We look at an assignment that is realized (i.e., a deterministic assignment).

The expression "ex-post" comes from Latin and means "after the event."

For instance, suppose that a random assignment β consists of obtaining the assignments μ, μ', and μ'', each one occurring with some probability. Evaluating β *ex-ante* consists of evaluating *together* the three assignments μ, μ', and μ'' *with their probabilities*. This contrasts with the *ex-post* evaluation, where we look at one assignment only, the one that is realized.

So in this section we will distinguish between two types of efficiency: **ex-ante efficiency** and **ex-post efficiency**. The efficiency we saw in section 11.1.2 corresponds to the ex-post efficiency. The definition for the ex-ante efficiency is very similar, except that now we have to use random assignment instead. So we will have that a random assignment β is ex-ante efficient if there does not exist another random assignment β' such that all individuals prefer β' to β.

How do we check that an individual prefers a random assignment to another one? The most common and compelling way is to use the stochastic dominance relation we studied in section 12.1.3, so the formal definition is the following.

Definition 12.2   A random assignment β is **ex-ante efficient** if there is no other random assignment β' such that, for each individual, β' stochastically dominates β.

Some authors use the terminology **ordinally efficient** instead of ex-ante efficient to emphasize the fact that we are working with *orderings* and not utilities or numerical payoffs (in which case we would talk about *cardinal efficiency*).

Now consider the Random Serial Dictatorship mechanism. This is a mechanism that randomizes over deterministic Serial Dictatorships (as many as there are different orderings of the individuals/dictators). We know that the (deterministic) Serial Dictatorship algorithm yields (ex-post) efficient outcomes. So, if we randomize, we would expect also to have an ex-ante efficient random assignment. Surprisingly, we do not. To see this, consider the following example proposed by Anna Bogomolnaia and Hervé Moulin.[2]

Example 12.5   We consider four individuals (Alice, Bob, Carol, and Denis) and four objects (an apple, an orange, a pear, and a strawberry). Alice and Bob have the same preferences and Carol and Denis have the same preferences over those four objects.

P_{Alice}	P_{Bob}	P_{Carol}	P_{Denis}
apple	apple	orange	orange
orange	orange	apple	apple
pear	pear	strawberry	strawberry
strawberry	strawberry	pear	pear

In total, there are $4! = 24$ different orderings of these four individuals, each one occurring with the same probability. Among those 24 orderings, there are 6 orderings where Alice is the first individual, 6 where she is the second, 6 where she is the third, and 6 where she is the last. For each of these cases, we count how many times she gets the apple.

- Alice is first. Then she gets the apple for sure.
- Alice is second. She can get the apple only if Bob is not first. She is ranked second six times, but of those six times there are two times where Bob is first, two times where Carol is first, and two times where Denis is first. So out of the six times where Alice is second, she can get the apple only four times (when Carol and Denis are ranked first, they both pick the orange, so Alice can take the apple).

- Alice is third. If Bob is first or second, then she cannot get the apple. So we need both Carol and Denis to be before Alice. But in this case the first will pick the orange and the second will pick the apple. So Alice will never get the apple if she is third.
- Alice is fourth. It is as in the previous case: she will never obtain the apple.

So, out of 24 orderings, Alice can get the apple 6 times (when she is first) or 4 times (when she is second). So she obtains it 10 times out of 24, which is 5/12. Since Alice and Bob have identical preferences, we deduce that Bob also has a probability of 5/12 of obtaining the apple. Doing the same for the other objects and for all individuals, we obtain the following random assignments, which we call β.

$$
\begin{array}{c}
\phantom{\text{Alice} \to} \begin{array}{cccc} \text{apple} & \text{orange} & \text{pear} & \text{strawberry} \\ \downarrow & \downarrow & \downarrow & \downarrow \end{array} \\
\begin{array}{c} \text{Alice} \to \\ \text{Bob} \to \\ \text{Carol} \to \\ \text{Denis} \to \end{array}
\begin{pmatrix} 5/12 & 1/12 & 5/12 & 1/12 \\ 5/12 & 1/12 & 5/12 & 1/12 \\ 1/12 & 5/12 & 1/12 & 5/12 \\ 1/12 & 5/12 & 1/12 & 5/12 \end{pmatrix}
\end{array}
\quad (12.2)
$$

Consider instead the following random assignment, which we call β'.

$$
\begin{array}{c}
\phantom{\text{Alice} \to} \begin{array}{cccc} \text{apple} & \text{orange} & \text{pear} & \text{strawberry} \\ \downarrow & \downarrow & \downarrow & \downarrow \end{array} \\
\begin{array}{c} \text{Alice} \to \\ \text{Bob} \to \\ \text{Carol} \to \\ \text{Denis} \to \end{array}
\begin{pmatrix} 1/2 & 0 & 1/2 & 0 \\ 1/2 & 0 & 1/2 & 0 \\ 0 & 1/2 & 0 & 1/12 \\ 0 & 1/2 & 0 & 1/12 \end{pmatrix}
\end{array}
\quad (12.3)
$$

Take, for instance, Carol. The probability of getting her favorite object (the orange) under β' is higher than under β (1/2 vs. 5/12). The probability that she gets either the orange or the apple (the two most preferred objects) is the same under β' as under β (1/2 + 0 vs. 5/12 + 1/12). Finally, the probability that she gets either the apple, the orange, or the strawberry (the three most preferred objects) is 1 under β' but only 11/12 under β (5/12 + 1/12 + 5/12). That satisfies equation (12.1), so for Carol β' stochastically dominates β.

The calculus for Alice, Bob, and Denis is similar. So β is not ex-ante efficient!

The fact that the Random Serial Dictatorship is not ex-ante efficient is very surprising. This can be annoying, because it is a widely used solution to allocate objects on the ground that it is a very simple procedure in that it is not only fair (each ordering has the same probability) but also *seems* to be (ex-ante) efficient.

Bogomolnaia and Moulin have proposed a way to fix this problem by defining a new algorithm, called the **Probabilistic Serial algorithm**, which works as follows. We will imagine that individuals will race to "eat" each object in order of their preferences. That is, each individual will start eating his or her most preferred object. Once an individual's most preferred object is completely eaten, he or she will start eating the next most preferred object that is still available (i.e., not yet completely eaten).

The key aspect of this procedure is the following: several individuals can eat the same object at the same time. In the simple version we present here, we will assume that the eating speed is the same for each individual. For instance, if for a problem there are five individuals who have the same object as their most preferred object, these five individuals will start eating that object at the beginning of the algorithm. Since the eating speed is the same, they will each eat only 1/5 of the first object.

Bogomolnaia and Moulin's idea is to set the share of an object that is eaten by an individual as the probability that that individual is assigned that object. That way we construct a random assignment.

The Probabilistic Serial algorithm looks quite different from the other algorithms we have seen. These other algorithms are described in steps, which is not the case for the Probabilistic Serial algorithm: it runs over a time interval. The length of the time interval of the algorithm is irrelevant; it could be one hour, one day, or whatever. What matters is that

- the "size" of each object (i.e., the amount to be eaten) is the same and that
- each individual needs the whole time interval to entirely eat one object.

Algorithm 12.1   Probabilistic Serial

When the algorithm starts, each individual starts eating her most preferred object.
   When the object an individual is eating is entirely eaten, then:

- The fraction of the object that has been eaten will be the probability that the individual is assigned that object.
- The individual starts eating the most preferred object among the objects that are not yet entirely eaten.

At the beginning, each person eats his or her most preferred object. So Alice and Bob both eat, at the same time, the apple, and Carol and Denis eat the orange.
   Alice and Bob eat at the same speed, so each will manage to eat one half of the apple. So the probability that Alice is assigned the apple is 1/2, and similarly for Bob.

Parallel to Alice and Bob's feast, Carol and Denis eat the orange. The situation is similar: Carol and Denis each obtain the orange with a probability 1/2.

So Alice and Bob on one side and Carol and Denis on the other side finish eating the first fruit at the same time. Hence, at the same time, they will start eating the most preferred fruit that is still available.

For both Alice and Bob, the next fruit according to their preferences is the orange. But the orange is gone; it is already completely eaten by Carol and Denis. So they go for the next fruit, the pear. For Carol and Denis, the most preferred fruit that is still available is the strawberry.

So when Alice and Bob eat the pear, Carol and Denis eat the strawberry. We obtain that Alice eats only half of the pear; the other half is eaten by Bob. Similarly, Carol and Denis each only manage to eat one half of the strawberry. So Alice and Bob will each get the pear with probability 1/2 and Carol and Denis will each get the strawberry with probability 1/2. We obtain the assignment β' described in equation (12.3).

It may happen that when an individual starts eating an object this object has already been partially eaten. The next example illustrates such situations.

Example 12.6 We consider three individuals (Alice, Bob, and Carol) and three objects (an apple, an orange, and a pear). Their preferences are as follows.

P_{Alice}	P_{Bob}	P_{Carol}
apple	apple	orange
pear	orange	pear
orange	pear	apple

At the beginning, Alice and Bob start eating the apple and Carol eats the apple. Recall that all individuals eat at the same speed, so the first fruit that is eaten is the one with the most individuals on it, the apple. Alice and Bob have each eaten one half, so they will be assigned the apple with probability 1/2.

Then Alice starts eating the pear and Bob eats the orange. Note, however, that when Bob starts eating the orange, Carol has already eaten some of it. How much? Since they all eat at the same speed, when Bob is done eating his half apple, Carol has only eaten half of the orange.

Therefore, when Bob starts eating the orange (with Carol), there is only half of it left. Now they start eating it, and since they eat at the same speed, each will get one half of this second half, namely 1/4. Earlier Carol ate 1/2 of the orange, so in total Carol ate $1/2 + 1/4 = 3/4$ of the orange, and Bob ate 1/4 of the orange.

When Bob or Carol eats only 1/4 of the orange, Alice, because she eats at the same speed, only eats 1/4 of the pear.

So far, we therefore have the following.

	apple	orange	pear
Alice	1/2	0	1/4
Bob	1/2	1/4	0
Carol	0	3/4	0

Now there is only the pear left. Alice is still eating it and is joined by Bob and Carol. How much of the pear is left? What Alice has not yet eaten, 3/4.

They all eat at the same speed, so each will get one third of 3/4 of the pear, or 1/4. For Alice, we add this to what she already ate, so she ate in total $1/4 + 1/4 = 1/2$ of the pear. So we obtain the following random assignment:

$$\begin{array}{c} \\ \text{Alice} \to \\ \text{Bob} \to \\ \text{Carol} \to \end{array} \begin{array}{c} \text{apple} \\ \downarrow \\ \left(\begin{array}{c} 1/2 \\ 1/2 \\ 0 \end{array}\right. \end{array} \begin{array}{c} \text{orange} \\ \downarrow \\ 0 \\ 1/4 \\ 3/4 \end{array} \begin{array}{c} \text{pear} \\ \downarrow \\ \left.\begin{array}{c} 1/2 \\ 1/4 \\ 1/4 \end{array}\right) \end{array} \qquad (12.4)$$

Bogomolnaia and Moulin defined a more general version of the probabilistic serial mechanism we just presented. In the general version they proposed, individuals may not only have different eating speeds, but also the speed at which someone eats may change over time (e.g., someone eats very fast at the beginning and then slows down). In other words, each individual has an *eating speed function* that indicates, at any time, the speed at which they eat. The remarkable result Bogomolnaia and Moulin obtain is the following.

Result 12.3 (*Bogomolnaia and Moulin*) For any problem and for any eating speed function, the random assignment calculated with the Probabilistic Serial algorithm is ex-ante efficient.

Conversely, for any ex-ante efficient random assignment, there exists an eating speed function for each individual such that the Probabilistic Serial algorithm yields this random assignment.

So, is the Probabilistic Serial mechanism the "right" random mechanism? It turns out to be not quite so. When studying the Probabilistic Serial mechanism, Bogomolnaia and Moulin highlighted an inherent conflict between two properties: ex-ante efficiency and strategyproofness. To be precise, the negative result they obtain is using a third property: **equal treatment of equals**. This property states that if two individuals have exactly the same preferences over objects, then they should obtain the same random assignment. For deterministic assignments, it is relatively easy to see that this property does not make much sense. Consider, for

Probabilistic Assignments

instance, example 12.5. For any order of the individuals, it will not be possible to ensure that Alice and Bob, who have the same preferences, will get the same object. However, when considering *random* assignments, we may want to give equal chances to both individuals; that is, to give them the same probabilities over the objects they may obtain. This property then captures some notion of fairness that random assignments should satisfy.

Result 12.4 (Bogomolnaia and Moulin)   Whenever there are at least four individuals, there is no random assignment mechanism that is always ex-ante efficient, strategyproof, and that satisfies the equal treatment of equals property.

We can see that result 12.4 states an impossible result when there are at least four individuals. When there are three individuals, Bogomolnaia and Moulin show that the mechanism that satisfies the three properties listed in result 12.4 is the Random Serial Dictatorship.

13 School Choice

The medical match we saw in chapter 10 was the first example where matching theory can help us understand how a market works and how to improve it. A more recent application that has opened the way to numerous developments is the assignment of children to schools. In the economic literature, "school choice" refers to the idea that parents may have a say in the school assignment of their children. In some regions or countries, parents have no influence in the selection of the school their children will attend. This is the case, for instance, where authorities draw a "zoning map" that specifies which school a student will attend as a function of his home address. But in many cities or countries across the globe, parents may have a say (beyond deciding where to live) in the school assignment. Of course, some schools are more popular and thus may be more in demand. The question then is how to accommodate parents' preferences with school district policies.

13.1 The Many-to-One Assignment Model

At the outset, the so-called school choice model is very close to the many-to-one matching model we saw in chapter 10. We just have to replace doctors and hospitals by *students* and *schools*, respectively. But there is one major difference between the medical match and the school choice model that will guide much of our modeling choices and the question we will address: schools in a school choice problem do *not* have preferences. So, the school choice problem is more an assignment problem. We will see that there are some subtle differences with respect to the assignment model we saw in chapter 11, though.

13.1.1 Preferences versus Priorities
In a school choice model, schools do *not* have preferences over students. When economists say that an agent has *preferences* over some alternatives, it is implicitly assumed that this agent *enjoys* what he obtains or consumes. Schools in

the present context are not assumed to have preferences; they are mere institutions (or "objects") offering a service to be consumed. This implies that any welfare analysis in a school choice problem only considers students' welfare (or their parents' welfare). For this reason, the school choice model is often referred to in the literature as an **assignment problem** instead of a matching problem. Having said that, from a mathematical point of view, there are not so many differences between the many-to-one *matching* model and the many-to-one *assignment* model. However, the assignment model will make some modeling choices more natural.

Schools do not have preferences over students, but they nevertheless rank them. To distinguish these rankings over students from the concept of preferences, we will call such rankings **priorities** (or *priority lists*). In real life, those priorities are often the outcome of administrative (or political) decisions. For instance, a common practice is that students who live close to a school have a higher priority for enrollment in that school than the students who live farther away. One may think that the school "prefers" students who live closer, although it is better to say that the students who live closer have a *higher priority* than the students who do not. Other criteria can be the presence of siblings in the school, being part of some specific social group, and other factors.

Regarding schools' priority rankings, a common assumption is that those rankings are over students and not groups of students. As with the medical match, this may create a problem because since schools can enroll more than one student, how do we proceed when a school is considering groups of students? In other words, how do we construct priorities over groups of students from the priority rankings over students? The solution consists of saying that schools' priority rankings are **responsive**. By this term we basically mean that if a school has already enrolled a group of students and it has to pick one additional student from a given set of students, then the school will pick the student with the highest priority. In other words, the choice between two students does not depend on who the other students are who are already enrolled.

In chapter 10, the use of responsive preferences was an assumption that we made to simplify the analysis, but there is no reason why hospitals must have such preferences. For the case of school choice, responsive preferences are much easier to justify. By saying that schools have a ranking of *priorities* over students, we are implicitly saying that schools do not consider that there is some sort of complementarity between students or any other types of group effects. In other words, students are seen as being independent from each other. Each student has a priority order, and that order is not affected by who the other students already enrolled in a school are. This notion of independence is captured by a concept economists call *separability*, which is in fact the notion of responsiveness in our context.[1]

13.1.2 The Model
To sum up, a **school choice problem** is given by a collection (I, S, q, P, π) that consists of:

1. a set of **students**, $I = \{i_1, \ldots, i_n\}$;
2. a set of **schools**, $S = \{s_1, \ldots, s_m\}$;
3. a **capacity** vector, $q = (q_{s_1}, \ldots, q_{s_m})$, which specifies, for each school, the maximum number of students the school can enroll;
4. a profile of strict **student preferences**, $P = (P_{i_1}, \ldots, P_{i_n})$, similar to the preferences that doctors have over hospitals we saw in chapter 10; and
5. a strict **priority structure** of the schools over students, $\pi = (\pi_{s_1}, \ldots, \pi_{s_m})$, where π_s denotes the priority ranking over students for school s.

Schools' priorities differ from students' preferences on one additional point. For a student i, her preference P_i consists of an ordering over the set $S \cup \{i\}$. That is, a student is allowed to consider some schools unacceptable. It is often argued in the literature that, for a student, being matched to herself is capturing the idea of an outside option such as attending a private school or studying at home. Schools' priorities, however, are orderings over *the whole set of students*. The motivation behind this assumption is that school choice refers to the problem of assigning students to *public schools* and, except in a few cases, the vast majority of schools are supposed to be able to enroll *any* student.

13.1.3 Assignments
We can now define what an assignment is in this many-to-one assignment problem. The definition is similar to the one we have seen for the basic one-to-one matching problem, with the exception that now we have to deal with schools' capacities. That is, the only additional requirement we will add is that a school cannot enroll more students than its capacity. Formally, an assignment is defined as a mapping μ from the set of schools and students, $S \cup I$, to the set of all possible sets of students and schools, $\mu : I \cup S \to 2^I \cup S$, such that for any $i \in I$ and any $s \in S$, we have:

(a) $\mu(i) \in S \cup \{i\}$. Each student must be matched to a school or to herself.

(b) $\mu(s) \in 2^I$. Each school is matched to a subset of students, where 2^I denotes the collection of all possible sets of students.[2] An equivalent way to write it is $\mu(s) \subseteq I$.

(c) $\mu(i) = s$ if and only if $i \in \mu(s)$. A student i is matched to a school s only if that student is in the list of students enrolled at school s, $\mu(s)$.

(d) $|\mu(s)| \leq q_s$. If $\mu(s)$ denotes the set of students matched to school s, $|\mu(s)|$ simply denotes the number of such students. This criterion states that the school s cannot be matched to more than q_s students, the capacity of school s.

13.1.4 Stability and Efficiency

The concept of stability for the school choice model is nearly identical to the one used for the medical match. To emphasize that schools do not have *preferences* (but rather priority rankings), the vocabulary changes slightly. Stability in a school choice problem is the conjunction of three requirements, given in definition 13.1.

Definition 13.1   An assignment μ is stable if

(a) it is **individually rational**, meaning that for all students $i \in I$, $\mu(i)$ is weakly preferred to the option of being unmatched;

(b) it is **nonwasteful**: if a student prefers a school to her assignment, then that school must have filled its capacity. Formally, for all $i \in I$ and all $s \in S$, $sP_i\mu(i)$ implies $|\mu(s)| = q_s$; and

(c) there is no **justified envy**: if a student i prefers a school s to his assignment, then all students matched to school s must have a higher priority than student i. Formally, for all $i,j \in I$ with $\mu(j) = s \in S$, $sP_i\mu(i)$ implies $j\pi_s i$.

By *weakly preferred* (condition (a) in the definition) we mean that it is either preferred or indifferent. As it is written, we do not know the assignment of student i. It could be a school or the student herself. Not knowing that, we can only require that the student's assignment be weakly preferred. An assignment is *not* individually rational if we have the opposite: a student who strictly prefers being unassigned rather than her assignment. Condition (c) is equivalent to the non-blocking condition for the definition of a matching we saw in the many-to-one matching model of chapter 10. If for an assignment μ there is a student i, a student j, and a school s such that $\mu(i) \neq s$, $\mu(j) = s$, $sP_i\mu(i)$, and $i\pi_s j$, then student i and school s **block** the assignment μ.

Another relevant property for assignment problems is *efficiency*.

Definition 13.2   An assignment μ is **efficient** if there is no other assignment μ' such that:

- All students *weakly prefer* μ' to μ, meaning all students are either indifferent between μ' and μ or prefer μ' to μ.
- There is at least one student who strictly prefers μ' to μ, meaning at least one student who is not assigned to the same school under μ' and μ and prefers the school she is assigned to under μ'.

There is a major difference between stability and efficiency which we illustrate in example 13.1. Stability relates to both the students' preferences *and* the schools' priorities, whereas efficiency only takes into account students' preferences.

Example 13.1   We consider three schools, called s_1, s_2, and s_3. Schools s_1 and s_3 have a capacity of one, and school s_2 can accommodate two students. So $q_{s_1} = 1$,

School Choice

$q_{s_2} = 2$, and $q_{s_3} = 1$. There are four students, called i_1, i_2, i_3, and i_4. The preferences of the students and the schools' priority lists are given in the following table.

P_{i_1}	P_{i_2}	P_{i_3}	P_{i_4}
s_2	s_1	s_1	s_2
s_1	s_2	s_2	s_3
s_3	s_3	s_3	s_1

π_{s_1}	π_{s_2}	π_{s_3}
i_1	i_3	i_4
i_2	i_4	i_1
i_3	i_1	i_2
i_4	i_2	i_3

Consider the following two assignments:

$$\mu = \{(i_1, s_1), (i_2, s_3), (i_3, s_2), (i_4, s_2)\},$$

$$\mu' = \{(i_1, s_2), (i_2, s_3), (i_3, s_1), (i_4, s_2)\}.$$

For instance, under μ, student i_1 is assigned to school s_1, students i_3 and i_4 are both assigned to school s_2, and student i_2 is assigned to school s_3.

The Assignment μ Is Not Efficient
The only differences between μ and μ' are the assignments of students i_1 and i_3. We see that both students i_1 and i_3 prefer the school they are assigned to under μ' to the school they obtain under μ. The other students are indifferent. Since there is *at least one student* who prefers μ' to μ (we have two, i_1 and i_2) and the other students are either indifferent or better off under μ', we deduce from the definition of efficiency that the assignment μ is *not* efficient.

The Assignment μ' Is Efficient
Observe that under μ' students i_1, i_3, and i_4 are assigned to their most preferred school. So the only way to find another assignment that can make at least one student better off is to find an assignment where student i_2 is better off. Therefore it has to be an assignment where i_2 is assigned to either s_2 or s_1. For either case, we must unassign another student:

(a) i_3 if we assign i_2 to s_1;

(b) i_1 or i_4 if we assign i_2 to s_2.

In case (a), the student reassigned, i_3, is necessarily worse off. In case (b), whether we reassign i_1 or i_4, the reassigned student is necessarily worse off. Therefore, there is no other assignment that can make all students either indifferent or better off. So we can conclude that μ' is an efficient assignment.

The Assignment μ Is Stable
To see this, note that only students i_1, i_2, and i_3 would like to have another school (i_4 is assigned to her most preferred school). Student i_1 would like to be assigned to s_2 instead of s_1. But school s_2 is assigned to i_3 and i_4, which is the set of students with the highest priorities, so student i_1 cannot block with school s_2. Students i_2 and i_3 would prefer to be with school s_1, but that school will "refuse" them, as it has the student with the highest priority, i_1. Student i_2 could also be better off with school s_2, but then we have the same situation as for student i_1: school s_2 will "refuse" the student. Hence, there is no blocking pair for μ; that is, μ is a stable assignment.

The Assignment μ' Is Not Stable
An assignment is not stable if either there is a blocking pair, it is not individually rational, or it is wasteful. It is not difficult to see that μ' is individually rational and not wasteful. So if it is not stable, it must be that there is a student-school pair that blocks the assignment. Observe that students i_1, i_3, and i_4 are assigned to their most preferred school under μ'. So clearly these students do not want to block, and thus the only student who could block the assignment is student i_2. Student i_2 can be better off with either s_2 or s_1. School s_1 is assigned to i_3 under μ'. Since i_2 has a higher priority at s_1 than i_3 does, we have that the pair (i_2, s_1) is a blocking pair! So μ' is not stable.

The existence of one blocking pair is sufficient to declare that an assignment is not stable. But as an exercise we can also look at the pair (i_2, s_2). School s_2 is assigned i_1 and i_4. To be able to block with student i_2, school s_2 must get rid of a student with a lower priority than i_2. This is indeed what responsiveness requires. If school s_2 replaces i_1 with i_2, then the set $\{i_2, i_4\}$ has a higher priority than the set $\{i_1, i_4\}$ if and only if i_2 has a higher priority than i_1. This is clearly not the case. The same occurs with student i_4: she has higher priority at school s_2 than student i_2, so student i_2 cannot block with s_2.

Stability and efficiency are among the two most important properties market designers look at. It turns out, however, that when designing an assignment procedure, we must choose which of these two properties we want; we cannot have both. In example 13.1, the assignment μ is stable (and is also the unique stable assignment for that problem). Yet μ is not efficient.

Result 13.1 It may happen that, for some specific preferences and priorities, a stable assignment is also efficient. But this is not true in general: it is impossible to guarantee obtaining at the same time efficient *and* stable assignments.

Haluk Ergin identified conditions on schools' priorities that ensure that the student-optimal assignment is also efficient.[3] However, these conditions are very demanding and are unlikely to be met in practice (very roughly, schools' priorities have to be very similar, limiting the way two students can be ordered differently by any pair of schools). Reconciling efficiency and stability, although desirable, seems to be an impossible goal to achieve.

If stability and efficiency are both desirable properties, we would like to be able to pinpoint an efficient *and* stable assignment whenever it exists. In example 14.2, such an assignment exists: μ' is both stable *and* efficient. Also, if there is a mechanism that does produce such an assignment, we would like it to be strategyproof. It makes participants' lives easier, as they do not have to think about which strategy is the best (i.e., which preference ordering to submit to the mechanism). Unfortunately, the following result by Onur Kesten tells us that this is an impossible task.[4]

Result 13.2 (Kesten) There is no efficient and strategyproof mechanism that selects the efficient and stable matching whenever it exists.

Result 13.2 reads as follows. Suppose we want to find a mechanism that always produces an efficient assignment. This is not difficult, and we will see in section 13.2 how to achieve this. We know from result 13.1 that this assignment is not necessarily stable. But what if the problem is such that there is an assignment that is both efficient and stable? It would be great if we could select this assignment. Result 13.2 says that there is no strategyproof mechanism that achieves this.

13.2 Competing Algorithms

The literature on school choice started with the seminal work of Atila Abdulkadiroğlu and Tayfun Sönmez.[5] Their work was motivated by the situation in Boston. Parents in Boston had to submit a preference list over schools to a central clearinghouse, and an algorithm was used to match students to schools. One key feature of the Boston procedure was that the algorithm used was *not* the Deferred Acceptance algorithm but instead the *Immediate Acceptance* algorithm that we will study in this section.[6] This algorithm is very similar to the Deferred Acceptance algorithm, the only difference being in the way the acceptance/rejection decisions are made. But we do not have to restrict ourselves to these two algorithms. Since the school choice model is fundamentally an assignment model, we can also consider using the Top Trading Cycle algorithm that we studied in chapter 11.

13.2.1 The Role of Each Side of the Market
In the medical matching model of chapter 10 or the simple one-to-one matching model of chapter 9, we considered two versions of the Deferred Acceptance algorithm, one in which doctors make proposals to hospitals and one in which

the hospitals make offers to candidates, and it is up to the market designer or the policy maker to choose which version to run.

In the school choice model, the question of which side makes offers has little bite. In both the Deferred Acceptance and Immediate Acceptance algorithms, agents on the proposing side make proposals starting from the top of their rank order lists. Proceeding this way amounts to saying that the agents on the proposing side will try to be matched to the highest-ranked partner(s). In a school choice model, students have *preferences*, so if students are proposing, the algorithm will try to match them to their most preferred choice. But, in this model, schools do not have preferences. In other words, schools do not really care about which students they enroll, and thus the idea of maximizing a school's payoff does not make sense, because schools do not have payoffs.[7]

For the Top Trading Cycle algorithm, we can also consider two cases. The first case would be one where students are assigned to the school they point to, and the second case would be the opposite, where schools enroll the students they point to (schools would use their priority rankings to point toward students). The Top Trading Cycle algorithm produces an assignment that is efficient for the agents that point in the algorithm. Choosing which version to use for this algorithm (students point or schools point) is thus a problem similar to the one for the Deferred Acceptance or Immediate Acceptance algorithms. Since schools are mere objects to be consumed, and each school's ranking over students only reflects a *priority* order and not a preference list, there is little sense in maximizing schools' priorities. So, in a school choice model, the Top Trading Cycle algorithm where students point to schools is the more relevant algorithm to use.

13.2.2 The Deferred Acceptance Algorithm

The Deferred Acceptance algorithm in a school choice context is identical to the one we have seen for the medical match, except that now we do not need to check whether a student applying to a school is acceptable for that school; by assumption, for each school, all students are acceptable.

Algorithm 13.1 Deferred Acceptance for School Choice

Step 1 Each student i applies to the school that is ranked first in her preference list (if there is no such school, then i becomes unassigned).

Each school s assigns students, one at a time, up to its capacity from the students applying to s, following the priority order π_s. That is, school s first admits the student with the highest priority, then the student with the second-highest priority, and so on until either school s has enrolled q_s students or it has enrolled all the students who applied to s. The remaining students are rejected.

Step k, $k \geq 2$ Each student rejected in the previous step applies to the most preferred school among those she has not yet proposed to and that are acceptable (if there is no such school, then the student becomes unassigned).

Each school receiving applications considers *the set of students it accepted at the previous step together with the set of new applicants*. From this larger set, the school accepts students up to its capacity, one at a time, following its priority order. The remaining students are rejected.

End The algorithm stops when no student is rejected or all schools have filled their capacities. Any remaining student remains unassigned.

The assignment we obtain with this algorithm is called the **student-optimal assignment**. From chapter 10, we can deduce the following result.

Result 13.3 (Properties with Deferred Acceptance for School Choice)

- The Deferred Acceptance algorithm produces a student's most preferred stable assignment.
- An assignment mechanism that uses the Deferred Acceptance algorithm to assign students to schools is strategyproof (for the proposing side, the students).

Note that from result 13.1 we know that the student-optimal assignment is not necessarily efficient.

Example 13.2   We consider four students, Alice, Bob, Carol, and Denis, and three schools, A, B, and C. The capacity of school A is 2 (i.e., it can be matched to at most two students), and the capacity of schools B and C is 1 (at most one student can attend these schools).

The preferences of the students over the schools and the priorities of the schools over the students are described in the following table.

P_{Alice}	P_{Bob}	P_{Carol}	P_{Denis}
A	A	B	A
B	B	A	C
C	C	C	B

π_A	π_B	π_C
Alice	Alice	Bob
Carol	Bob	Carol
Denis	Denis	Denis
Bob	Carol	Alice

Step 1 Alice, Bob, and Denis propose to school A. School A can admit at most two students, so it starts accepting the students who applied to it following the order given by its priority order. Alice is the top priority student, so she is accepted. The student with the next highest priority and is applying to A is Denis. So Denis is accepted, and thus Bob is rejected because school A has filled its capacity with Alice and Denis. The other student, Carol, proposed to school B. She is the only applicant, so she is accepted.

So, at the end of the first step, Alice and Denis are temporarily accepted at school A, Carol is temporarily accepted at school B, and Bob is not assigned to any school.

Step 2 Bob applies to school B. School B now has two applicants: Carol (from step 1) and Bob. Bob has higher priority than Carol, and since school B has only one seat available, Bob is now temporarily accepted at school B and Carol is rejected.

Step 3 Carol (the only student rejected in step 2) applies to school A. School A now has three applicants: Alice and Denis (from step 1) and Carol. The two students with the highest priorities are Alice and Carol, so Denis is rejected.

Step 4 Denis applies to school C. He is the only applicant, so he is accepted. No student is rejected so the algorithm stops.

So the student-optimal assignment is

$$\mu(\text{Alice}) = A, \quad \mu(\text{Carol}) = A,$$
$$\mu(\text{Bob}) = B, \quad \mu(\text{Denis}) = C.$$

13.2.3 The Immediate Acceptance Algorithm

The Immediate Acceptance algorithm, although similar to the Deferred Acceptance algorithm, introduces a twist at steps 2, 3, ... The crucial difference is that in the Immediate Acceptance algorithm the matches that are realized at the various steps are *not* temporary; they are final. In other words, the final acceptance here is *immediate* and not deferred to the end, when the algorithm stops. Apart from this, the two algorithms are identical.

Algorithm 13.2 Immediate Acceptance

Step 1 Each student i applies to the school that is ranked first in her preference list (if there is no such school, then i remains unassigned).

Each school is assigned, up to its capacity, to the students applying to it, one at a time, following its priority order. That is, school s first admits the student with the highest priority according to π_s among the students who applied to it. It then accepts, among the students who applied to it, the student with the second-highest priority in π_s. The acceptance decision is repeated until either school s has enrolled q_s students or it has enrolled all the students who applied to it. The remaining students are rejected.

Step k, $k \geq 2$ Each student i that has been rejected in the previous step applies to the most preferred school among the schools to which the student has not yet made any proposal. (If there is no such school, then i remains unassigned.)

Each school is assigned to the following students:

(a) All students that were assigned to that school at a previous step are assigned again to that school. The **remaining capacity** is the school's capacity minus the number of such students.

(b) Each school is assigned, up to its *remaining capacity*, to the students applying to it, one at a time, following its priority order. The remaining students are rejected.

End The algorithm stops when no student is rejected or all schools have filled their capacities. Any remaining student remains unassigned.

The only difference between the Deferred Acceptance and the Immediate Acceptance algorithms is how a school treats students it accepted at a previous step. In the Deferred Acceptance algorithm, the acceptance is reconsidered, while it is not in the Immediate Acceptance algorithm. In the first step, there is no previous step, so there are no students to reconsider, and thus the first step is the same for both algorithms.

Under the Immediate Acceptance algorithm, if a student i is matched to a school s, say, in the first step, then we can immediately deduce that at the end of the algorithm student i will be matched to s. With the deferred acceptance algorithm, this is not the case. It could be the case that student i is rejected at a later step by school s and ends up being matched to a less preferred school.

What are the properties satisfied by the Immediate Acceptance algorithm? First, the assignment is not necessarily stable. Example 13.2 illustrates this fact. The Immediate Acceptance algorithm does produce efficient assignments, though. However, we will see in the next section that students may benefit from misrepresenting their preferences.

Using the same preferences and priorities, we now compute the assignment using the Immediate Acceptance algorithm. Since step 1 of the Immediate Acceptance algorithm is *identical* to that of the Deferred Acceptance algorithm, we could start directly with step 2, using the result at the end of step 1 that we found with the Deferred Acceptance algorithm. For the sake of completeness, we run step 1 again.

Step 1 Alice, Bob, and Denis propose to school A. School A can admit at most two students, so it starts accepting the students who applied to it following the order given by its priority order. Alice is the top priority student, so she is accepted. The next student with the highest priority and is applying to A is Denis. So Denis is accepted, and Bob is rejected because school A has filled its capacity with Alice and Denis. The other student, Carol, proposed to school B. She is the only applicant, so she is accepted.

So at the end of the first step Alice and Denis are accepted at school A, Carol is accepted at school B, and Bob is not assigned to any school.

Step 2 Bob is the rejected student. He applies to school B. But school B already filled its capacity in step 1. This means that in step 2 the remaining capacity of school B is 0. Therefore school B can no longer accept any additional student and thus rejects Bob.

Step 3 Bob is the only rejected student from the previous step. He applies to his most preferred school among those he has not yet proposed to meaning he applies to school C. He is the only applicant, so he is accepted. No student is rejected, so the algorithm stops.

So the assignment under the Immediate Acceptance algorithm is

$$\mu'(\text{Alice}) = A, \qquad \mu'(\text{Carol}) = B,$$
$$\mu'(\text{Bob}) = C, \qquad \mu'(\text{Denis}) = A.$$

Notice that this assignment is *not* stable. To see this, observe that the pair (Bob, school B) forms a blocking pair: Bob prefers school B to his assignment (school C), and school C gives Bob a higher priority than Carol (who is assigned to it).

13.2.4 Top Trading Cycles

The Top Trading Cycle algorithm we use for school choice is very close to the algorithm we saw in chapter 11, except that now we have to take into account that each school can be assigned to many different students.

The key intuition to rewrite the algorithm for a many-to-one assignment problem is the following. In the Top Trading Cycle algorithm, notice that when an assignment is found at some step, then it is final for both the individual and the object: they are removed from the problem. For the many-to-one version with students and schools, we will proceed the same way. Whenever a student is assigned to a school, the student is removed. As for the school, we will not remove it but instead will decrease its capacity by one unit. It is when the updated capacity of a school becomes zero that the school is removed from the problem.

Apart from this (minor) complication, the description of the algorithm is almost identical to the one we saw in chapter 11.

School Choice

Algorithm 13.3   Top Trading Cycles for School Choice

Step 1 For each school $s \in S$, we first define q_s^1 as the **remaining capacity**, and we set $q_s^1 = q_s$.

For each student, we draw an arrow starting from the student to the school that is the most preferred by that student. If there is no such school, then the arrow goes to herself (we call that a *self-cycle*).

For each school, we draw an arrow starting from the school to the student who has the highest priority for the school.

There must be at least one cycle: if we start from a student in a cycle and we follow the arrows, we eventually visit that student again.

If a student belongs to a cycle, then we assign her to the school she is pointing to (i.e., the school at the end of the arrow starting from that student), and the student is removed from the problem. If the student is in a self-cycle, we remove the student from the problem, and the student is not assigned to any school.

If a school s is in a cycle, then the new remaining capacity q_s^2 is $q_s^1 - 1$. If a school s does not belong to any cycle, then the remaining capacity is unchanged: $q_s^2 = q_s^1$.

If the new remaining capacity of a school is zero, then that school is removed from the problem.

Step k, $k \geq 2$

Reminder: If a school s is still present at the beginning of step k, its remaining capacity is q_s^k.

For each student, we draw an arrow starting from the student to the school that is the most preferred by that student among the schools that are still present in the problem. If there is no such school, then the arrow goes to herself (a self-cycle).

For each school, we draw an arrow starting from the school to the student that has the highest priority for the school among the students remaining in the problem.

There must be at least one cycle. If a student belongs to a cycle, then we assign her to the school she is pointing to (i.e., the school at the end of the arrow starting from that student), and the student is removed from the problem. If the student is in a self-cycle, we remove the student from the problem, and the student is not assigned to any school.

If a school s is in a cycle, then the remaining capacity becomes q_s^{k+1}, setting $q_s^{k+1} = q_s^k - 1$. If a school s does not belong to any cycle, then the remaining capacity is unchanged; that is, $q_s^{k+1} = q_s^k$.

If the new remaining capacity of a school is zero, then that school is removed from the problem.

End The algorithm stops when all students or all schools have been removed. Any remaining student is assigned to herself.

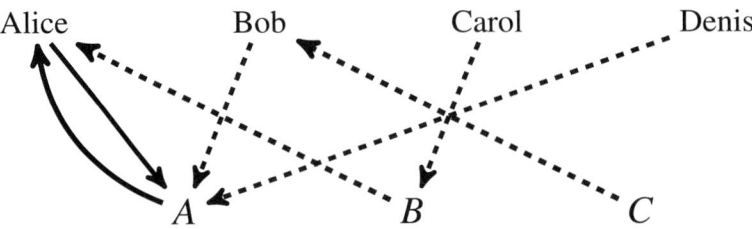

Figure 13.1.
Step 1

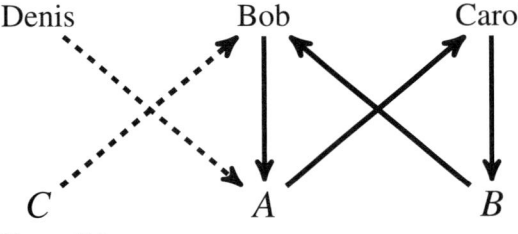

Figure 13.2.
Step 2

Example 13.2 continued Using the same preferences and priorities as in the first part of this example, we now compute the assignment using the Top Trading Cycle algorithm.

Step 1 Each student points to her most preferred school. So Alice, Bob, and Denis all point to school A, and Carol points to school B. Each school points to the student with the highest priority, so schools A and B both point to Alice, and school C points to Bob (see figure 13.1).

There is only one cycle, involving Alice and school A. The cycle is depicted with the solid arrows (the dashed arrows are not part of any cycle). So Alice is assigned to school A and is removed from the problem. School A has a total capacity of 2, so now that Alice is using one of its seats, the remaining capacity of school A is 1.

Step 2 Each remaining student points to her most preferred school. So Bob and Denis point to school A, and Carol points to school B. Each school points to the student with the highest priority among the remaining students, so school A points to Carol and both schools B and C point to Bob (see figure 13.2).

We now have the following cycle: Bob $\to A \to$ Carol $\to B \to$ Bob ... (depicted with the solid arrows). This implies that Bob is assigned to school A and Carol to school B. These two students are withdrawn from the problem. School A has now filled its capacity and is thus withdrawn from the problem. School B also has filled its capacity and is thus withdrawn.

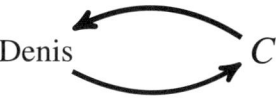

Figure 13.3.
Step 3

Step 3 Only Denis and school C remain in the problem. Denis then points to school C, which points to Denis (see figure 13.3).

We have an obvious cycle, so Denis is assigned to school C, and Denis and school C are both removed from the problem. There are no students left, so the algorithm stops.

So the assignment under the Top Trading Cycle algorithm is

$\mu''(\text{Alice}) = A, \quad \mu''(\text{Carol}) = B,$

$\mu''(\text{Bob}) = A, \quad \mu''(\text{Denis}) = C.$

The Top Trading Cycle algorithm we have presented inherits the same properties we saw in chapter 11.

Result 13.4 (Properties with Top Trading Cycle for School Choice)

• The assignment we obtain with this algorithm is efficient for any preferences, priority lists, and school capacities.

• The assignment mechanism that uses the Top Trading Cycle algorithm is strategyproof (for the students).

Note, however, that the Top Trading Cycle algorithm does not necessarily produce stable assignments. It could be that, for some specific situations, the assignment computed by the algorithm is stable. But this is not true in general.

There is an equivalence between the Top Trading Cycle algorithm we have seen in this chapter and the one we saw in chapter 11. Instead of having schools pointing to students, consider an assignment model where there are only students and where some students own one or more seats at a school in the following way. If a student has the highest priority at a school, then that student owns a seat at that school. With this transformation done, we can run the Top Trading Cycle algorithm of chapter 11. As soon as a student "who owns one or more schools" is assigned to a school by the algorithm, we consider each of the schools she owns. If a school has not filled its capacity, then the next owner of that school will be the next student in the priority ordering of that school among the students who are not yet assigned to a school. Once the school ownerships are updated, we again start drawing the arrows and look for cycles.

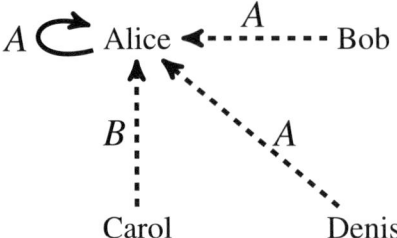

Figure 13.4.
Step 1

To illustrate the relation between the two models, let us consider (again) example 13.2. Alice has the highest priority for schools A and B. So we can say that she "owns" those two schools. School C is owned by Bob, and Carol and Denis do not have any endowments. If we run the first step of the Top Trading Cycle algorithm of chapter 11, we then have:

- Alice points to herself.
- Bob points to Alice (she owns Bob's most preferred school, A).
- Carol points to Alice (she owns Carol's most preferred school, B).
- Denis points to Alice (he owns Denis' most preferred school, A).

The graph is depicted in figure 13.4. Next to each arrow there is the school that the student is asking for. For instance, Carol points to Alice because she wants school B, and Bob points to Alice but because he wants school A.

The only cycle is the one involving Alice (with the solid arrow). So Alice is assigned school A and is removed from the problem. There is now only one seat left at school A. Now that Alice is removed, who owns schools A and B? The next student in school A's priority ordering is Carol, so she now owns school A. For school B, the next owner is Bob. So the students' most preferred schools (among the available schools) are as follows.

Student	Most preferred school	Owner
Bob	A	Carol
Carol	B	Bob
Denis	C	Bob

So now Bob points to Carol, Carol points to Bob, and Denis points to Bob, as shown in figure 13.5.

There is a cycle, involving Bob and Carol (with the solid arrows). So Bob is assigned school A, and Carol is assigned school B. Both Bob and Carol are removed from the problem. Both schools A and B have now filled their capacities, so they

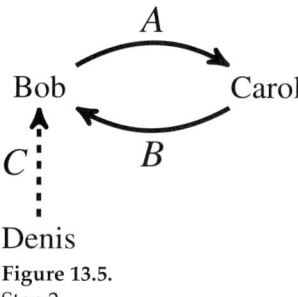

Figure 13.5.
Step 2

Figure 13.6.
Step 3

also are removed from the problem. Only school C and Denis remain. The new owner of school C is now Denis, and it is his most preferred school among the schools that are still available. So Denis points to himself (see figure 13.6).

Denis is in a self-cycle and is thus assigned school C. So the final assignment is

$\mu'''(\text{Alice}) = A, \qquad \mu'''(\text{Carol}) = B,$
$\mu'''(\text{Bob}) = A, \qquad \mu'''(\text{Denis}) = C.$

One can easily compare the assignment μ''' with assignment μ''. They are both calculated with different versions or interpretations of the Top Trading Cycle algorithm, yet they give the same outcome.

The Top Trading Cycle algorithm is a little bit equivalent to a situation where, for each school, we print coupons with the ranks in the priority ordering and those coupons are sent to the corresponding students. For instance, in example 13.2, Alice receives the following three coupons:

- school A, ♯1;
- school B, ♯1; and
- school B, ♯4.

We then let the students trade those coupons, first only allowing students who have rank priority ♯1 to trade their coupons. Once a student gets a new coupon, he takes a seat at that school and exits the market, and, for that school, we allow the student with the next highest priority ♯ among the students who are still in the market to start trading her coupon. For instance, if Alice gives the coupon "school B, ♯1" to, say, Bob, then Bob can claim a seat at school A, because the paper does not say to which student it corresponds; the right belongs to the bearer of the coupon.

13.3   The Problem with the Immediate Acceptance Algorithm

We know that, in a school choice problem, if we use the Deferred Acceptance or Top Trading Cycle algorithms, it is a dominant strategy for the students to reveal their true preferences. But what about the Immediate Acceptance algorithm? To answer this question, we use the problem developed in example 13.2.

In this particular example, we can see that the assignments we obtain with the Immediate Acceptance and Deferred Acceptance algorithms are different, even though we used the same preferences and priorities! To understand what happened, let us analyze in detail the execution of the Immediate Acceptance algorithm.

To begin with, observe that in the assignment computed with the Immediate Acceptance algorithm there is justified envy:

- Bob prefers school B to school C (his assignment), and,
- at school B, Bob has a higher priority than Carol (the student assigned to school B).

The problem for Bob is that he applied "too late." He applied to school B in step 2, but that school already filled its capacity in step 1 (with Carol). Bob has a higher priority than Carol at that school, but the school cannot renege on its acceptance decisions from the previous step.

Can Bob do something to prevent such a bad outcome? The answer is yes. If Bob submits a preference ordering with school B in the first position, he will end up being assigned to that school. To see this, observe that in this case (Bob putting school B as his top choice), in step 1, school B receives two proposals: from Bob and from Carol. Since Bob has a higher priority, he is accepted (and Carol is rejected).

To sum up, when using the Immediate Acceptance algorithm, if Bob submits

(false preferences) $P'_{Bob} = B, A, C$ $\Rightarrow$ Bob is assigned school B,

(true preferences) $P_{Bob} = A, B, C$ $\Rightarrow$ Bob is assigned school C.

This strategy that Bob would use to get a better school is very intuitive: put as a first choice a school for which he has "good chances" to be admitted. Unlike strategic manipulations in the Deferred Acceptance algorithm (for the side receiving proposals), this type of manipulation is relatively easy to undertake under the

Immediate Acceptance algorithm. Indeed, the student does not need to have a complete picture of the priorities at each school and the preferences submitted by the other students. The only information that is needed is the following:

If I put a school as my top choice, how many students with higher priority than me also put that school as their top choice?

If the number of such students is less than the capacity of the school, then I am sure to be assigned to that school if I declare it as my top choice. This type of manipulation is relatively well documented, essentially because the Immediate Acceptance algorithm turns out to be an algorithm used by many school districts to assign students to schools. For instance, in a conference organized by the Federal Reserve Bank of Chicago in 1994 called Midwest Approaches to School Reform, Robert Meyer and Steven Glazerman report,[8]

It may be optimal for some families to be strategic in listing their school choices. For example, if a parent thinks that their favorite school is oversubscribed and they have a close second favorite, they may try to avoid "wasting" their first choice on a very popular school and instead list their number two school first.

In a meeting of the West Zone Parents Group of the city of Boston, it was said:

One school choice strategy is to find a school you like that is undersubscribed and put it as a top choice, or, find a school that you like that is popular and put it as a first choice and find a school that is less popular for a "safe" second choice.

Parents who are strategic would then end up with a school that is their first choice in their submitted preferences but not necessarily their true top choice. For nonstrategic parents, most of them would easily end up in a not so desired school like Bob in our example when he is assigned to school C.

13.4 Applications

Since the seminal work of Atila Abdulkadiroğlu and Tayfun Sönmez, many school districts have modified their assignment procedures, with most of them adopting the Deferred Acceptance algorithm. We discuss in this chapter two of the earliest (and most famous) implementations of that algorithm.

13.4.1 The Boston School Match

Shortly after Abdulkadiroğlu and Sönmez published their work on school assignment, the *Boston Globe* newspaper contacted them and other scholars for their comments on the Boston Public Schools procedure for assigning students to schools.

In its 2003 article, the *Boston Globe* pointed out the flaws of the procedure used at that time in Boston, which was using the Immediate Acceptance algorithm.[9]

In the fall of 2003, Abdulkadiroğlu and Sönmez, together with Parag Pathak and Alvin Roth, were asked to study in detail the mechanism used in Boston and make some recommendations for possible improvements.[10] This was the first opportunity for a large-scale test of the assignment theory we have studied in this chapter. Boston's school district consists of over 60,000 students between kindergarten and twelfth grade in almost 140 schools. Assignments essentially take place in grades K, 1, 6, and 9. Each year, there are on average about 4000 students entering each of those grades.

The case of Boston is a perfect example of a school choice problem, because students' priorities regarding schools are set not by the schools but by the central administration. In Boston, the priorities for each school are constructed as follows, in this order:

1. The students with the highest priority at a school are those who have an older sibling attending that school.

2. Next are the students who live within walking distance of the school (the *walk zone*, and these zones are defined by the Boston Public Schools). Those students have priority over half of the seats offered by that school. For instance, if a school can admit 100 students, the students in the walk zone have higher priority than the other students for only 50 seats. For the remaining 50 seats, there is no higher priority granted to the student living near the school compared to the other students.

3. Last are all the other students.

Clearly, these categories are not sufficient to construct a strict ordering of the students, for there are many students in each category. To obtain a priority ordering, each student is assigned a random number, and within each category students are ordered according to the random number they have been assigned.

The Immediate Acceptance algorithm forces parents to play a difficult game. They have to be strategic when choosing which school to put as their top choice. It quickly became obvious that this algorithm had to be replaced. But the question was: with which algorithm?

The first criterion for which a revamped mechanism was asked for was strategyproofness. The rationale for having a strategyproof assignment mechanism was that it would "level the playing field" by giving equal chances to parents who have different levels of sophistication about the mechanism. In Boston, some parents had a good understanding of the algorithm that was used (the Immediate Acceptance algorithm) and were able to optimize their submitted preference lists.

Other parents had more difficulties strategizing, so it was more difficult for them to obtain one of their most preferred schools. A strategyproof mechanism is thus desirable because it does not give more chances to the parents who are able to strategize efficiently.

The next question then was which property for the assignment should be considered: stability or efficiency? Answering this question will immediately tell us which algorithm we have to use: the Deferred Acceptance algorithm if we want stable assignments, or the Top Trading Cycle algorithm if we want efficient assignments.

It turns out that choosing between stability and efficiency sheds light on how we can "interpret" schools' priority orderings. We have indeed seen that the Top Trading Cycle algorithm implicitly assumes that students *trade* their priorities. If a priority rank at a school can be considered a good that is owned by a student (and can then be traded), then an efficient assignment is the natural choice. If, however, we consider that priorities are not tradable (or, put more crudely, that students have no ownership of their enrollments), then stability is a better choice. In December 2004, Boston Public Schools announced that it planned to change the system. Although the task force in charge initially advocated for the Top Trading Cycle algorithm, it was the Deferred Acceptance algorithm that was eventually adopted, in 2007.

13.4.2 The New York City School Match

At about the same time Boston was reforming its procedure to assign students to schools, the New York City Department of Education (NYCDOE) also underwent an overhaul of its assignment system.

The case of New York City is quite different from that of Boston for several reasons. First, the problem is of a much larger scale. There are over a million students attending public schools in New York, with about 90,000 students entering one of the 500 different academic programs offered by public high schools. Unlike Boston, the reform of school choice in New York did not start with the *Boston Globe* article (or Abdulkadiroğlu and Sönmez's publication). Instead, several people at the NYCDOE were aware of the National Resident Matching Program (see chapter 10) and wondered if it could be adapted to New York City.

The procedure in place in New York was quite different from that of Boston. Without entering into much detail, the assignment mechanism in New York was decentralized. Students would apply to schools, and the schools had to decide which students to admit, reject, or place on a waiting list. Students were restricted in the number of applications they could send (they also had to establish a preference ordering over schools that could be observed by the schools). Some schools also faced constraints that others did not. For instance, schools offering *unscreened programs* admitted students by lottery, whereas schools that had the status of *zone schools* had to give priority to students from the neighborhood, and schools offering *screened programs* evaluated students individually.

Students would receive decision letters from schools, and successful applicants could accept at most one offer (and be on one waiting list). After this, schools with vacant positions would make new offers to students. There were three such rounds of processing. This clearly was not enough, as about one-third of the students were ending up without any school and had to be assigned to a school by the authorities (and of course that school was not on their preference list). In other words, the assignment procedure in New York City suffered from *congestion* (see section 1.3.1).

The fact that many schools could decide which students to admit had a profound impact on the design of the new system. Some schools were not only choosing which students to accept or reject but were also strategically concealing their capacities from the central administration. By not revealing their exact capacities, schools can reserve places that would be assigned later (and thus have more choice over which students to admit). If the assignment is stable, the incentives for schools to conceal their capacities are minimized. Also, unlike in Boston, many schools in New York City have special academic programs that target students with specific needs and skills. In other words, unlike in Boston, many schools in New York City do have *preferences* among students. Finally, the fact that many schools were strategic in their decisions (trying to game the system) convinced the team of market designers, made up of Abdulkadiroğlu, Pathak, and Roth (with the help of Sönmez), that the situation in New York was a *matching* problem and not an *assignment* problem.[11] The experience and the lessons from the medical match (see chapter 10) made it clear that stability of the matching was the key property that was needed in New York.

Once it was clear that an algorithm that produces stable matching was needed, there came another question: should we use the student proposing or the school proposing version of the Deferred Acceptance algorithm? The choice of the student proposing version was quickly considered the best option for several reasons:

- It is a dominant strategy for the students to reveal their true preferences and produces the *student-optimal matching* (i.e., the stable matching that is the most preferred among stable matchings by all the students).

- It turns out that in a many-to-one matching problem there is no algorithm that produces stable matchings such that it is a dominant strategy for the schools to reveal their true preferences and their true capacities.

In the first year of operation of the new matching mechanism, over 70,000 students managed to be matched to a school on their initial preference list (20,000 more than under the previous system). Students who would not be assigned to a school (and did not withdraw from the New York City public schools) could submit a secondary preference list containing schools that still had vacancies. At the end of the matching process, there were about 3000 students who were administratively matched to a school that was not on their preference list (compared to 30,000 students under the previous system).

14 School Choice: Further Developments

One might think that simply applying the assignment model to the school choice problem would be enough. In fact, this is just the beginning; there is much more to do. The use of the two-sided matching literature to address the problem of assigning students to schools unveiled an impressive list of new questions and problems that had to be addressed. This chapter reviews some of these new problems that arose when the assignment model started to be applied in real life.

14.1 Weak Priorities

Until now, we have assumed that schools' priorities over students are strict, that at any school no pair of students have the same priority. In many matching or assignment problems, this assumption is innocuous. For school choice problems, this assumption is difficult to justify. It has to be relaxed.

14.1.1 The Problem

In chapter 13, we saw that in the school choice model schools do not have preferences over students but rather priorities. Those priorities are in general the outcome of administrative decisions, which simply follow some criteria set by policy makers. For instance, a common criterion for establishing priorities is the sibling rule: students with a sibling already attending a school have a higher priority (at that school) than students without a sibling. Another common criterion is proximity (students living close to the school have a higher priority).

If we look carefully, such criteria do not generally allow for a fine sorting of students. Even if policy makers decide that priorities should depend on many different criteria (e.g., grades, ethnicity, family income), it is very likely that many students will end up having the *same priority*. A school priority order thus typically consists of a collection of ordered tiers. Students in the first tier have the same priority and have a higher priority than any other student from any other tier. Students in the second tier have the same priority, a lower priority than any student from the first tier and a higher priority than students from any lower tier. We call such priorities **weak priorities**. An example of a weak priority is the following:

$\bar{\pi}_s =$ [Alice, Bob], Carol, [Denis, Erin, Fred], Gilda.

The strict priority ordering of a school s was denoted π_s in chapter 13. When dealing with weak priorities, we will add a "bar," so $\bar{\pi}_s$ is the weak priority ordering for school s. Note that any strict priority ordering is also a weak priority ordering where tiers only contain one student. Students who share the same priority are in brackets. So, for school s, the weak priority $\bar{\pi}_s$ establishes that Alice and Bob are in the same tier, which is also the highest: they have a higher priority than any other student, but Alice does not have a higher priority than Bob (and Bob does not have a higher priority than Alice). Carol is in the second tier and is the only student in that tier. She has a lower priority than Bob or Alice but has a higher priority than Denis, Erin, Fred, or Gilda. Denis, Erin, and Fred are in the same tier: none of them has a higher priority than the two other students in that tier. However, they have a lower priority than Alice, Bob, and Carol, and a higher priority than Gilda.

If schools do not have strict priorities, we cannot run the Deferred Acceptance algorithm, the Immediate Acceptance algorithm, or the Top Trading Cycle algorithm. Such algorithm are only defined when schools have strict priority orderings over students. To see this, suppose that for some school two students, say Alice and Bob, have the same priority and are the only two students in their category. Suppose also that this school has only one seat and that it is the most preferred school for both Alice and Bob. If we run the Deferred Acceptance algorithm, both Alice and Bob make a proposal to that school. This school can only accept one student, so it has to reject Bob's proposal or Alice's. But which one? One way to proceed would be to break ties in schools' priorities, for instance by flipping a coin. This solves our problem, but it creates a new one.

14.1.2 Efficiency Loss

We have seen that when schools' priorities are strict, the assignment obtained with the Deferred Acceptance algorithm (with student proposing) gives students' most preferred stable assignment. If we break ties in schools' priorities, this property may be lost. Before presenting an example to illustrate this negative result, it is important to note that stability is defined using schools' *original priorities*, the weak priorities. This is because the strict priorities introduced by breaking ties randomly are artificially obtained (and may change with another random draw). So in our context an assignment will not be stable if it is any of the following:

- **not individually rational**: a student is assigned to an unacceptable school;
- **wasteful**: a student is not assigned to any school, and there is an acceptable school that has not filled its capacity; and

School Choice: Further Developments

- **there is justified envy**: a student prefers a school to her assignment, and a student with a *strictly* lower priority is assigned to that school.

For individual rationality and wastefulness, we do not need to differentiate between weak and strict priorities; only the definition of justified envy is sensitive to that distinction. The next example illustrates how justified envy works with weak priorities.

Example 14.1 Alice is assigned to a school that is *not* her most preferred school. Her most preferred school is school s. The priority ordering at school s is

$\bar{\pi}_s = $ [Alice, Bob], Carol, Denis.

Suppose that the tie between Alice and Bob is broken in such a way that Bob now has a higher priority than Alice. If Bob is assigned to school s, then we cannot say that Alice has justified envy because under the original priority ordering $\bar{\pi}_s$ they both have the same priority. The strict priority that Bob now has over Alice is artificial.

If, however, Carol or Denis are assigned to school s, then we can say that Alice has justified envy. Justified envy can thus only be invoked when it does not depend on the way we broke ties within each tier of a school's priority ordering.

14.1.3 The Student-Optimal Assignment with Weak Priorities

Breaking ties within each tier in a weak priority ordering is a necessary step if we want to use, say, the Deferred Acceptance algorithm. But, as the next example shows, this comes with a cost.

Example 14.2 We consider three students, Alice, Bob, and Carol, and three schools, s_1, s_2, and s_3. The capacity of each school is 1. The students' preferences and the weak priorities of these schools are given in the following table.

P_{Alice}	P_{Bob}	P_{Carol}
s_2	s_3	s_2
s_1	s_2	s_3
s_3	s_1	s_1

$\bar{\pi}_{s_1}$	$\bar{\pi}_{s_2}$	$\bar{\pi}_{s_3}$
Alice	Bob	Carol
[Bob, Carol]	[Alice, Carol]	[Alice, Bob]

So, for instance, for school s_1, Alice has the highest priority, while Bob and Carol are both in the second tier and thus have the same priority.

Suppose that ties are broken such that Alice always has the highest priority (if she has the same weak priority as another student) and Carol always has the lowest priority (if she has the same weak priority as another student). So we obtain the following strict priority orderings.

π_{s_1}	π_{s_2}	π_{s_3}
Alice	Bob	Carol
Bob	Alice	Alice
Carol	Carol	Bob

The student-optimal assignment with the priorities π_{s_1}, π_{s_2}, and π_{s_3} gives the assignment

$\mu(\text{Alice}) = s_1, \quad \mu(\text{Bob}) = s_2, \quad \text{and} \quad \mu(\text{Carol}) = s_3.$

Now consider the following assignment:

$\mu'(\text{Alice}) = s_1, \quad \mu'(\text{Bob}) = s_3, \quad \text{and} \quad \mu'(\text{Carol}) = s_2.$

It is not difficult to see that the assignment μ' is also stable. First, Bob and Carol are assigned to their most preferred school under μ'. So the only student who may have justified envy is Alice, who is assigned to her second most preferred school. Her most preferred school is s_2, which is assigned to Carol under μ'. Both Alice and Carol are in the same tier in $\bar{\pi}_{s_2}$, so Alice does not have justified envy (we would need Carol to be in a lower tier than Alice). So μ' is stable.

We now compare μ and μ'. Alice is assigned to the same school under μ and μ', but both Bob and Carol obtain a preferred school under μ'. So the assignment μ is not efficient (see definition 13.2).

The conclusion from example 14.2 is very simple. Breaking ties arbitrarily can create an efficiency loss because the assignment we obtain when running the Deferred Acceptance algorithm with student proposing may not be the student-optimal assignment. This is worrisome because we already know that stability and efficiency are, in general, not compatible (see result 13.1). Here we have an *additional* source of inefficiency that is introduced by the presence of weak priorities.

14.1.4 Restoring Efficiency

If we want to produce stable assignments in a school choice problem, we have two potential sources of inefficiency:

1. Stability and efficiency are, in general, not compatible.

School Choice: Further Developments

2. Weak priority orderings *may* generate (when we break ties within tiers) an assignment that is not the student-optimal matching. That is, there might be another stable assignment that all students prefer.

We know that restoring efficiency in the first case is hopeless unless we are willing to abandon strategyproofness (see results 13.1 and 13.2). But what about the second source of inefficiency?

From the definition of an efficient assignment, if a stable assignment μ is not efficient, then there exists another assignment μ' such that we can separate students into two groups:

- *Group 1*: the students that have a different assignment between μ' and μ. Those students are assigned under μ' to a school they prefer to the one they have under μ.
- *Group 2*: the students who are assigned to the same school under μ and μ'.

The inefficiency of μ implies that group 1 is not empty. Now take any student in group 1, say Janis. Without loss of generality, let us call s_1 the school Janis is assigned to under μ and s_2 her school under μ'.

We now assume that μ is a stable assignment. If Janis prefers s_2 to s_1, then it must be that under μ school s_2 is full (otherwise μ would be a wasteful assignment and thus not stable). Since under μ' Alice is assigned to s_2, there must be a student who was assigned to s_2 under μ, say John, and who is not assigned to s_2 under μ'. So John has a different assignment under μ than under μ' and is thus in group 1 as well. We can repeat this reasoning for all students in group 1. We then have the following observation:

Each student in group 1 is taking under μ' the seat that another student of group 1 had under μ.

We can then imagine the following scenario to obtain μ'. Take all the students in group 1, and give to each of them a paper that contains the name of the school they are assigned to under μ. Now let the students exchange their papers, and once they have done so, assign them to the school written on the paper they received. Of course, students will be willing to exchange their papers only if another student's paper corresponds to a school they prefer. In our example, it would be Janis receiving John's paper.

Our scenario conveys the following message. Obtaining μ' from μ is equivalent to having students in group 1 trading assignments with each other. Since μ' is a preferred assignment for all the students in group 1, the trading that leads to μ' is thus an improving assignment. But we know how we can get students to trade their assignments in an efficient way! The Top Trading Cycle algorithm does just that.

It is not difficult to see that to go from μ to μ' we also have cycles like the one in the Top Trading Cycle algorithm. To see this, again take our example, but now consider John. John is going to another school, so he must have taken a paper from another student. If it is Janis, then we have a cycle involving Janis and John: Janis gives her paper to John, and John gives his to Janis. Suppose instead that John takes his paper from another student, say Nina. If Nina takes her paper from Janis, then we have a cycle involving Janis, John, and Nina. If not, then we have a fourth student to consider. The number of students in group 1 is finite, so at some point we eventually have a student taking her paper from a student we already put in the collection (Janis, John, Nina, ...).

The cycles we "constructed" were relatively easy because we already knew which assignment we wanted to construct from the initial assignment, μ. What if we do not have a "target" assignment (like μ' in our discussion)? In principle, this is not a problem; we simply start from a stable assignment and consider that, for each student, her assignment is her endowment. Then we could simply run the Top Trading Cycle algorithm with private endowment (see section 11.2.2). But there is a risk in doing that, for we can end up with an unstable assignment. To avoid this, we will have to limit trades between students in the following way. A student i can take the seat of another student j at a school s in a cycle only if for school s student i has a weakly higher priority than student j.

To summarize, the only cycles we consider, which are called **improvement cycles**, are the cycles of students $(i_1, i_2, \ldots, i_k, i_{k+1})$ with $i_{k+1} = i_1$ such that, given an initial stable assignment μ:

(a) Each student in the cycle is assigned to a school under μ, meaning $\mu(i_h) \in S$ for $h = 1, \ldots, k$.

(b) Each student in the cycle prefers the school to which the next student in the cycle is assigned under μ; that is, for $h = 1, 2, \ldots, k$, student i_h prefers the school of student i_{h+1}, $\mu(i_{h+1})$, to her school, $\mu(i_h)$.

(c) Each student in the cycle is, for the school assigned to the next student in the cycle, among the highest-priority students among all students who prefer that school to their assignment. So, for $h = 1, 2, \ldots, k$, if s is the school of student i_{h+1} and A_s is the set of all students who prefer s to their school under μ, then i_h is the student in A_s such that there is no other student in A_s who has a *strictly* higher priority at school s than student i_h.

When there is an improvement cycle, we can thus construct the assignment μ':

$$\mu'(i) = \begin{cases} \mu(i) & \text{if } i \notin (i_1, i_2, \ldots, i_k) \\ \mu(i_{h+1}) & \text{if } i = i_h \end{cases} \qquad (14.1)$$

School Choice: Further Developments

Points (a) and (b) simply capture the idea that in a cycle each student is better off with the school she is trying to obtain. Point (c) is the condition that will guarantee the assignment μ' is still stable.

Example 14.2   We start from the stable assignment μ,

$$\mu(\text{Alice}) = s_1, \quad \mu(\text{Bob}) = s_2, \quad \text{and} \quad \mu(\text{Carol}) = s_3,$$

and we consider stable improvement cycles. To this end, we first select for each student the schools they can consider that satisfy conditions (b) and (c)—condition (a) is trivially satisfied.

For each student, we first see what school they would like to get in an improvement cycle—condition (b).

- *Alice*: School s_2 is the only school that she prefers to her assignment.
- *Bob*: School s_3 is the only school that he prefers to his assignment.
- *Carol*: School s_2 is the only school that she prefers to her assignment.

We now consider condition (c). There is only one student who wants school s_3, Bob. So for him condition (c) is trivially satisfied. For school s_2, we have two students who want that school, Alice and Carol. They both have the same priority, so in principle they can both be part of a cycle. If one of them, say Alice, had a *strictly* lower priority than the other at school s_2, then Alice would not be allowed to claim school s_2.

We are now ready to construct the cycles. We have:

- Alice points to Bob (she wants school s_2, Bob's assignment under μ);
- Bob points to Carol (he wants school s_3);
- Carol points to Bob (she wants school s_2).

The graph is depicted in figure 14.1, where next to the arrow is the name of the school the student is trying to get.

We can see that there is a cycle involving Bob and Carol. By construction, our cycle satisfies conditions (a), (b), and (c). So we can construct the new assignment

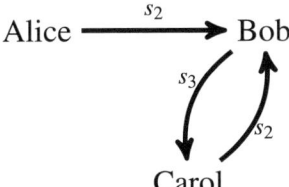

Figure 14.1.
A stable improvement cycle

using equation (14.1) with (Bob, Carol, Bob) as the cycle; that is, Bob takes Carol's school and Carol takes Bob's school. We can see that we obtain the matching μ', which we already know is stable *and* efficient.

Since each student in the cycle is better off with the new assignment μ', we say that μ' **Pareto dominates** μ. The following result, due to Aytek Erdil and Haluk Ergin, summarizes the discussion we just had.[1]

Result 14.1 (Erdil and Ergin) If μ is a stable assignment and μ is Pareto dominated by another stable assignment μ', then there exists a stable improvement cycle.

This result is important because it indirectly tells us how to find a stable and efficient matching. Start from a stable matching and then run the Top Trading Cycle algorithm with private endowments, where students are restricted to considering only schools that satisfy requirement (c) of the definition of an improvement cycle.

Using data from the New York City High School match, Atila Abdulkadiroğlu, Parag Pathak, and Alvin Roth find that for the years 2003–2007 they can improve on average the assignment of about 1700 students (around 2.5% of the number of students assigned to a school). So, the lack of efficiency due to weak priority orderings can be substantial. Fortunately, this can be solved using Erdil and Ergin's improvement cycles.

There is one question that remains, though. Is it possible to obtain a stable and efficient matching while still maintaining strategyproofness? The following example, due to Erdil and Ergin, shows that the answer is negative: we cannot hope for a mechanism that selects an efficient assignment when there are weak priorities. A key aspect of this example is that it relies not on stable improvement cycles but only on the fact that, when schools have weak priorities, the stable and efficient matching is not unique.

Example 14.3 We consider three students, Alice, Bob, and Carol, and three schools, s_1, s_2, and s_3. Each school can be assigned to at most one student (i.e., the capacity of each school is 1). The students' preferences and the weak priorities of these schools are given in the following table.

P_{Alice}	P_{Bob}	P_{Carol}
s_2	s_2	s_1
s_3	s_3	s_2
s_1	s_1	s_3

$\bar{\pi}_{s_1}$	$\bar{\pi}_{s_2}$	$\bar{\pi}_{s_3}$
Alice	Carol	Carol
Bob	[Alice, Bob]	Bob
Carol		Alice

There are two assignments that are stable in this problem (this is due to the fact that Alice and Bob have the same priority at school s_2):

$\mu(\text{Alice}) = s_2$, $\quad \mu(\text{Bob}) = s_3$, $\quad$ and $\quad \mu(\text{Carol}) = s_1$.

$\mu'(\text{Alice}) = s_3$, $\quad \mu'(\text{Bob}) = s_2$, $\quad$ and $\quad \mu'(\text{Carol}) = s_1$.

Now consider the following two misrepresentations for Alice and Bob:

$P'_{\text{Alice}} = s_2, s_1, s_3,$

$P'_{\text{Bob}} = s_2, s_1, s_3.$

If Alice submits the preference ordering P'_{Alice} instead of P_{Alice} (and Bob and Carol are truthful, submitting P_{Bob} and P_{Carol}), then there is a unique stable assignment, μ.

If Bob submits the preference ordering P'_{Bob} instead of P_{Bob} (and Alice and Carol are truthful, submitting P_{Alice} and P_{Carol}), then there is a unique stable assignment, μ'.

Now consider the initial problem with the true preferences of Alice, Bob, and Carol (P_{Alice}, P_{Bob}, and P_{Carol}), and suppose that there exists an assignment mechanism that is efficient and stable. With these preferences, that mechanism must then output either μ or μ', the only two assignments that are efficient and stable.

If that mechanism outputs μ, then Bob has an incentive to misrepresent his preferences by submitting P'_{Bob} instead of P_{Bob}: with the profile (P_{Alice}, P'_{Bob}, P_{Carol}) there is a unique stable and efficient assignment, μ'. Bob is better off not being truthful, because he prefers $\mu'(\text{Bob}) = s_2$ to $\mu(\text{Bob}) = s_3$.

So, in order to maintain strategyproofness, this mechanism must select μ' when the students submit the preferences P_{Alice}, P_{Bob}, and P_{Carol}. But if the mechanism produces μ' with these preferences, then it is now Alice who has an incentive to submit P'_{Alice} instead of P_{Alice}: with the profile (P'_{Alice}, P_{Bob}, P_{Carol}), there is a unique efficient and stable assignment, μ, that Alice prefers to μ'. So, to maintain strategyproofness, the mechanism cannot select μ'.

To sum up, there are only two efficient and stable assignments, but none of them can be selected if we want to maintain strategyproofness.

14.1.5 How to Break Ties If You Must

As we saw in section 14.1.1, when schools have weak priority orderings over students, we need to break ties within each tier of a priority ordering (so that we obtain a strict priority ordering) if we want to run an algorithm like Deferred Acceptance or Top Trading Cycles.

How should we break ties? There are two different ways:

- *Multiple tie-breaking*: Each school has its own tie-breaking.

This means the following. Suppose that Alice and Bob are two students who belong in the same tier for two different schools. Multiple tie-breaking means that in one school the tie may be broken in favor of Alice and for the other school in favor of Bob.

- *Single tie-breaking*: The tie-breaking rule is the same for all schools. Consider again the case of Alice and Bob. With a single tie-breaking rule, if the tie is broken in favor of Alice in one school, then it is also broken in favor of her in the other school.

A simple way to break ties is to assign a random number to each student. If two students belong to the same tier in a priority ordering, then the student with the lowest random number obtains a strictly higher priority order than the other student.[2] Suppose that Alice, Bob, and Carol are in the same tier for some school, and they are assigned the numbers 347, 42, and 728, respectively. After breaking ties, we have that Bob is ranked above Alice, who is ranked above Carol.

A multiple tie-breaking simply amounts to drawing different random numbers for each school. So Bob can have the number 42 for one school and the number 827 for another school. With a single tie-breaking rule, each student has a unique random number, and that number is used for all schools.

One may argue that a multiple tie-breaking rule is fairer. If a student has a bad random draw for one school (i.e., she is the last of her tier after breaking ties), then it may be difficult for that student to be assigned to that school under a stable assignment. To see this, suppose that a school s has the following weak priority ordering (and the capacity of school s is 1):

$\bar{\pi}_s =$ Alice, [Bob, Carol, Denis], Erin.

First consider the case where the tie between Bob, Carol, and Denis is broken in favor of Bob. So we obtain the priority ordering

$\bar{\pi}_s =$ Alice, Bob, ...

To obtain a stable assignment where Bob is assigned to school s, we only need that Alice not be assigned to a school that is less preferred (for her) than school s.

Now consider the case where Bob is the last of his tier, so we obtain, for instance, the priority order

$\bar{\pi}_s =$ Alice, Carol, Denis, Bob, ...

(Another possibility is where Denis is ranked above Carol.) If we want to obtain a stable assignment where Bob is assigned to school s, we now have more constraints: we need not only that Alice be assigned to a school she prefers to school s but also that both Carol and Denis be assigned to a school they prefer to school s.

To sum up, having a high random number decreases the chances of being assigned to a school you like. We would thus argue that, if we want to give equal chances to each student, a multiple tie-breaking rule is better than a single tie-breaking rule.

How to break ties was one of the important questions that was raised when reforming the New York City High School match. During the design of the new school assignment procedures, policy makers in New York initially believed that a multiple tie-breaking rule was a better option. Abdulkadiroğlu, Pathak, and Roth, who were involved in the design of the new mechanism for the city, initially suggested by means of examples and computer simulations that a single tie-breaking rule would have superior welfare properties. Their intuition was also confirmed with the following result.[3]

Result 14.2 (Abdulkadiroğlu, Pathak, and Roth) For any school choice problem, if an assignment can be produced with the Deferred Acceptance algorithm using a multiple tie-breaking rule but cannot be produced when using a single tie-breaking rule, then that assignment is not a student-optimal assignment.

This result reads as follows. When we break ties in schools' priority orderings, we already know that the assignment we obtain with the Deferred Acceptance, although stable, is not necessarily efficient. In example 14.2, the assignment μ is not efficient. So μ is not the *student-optimal assignment*. An assignment is the student-optimal assignment if it is stable and if there is no other stable assignment that Pareto dominates it.

The intuition behind result 14.2 is the following. Consider any school choice problem where schools have weak priority orderings, and let μ be a student-optimal assignment. So μ is stable and there is no other stable assignment that Pareto dominates μ. Abdulkadiroğlu, Pathak, and Roth show that in this case there exists a single tie-breaking rule such that, when we break ties using that rule and we run the Deferred Acceptance algorithm, we obtain the assignment μ. In other words, any student-optimal assignment can be obtained using a single tie-breaking rule. It follows that if an assignment cannot be reproduced with a single tie-breaking rule using the Deferred Acceptance algorithm (but it can be reproduced with a multiple tie-breaking rule), then that assignment cannot be a student-optimal assignment.

The question is, what do we lose if we use a single tie-breaking rule? It is obvious that a multiple tie-breaking rule can yield more assignments (one for each possible way we break ties) than a single tie-breaking rule. This follows from the observation that any single tie-breaking is a particular instance of a multiple tie-breaking: each school just happens to have drawn the same numbers for each student! (We are not saying that it is likely to be the case, just that it *can* happen.)

So suppose that there are some assignments that can be obtained using a multiple tie-breaking rule that *cannot* be obtained using a single tie-breaking rule. Result 14.2 says that such assignments cannot be student-optimal assignments. In other words, these assignments are not desirable (we could even say not interesting). So we do not lose anything by using a single tie-breaking rule.

To make the point and convince the policy makers in New York City, Abdulkadiroğlu, Pathak, and Roth ran simulations using the preference lists submitted by the students (or their parents). The comparison between the two tie-breaking rules is reported in table 14.1. Since ties are broken randomly, the researchers made 250 random draws to obtain robust estimates. The numbers in the table are averages over these 250 draws.

Table 14.1
Average number of students assigned per choice for grade 8 applicants in NYC in 2006–2007

Rank in choices	Single tie-breaking	Multiple tie-breaking	Student-optimal (efficient)
1	32,105.3	29,849.9	32,701.5
2	14,296.0	14,562.3	14,382.6
3	9279.4	9859.7	9208.6
4	6112.8	6653.3	5999.8
5	3988.2	4386.8	3883.4
6	2628.8	2919.1	2519.5
7	1732.7	1919.1	1654.6
8	1099.1	1212.2	1034.8
9	761.9	871.1	716.7
10	526.4	548.4	485.6
11	348.0	353.2	316.3
12	236.0	229.3	211.2
Unassigned	5613.4	5426.7	5613.4

Source: Abdulkadiroğlu, Pathak, and Roth (2009).

Table 14.1 reads as follows. The first column indicates the rank in the preferences of the students. The second column reports the average number of students being assigned to their *k*th choice when using a single tie-breaking. For instance, with 250 simulations, there are on average 32,105.3 students being assigned to their top choice, 14.296 students assigned to their second choice, and so on The third column is reporting the same types of numbers but now using a multiple tie-breaking. For instance, on average, there are 29,849.9 students being assigned to their top choice, 14,562.3 students assigned to their second choice, and so on. Finally, the last column gives the average number of students obtaining their *k*th choice but when we compute the student-optimal matching (using the stable improvement cycles we saw in section 14.1.4). The last row gives the average number of students who end up not being assigned to any school for the two versions of the tie-breaking rule and for the student-optimal assignment.

The statistics of table 14.1 give two messages. First, under the single tie-breaking, for almost any rank in the choices, the average number of students assigned to a school ranked above that rank is higher with the single tie-breaking rule. This is obvious for the top choice: on average, 32,105.3 students obtain their top choice with the single tie-breaking rule, while only 29,849.9 do so with the multiple tie-breaking rule. What about the second choice? With the single tie-breaking rule, on average there are $32,105.3 + 14,296 = 46,401.3$ students getting their top or their second most preferred choice. With the multiple tie-breaking rule, it is only $29,849.9 + 14,562.3 = 44,412.2$ students. We can continue this way, comparing the number of students who get either their first, second, or third choice and so on. It is only when we reach the seventh choice that the numbers of students getting their seventh choice or better are similar between the two methods.[4] So the single-tie breaking rule manages to assign more students a higher choice.

The second message of table 14.1 is that the single tie-breaking rule gives outcomes that are virtually identical to student-optimal assignment. We know from example 14.2 that when breaking ties the assignment obtained by running the Deferred Acceptance algorithm might not be the student-optimal assignment (i.e., it is Pareto dominated by another assignment that is also stable), and a student-optimal assignment can be calculated using Ergin and Erdil's stable improvement cycles algorithm. The simulations run with real data from New York City show that in fact we may not need to use the stable improvement cycles algorithm, as single tie-breaking generates few aggregate inefficiencies. Why is this so? The result of Erdil and Ergin (result 14.1) is a *theoretical* result, which means that it holds for *any* profile of students' preferences and schools' priorities, whereas the numbers in table 14.1 are obtained with *real* preferences. So it is perfectly possible that in theory we obtain a "negative" result but in real-life settings the conditions needed to obtain such negative results are not met.

14.2 Constrained Choice

In many assignment or matching markets, there are a large number of participants. The medical match, for instance, involves about 20,000 candidates (see chapter 10) and the New York City High School match involves each year about 90,000 students with an offering of more than 500 different academic programs.

In this context, it is not uncommon to see assignment or matching mechanisms in real life where participants are constrained in their choices: participants can submit preference lists over a limited number of options. Examples of such limitations are:

• For the New York City High School match, students can put in their preference lists at most 12 academic programs.

• In Spain, college admission decisions are made with a centralized matching procedure where students can submit a preference over at most eight academic programs.

• In France, the faculty hiring in universities is done through a centralized matching procedure (that uses an algorithm equivalent to the Deferred Acceptance algorithm). Until 2008, departments could only submit a preference list over at most five candidates.

There are many reasons why policy makers may want to curb the length of participants' preference lists. The first reason is nudging. Many people may have difficulties figuring out their preferences among five or more options. In the case of a large mechanism like school choice in New York, we cannot expect a student to submit preferences over 500 programs. One can thus argue that a constraint on the length of the preference list serves as some sort of guidance or incentive to focus on a small number of options in order to make the process easier.

Another reason is to force participants to focus only on options that matter. This is especially relevant in school assignment, where public authorities often have to deal with "no-shows." This occurs when a student is accepted for enrollment at a school, but when the academic year starts, the student is not coming (e.g., she went to a private school). Allowing participants to include too many options may lead some of them to put on their preference lists schools that they are not particularly interested in. When choice is limited, students would thus consider only the schools they are really interested in.

A third reason may be simply historical. Some matching mechanisms like the faculty match in France were implemented several decades ago when the matching process was not fully automated (e.g., preference lists were submitted on paper

School Choice: Further Developments

and then entered manually into a computer program). Limiting choices would then make the matching/assignment process faster and more convenient.

14.2.1 Issues

What is the impact of having constrained choices? Can we still have a stable or efficient assignment? Do participants still have incentives to report their preferences truthfully? If choice is constrained, is it better to have a harsh constraint (e.g., limiting choice to only, say, 3 or 4 options) or a loose one (e.g., limiting choice to, say, 10 or 12 options)?

One straightforward observation that we can make is that if participants are constrained in their choices, the mechanism used to assign students to schools cannot be strategyproof. This is because submitting one's true preferences is not possible whenever a student has more acceptable schools in her preference list than the number she is allowed to report. So, even if we use the Deferred Acceptance or the Top Trading Cycles algorithm, which are strategyproof when choices are not constrained, students have to be careful when they submit their preferences when choices are constrained.

But there is a silver lining. If the assignment mechanism is strategyproof when choice is not constrained, then in the constrained case students only have to care about *which* schools they put in their submitted preference lists. They do not have to care about how to order these schools.[5]

Result 14.3 (*Haeringer and Klijn*) If students cannot put more than k schools in their preferences and the algorithm used is the Deferred Acceptance or the Top Trading Cycle algorithm:

(a) If a student finds at most k schools acceptable, then she can do no better than submitting her true preferences.

(b) If a student finds more than k schools acceptable, then she can do no better than employing a strategy that selects k schools among the acceptable schools and ranks them according to her true preferences.

What result 14.3 does not say for case (b) is which strategy optimally chose the schools to put in the preferences. The choice of these schools obviously depends on the strategies employed by the other students and the schools' priorities. To see this, suppose that Alice's most preferred school is school s_1 and school s_2 is her second most preferred school, and assume that Alice can only put one school in her submitted preference list (for simplicity, assume that each school can accommodate at most one student). If there is a student with a higher priority at school s_1 who puts s_1 in his preference list, then Alice should not do the same because by doing so she will end up not being assigned to any school (that other student and school s_1 would be a blocking pair). If, however, no student with a higher priority at school s_1 puts school s_1 in his preference list, then Alice's best strategy is to put s_1 in her preference list.

What are the consequences of constraining choices? With empirical data, this is an extremely difficult question to answer because we would need to estimate students' true and complete preferences by observing their constrained choices and then rerun the Deferred Acceptance algorithm with those true preferences to see what changes. An easier exercise is to run a controlled experiment that mimics the constrained and unconstrained situations.

With Caterina Calsamiglia and Flip Klijn, we conducted such an experiment.[6] In this experiment, participants were given a list of schools, and each school was given a corresponding monetary payoff. They were told that they had to report a preference list over the schools and would be paid at the end of the experiment, with the amount corresponding to the school they managed to be assigned to. The experiment in fact consisted of various experiments, depending on the algorithm that was used (Immediate Acceptance, Deferred Acceptance, or Top Trading Cycles) and the structure of the monetary payoff. There was one case where the payoffs were correlated and another case where payoffs were random.[7]

The results obtained in the experiment show clearly that constraining choice yields worse outcomes. For the three algorithms (and the two versions of payoffs), participants obtained on average a significantly lower payoff with the constrained choice. The reason is relatively intuitive. When choice is constrained, the risk that a student ends up being unassigned becomes nonnegligible. A student who fears not being accepted by her most preferred school may prefer to focus on lower-ranked schools that are likely to be less demanded.

The experiment's design was aimed at reproducing current school choice mechanisms, so each participant in the experiment was assigned a "zone school," a school for which the participant has the highest priority. When choice is constrained, the zone schools become more attractive, as they constitute a safe choice. In the experiment, we observed that almost all participants put their zone schools in their submitted preference lists (the few who did not were those for which the zone school gave the lowest payoff). Consequently, for the constrained case, on average, 35% more students ended up being assigned to their zone school. In other words, constraining choices gives incentives to be more cautious, which in turn increases segregation (more students are assigned to their zone school).

Constraining choices, although it has certain benefits (e.g., it forces students to focus on meaningful choices), can thus have perverse effects. It forces students to be strategic (about which school they should put in their preference list), can reduce students' overall welfare, and can increase segregation across neighborhoods.

14.2.2 From Very Manipulable to Less Manipulable

We saw in section 14.2.1 that constraining choices can have a negative impact on the performance of assignment mechanisms. But perhaps more worrisome for policy makers is that it makes some mechanisms vulnerable to strategic manipulations, while their unconstrained versions are not.

When Boston decided in 2005 to change its school choice mechanism, replacing the Immediate Acceptance algorithm with the Deferred Acceptance mechanism, one of the objectives that the policy makers in Boston had was to have a strategyproof mechanism. As we have outlined, strategyproof mechanisms level the playing field between sophisticated and less sophisticated families (the former being more able to game the mechanism optimally). The idea of gaming a mechanism is in fact considered objectionable by many policy makers, as it not only gives advantage to those who are able to figure out how to play the mechanism but also can go against the initial motivations that led to the design of the mechanism. By being strategic, the outcomes that we obtain may differ from the objectives set by policy makers.

In an article published in 2013, Parag Pathak and Tayfun Sönmez reported how officials in England, Boston, and Chicago aimed at reducing gaming in school choice mechanisms. In 2003, England started a major reform of its school assignment mechanisms. In 2007, it was established that school choice mechanisms cannot give an advantage to a student in gaining admission to a school because that student listed that school first in her preferences. In other words, students should be accepted at schools according to their priorities and not according to the rank of schools in their preferences. Therefore, algorithms like the Immediate Acceptance algorithm were banned in England in 2007.

Chicago experienced a similar policy change in 2009. That year, after parents submitted preference lists but before assignments were announced, Chicago Public Schools asked parents to resubmit a preference list over schools. The Immediate Acceptance algorithm used in Chicago had perverse effects. According to Chicago authorities, "high-scoring kids were being rejected simply because of the order in which they listed their college prep preferences."

With these concerns in mind, one would expect that policy makers in Chicago or England would have replaced their deficient mechanisms with strategyproof ones. Surprisingly this was not the case: the mechanisms were replaced with other mechanisms that are still not strategyproof. Pathak and Sönmez undertook a careful scrutiny of those new mechanisms and compared them with the previous ones. One of their goals was to assess whether the new mechanisms were in line with policy makers' objective to reduce the chance of strategic misrepresentation of one's preferences. To this end, they proposed a new concept to compare the degree of manipulability of two mechanisms, where a mechanism is said to be *manipulable* if it is not strategyproof.

Definition 14.1 A mechanism F is **more manipulable** than a mechanism G if

(a) there is a profile of preferences for which a student can be better off by misrepresenting her preferences with the mechanism G, then with mechanism F there is also a student who can be better off by misrepresenting her preferences; and

(b) there is a problem (students, preferences, and schools) such that with mechanism F a student can be better off by misrepresenting her preferences but with mechanism G no student can be better off by misrepresenting her preferences (for the same problem).

Note that in condition (a) we do not say that the student who can benefit by misrepresenting her preferences in F is the same student who can do so under mechanism G. It could be another student. Another way to phrase definition 14.1 is the following. Condition (a) says that whenever we have a school problem in which a student can gain by not being truthful about her preferences under the mechanism G, then there is a student (not necessarily the same one) who can also gain by misrepresenting her preferences under F. Condition (b) says that there are some cases where a misrepresentation can benefit a student under the mechanism F, but if we take instead the mechanism G, then no student can gain by misrepresenting her preferences (with the same students, same students' preferences, and same schools' priorities).

Until 2009, Chicago used for its nine selective high schools the Immediate Acceptance algorithm but constrained students to list at most four schools in their submitted preference lists. In fall 2009, after having computed the assignment (but before disclosing it), officials from Chicago Public Schools (CPS) started to worry. The *Chicago Sun-Times* reported on November 12, 2009:[8]

Poring over data about eighth-graders who applied to the city's elite college preps, Chicago Public Schools officials discovered an alarming pattern.

High-scoring kids were being rejected simply because of the order in which they listed their college prep preferences.

"I couldn't believe it," schools CEO Ron Huberman said. "It's terrible."

CPS officials said Wednesday they have decided to let any eighth-grader who applied to a college prep for fall 2010 admission re-rank their preferences to better conform with a new selection system.

For those who know and understand what is at stake with the Immediate Acceptance algorithm, the Chicago case is not surprising. In section 13.3, we saw that putting highly demanded schools on top of one's preference list is a risky strategy.

To solve the "crisis," CPS officials decided to ask students to resubmit their preferences over schools, but this time using a different algorithm. Before describing

School Choice: Further Developments

the algorithm they used, it is important to explain how schools' priorities are constructed in Chicago. Unlike Boston, where priorities depend on sibling and walk zone priorities, Chicago schools' priorities are constructed using students, composite scores. This implies that all schools have the same priority ordering.

For Chicago officials, the smoking gun was the fact that students with a high score were rejected at some schools in favor of students with low scores. To avoid obtaining such outcomes again, they decided to let students with the highest scores choose their schools first, then students with the second-highest scores, and so on until all the schools filled their capacities. Proceeding this way is nothing but using the **Serial Dictatorship algorithm** we saw in chapter 11 (where the order of dictators here is given by students' composite scores). The algorithm thus changed, but Chicago maintained the constraint on the length of submitted preference lists: students were still obliged to rank no more than four schools. As we explained in section 14.2.1, because of a limitation on the number of choices, this version of the Serial Dictatorship mechanism is *not* strategyproof (although the unconstrained version is). Yet, as the following result shows, it is better than the previous mechanisms.

Result 14.4 (Pathak and Sönmez) For any $k > 1$, the Immediate Acceptance mechanism where students cannot put more than k schools in their submitted preferences is more manipulable than the Serial Dictatorship where students cannot put more than k schools in their submitted preferences.

In a more general perspective, Pathak and Sönmez showed that for the Deferred Acceptance and Serial Dictatorship mechanisms, lowering the constraint on the length of submitted preferences (i.e., allowing listing of more schools) always reduces manipulability. In other words, for any pair of constraints k and k' (with $k' > k$), the mechanism where students cannot list more than k schools is always more manipulable than the mechanism (with the same algorithm) where students cannot list more than k' schools.

Pathak and Sönmez gathered an impressive list of school districts that had changed their school choice mechanisms. Many of those school districts are in England, but the authors also identified changes (beyond Boston and Chicago) in Seattle, Denver, and Ghana. Other than Boston, none of those districts adopted a fully unconstrained (and strategyproof) mechanism. The authors nevertheless observed that all (except Seattle) moved to a less manipulable mechanism. A surprising fact in their data is that virtually all the school districts they list managed to decrease manipulability not by softening the constraint on the length of the submitted preference list but by changing the algorithm (for England, the new mechanism uses the Deferred Acceptance algorithm).

15 Course Allocation

In the previous chapters, all the matching or assignment problems we saw were one-to-one or many-to-one problems. For those cases, it was not too difficult to come up with solutions that were easy to implement and had good properties. The many-to-many case is different in that it is much more complex. A typical yet important case of the many-to-many problem is the assignment of students to courses. In many universities or colleges, students have to choose which course they want to enroll in, and courses usually have limited capacities. This chapter reviews several approaches that address that problem.

15.1 Preliminaries

In its simplest form, a course assignment problem is similar to a "classic" assignment problem. There is a set of students and a set of courses (where each course has a limited enrollment capacity), and the problem is to determine which student will be enrolled in which course. The set of all courses a student enrolls in is called a **schedule**. What differs from the matching or assignment problems we studied in the previous chapters is that now we have a *many-to-many* problem.

When studying the many-to-one matching or assignment problems, we have seen that the side of the market that can be assigned or matched to several individuals from the other side must have preferences or priorities over *sets* of individuals. However, we always made the assumption that those preferences or priorities over sets had a specific structure that made the models tractable and relatively easy to solve; that is, we assumed that the preferences or priorities were *responsive*. This is one problem that course assignment has. In many schools, some courses may be complements (e.g., it is best to have *both* courses A and B) or substitutes (e.g., it is best to have course C or D, but not both). In other words, students have to build schedules from what is proposed, and because of the complementarity or substitutability problem, we can no longer assume that students' preferences over *sets* of courses are responsive. The lack of responsiveness in students' preferences greatly complicates the analysis.

Another problem, perhaps more worrisome, is that the many-to-many assignment problem is not as easy to solve as the one-to-one or many-to-one problems for which we have identified strategyproof mechanisms that yield optimal outcomes (e.g., stability for the Deferred Acceptance algorithm or efficiency for the Top Trading Cycle algorithm). In other words, for the many-to-one or one-to-one problems, we have a choice between various mechanisms, depending on which property we seek. In the many-to-many case, we do not have much choice: the Serial Dictatorship algorithm is the only "relevant" algorithm.[1]

Result 15.1 (Pápai)   A many-to-many assignment mechanism is strategyproof, nonbossy, and Pareto efficient if, and only if, it is a Serial Dictatorship mechanism.

The nonbossiness property invoked in result 15.1 states the following. Suppose that using some mechanism where students report their preferences over courses, we obtain an assignment (i.e., a schedule for each student), and suppose now that a student modifies her submitted preference list over courses but that does not change her schedule. The mechanism is said to be nonbossy if, in this case, the schedules of the other students have not changed either. In other words, a mechanism is nonbossy if a student cannot change the assignment of other students without changing her own assignment. This property prevents individuals from gaming the mechanism in the following way. If the mechanism is bossy, then an individual, say Alice, may want to pay a bribe to another individual, say Bob, for altering his preferences in a way that benefits her but that does not change Bob's schedule.

In light of this negative result, it is interesting to study the mechanisms that are used in practice. Also, this chapter on course assignment will also be a nice opportunity to merge notions we saw in the chapters on assignments and matching problems with the chapters on auctions. We will indeed see that some of the solutions that have been proposed (and used) for the course assignment problem consist precisely of running auctions. Another solution, which we will see in section 15.4, will also partially go back to the traditional microeconomic approach of competitive markets with prices! In other words, although most of the market design questions we have seen so far are about problems for which the traditional competitive market analysis is not adequate, that does not mean that this traditional approach is completely useless.

15.2 Bidding for Courses

A common assignment mechanism used by many business schools consists of running an auction. This is the case at Columbia Business School, the Haas Business School at the University of California at Berkeley, Princeton University, and the

University of Michigan Business School. Although these schools do not use *exactly* the same mechanism, the basic structure is the same:

1. Each student is endowed with a *budget* of coins, tokens, points, or whatever.
2. Students submit bids for each of the courses in which they want to enroll.
3. All students' bids, *for all courses*, are ordered into a *single list*.

Bids are processed one at a time, starting with the highest-ranked bid. When considering a bid, the bidder is admitted for the corresponding course if the course has not filled its capacity. Also, since students may bid for more courses than they need, the bid of a student for a course is ignored if that student has already filled her schedule with other courses.

15.2.1 The Bidding and Allocation Process

This auction mechanism can be considered a first-price, sealed-bid auction. That is, students submit a bid only once, and they "pay" their bid for each course they are enrolled in. Observe, however, that the payment rule has little importance here when we consider a sealed-bid auction. Since students cannot revise their bid as in an ascending auction, the price they "pay" for a course cannot affect their bids for the other courses. In other words, the bids are just there to rank students for each course.

In fact, this auction can be seen as a variant of the Serial Dictatorship mechanism where the order of the dictators (the students) is given by the bids. Also, since students bid for several courses, students appear several times in the order. Finally, a dictator is restricted in her choices in that, when it is her turn, she can only pick the course that corresponds to the bid. For instance, if Alice has the third-highest bid and that bid was made for course X, when we consider this bid, she has to take pick X.

The following example, taken from an article by Tayfun Sönmez and M. Utku Ünver, illustrates this bidding mechanism.[2]

Example 15.1 There are four students: Alice, Bob, Carol, and Erik. Each student has to take two courses among courses X, Y, and Z. Courses X, Y, and Z have a capacity of three students, two students, and four students, respectively. The following table gives the bid of each student for each course.

	X	Y	Z
Alice	60	38	2
Bob	48	22	30
Carol	47	28	25
Erik	45	35	20

So, for instance, Alice bids 60 for course X, Carol 28 for course Y, and so on. We now proceed with the bids to allocate the courses.

- The highest bid is Alice's, for course X. So she is assigned that course.
- The second-highest bid is from Bob, for course X. So he is assigned that course.

If we proceed with our analogy with the Serial Dictatorship algorithm, we have that Bob is the second dictator. However, he cannot take just any course among courses X, Y, and Z. He has to pick the course corresponding to the bid that put him second in the order, which is course X.

At this stage, there is only one remaining seat available for course X.

- The third-highest bid is by Carol, for course X. So she is assigned that course. Course X is now full, and all the subsequent bids for course X will be ignored.
- The fourth-highest bid is from Erik. He is bidding for course X, but that bid cannot be processed, because course X is already full.
- The fifth-highest bid is from Alice, with a bid of 38 for course Y. So she is admitted to course Y.
- The sixth-highest bid is from Erik, for course Y. There is one seat left for that course, so Erik is enrolled. Course Y is now full.
- The seventh-highest bid is from Bob, for course Z. He is admitted.
- The eighth-highest bid is from Carol, for course Y. She is rejected because course Y is full.
- The ninth-highest bid is from Carol, for course Z. She is admitted.
- The tenth-highest bid is from Bob. His schedule is already full, so that bid is not considered.
- Continuing this way until all bids are processed, we obtain the following assignment μ:

$\mu(\text{Alice}) = \{X, Y\}, \quad \mu(\text{Bob}) = \{X, Z\},$

$\mu(\text{Carol}) = \{X, Z\}, \quad \mu(\text{Erik}) = \{Y, Z\}.$

In practice, it may happen that a student submits the same bid for different courses or that two students submit the same bid for the same course. In such cases, the usual procedure consists of breaking ties with a random draw.

15.2.2 Issues: Nonmarket Prices and Inefficiency

The rationale of the bidding mechanism we presented in the previous section is to elicit students' preferences in a way that allows them to express the intensity of their preferences over courses. To see this, suppose first that Janis bids, say, 41 and 40 for courses X and Y, respectively. So she is signaling that she prefers course X to course Y. By bidding higher for course X, her turn in the ordering of the dictators for course X will come before her turn for course Y. Since a higher bid means a higher rank in the dictators' order, she gives herself a higher chance of getting course X. Suppose now that she instead bids 70 and 30 for courses X and Y, respectively. With those bids, she is still expressing that she prefers course X to course Y but, keeping the bids of the other students fixed, she is now ranked higher in the dictators' ordering for course X. Therefore she increases the probability that she can enroll in course X. With bids 70 and 30, Janis is expressing that she cares more about being enrolled in course X than in course Y.

Interpreting bids as preferences (with intensities) is fine as long as the bidding mechanism is strategyproof. However, it is not difficult to see that students may have an incentive to misrepresent their preferences in such a bidding mechanism. For instance, if a student's most preferred course is not popular and always has some empty seats, that student can be better off by bidding very low for that course, saving tokens for more competitive courses. In that case, the ordering of bids for a student would not coincide with the preferences over courses. The bidding mechanism thus has a little bit of the flavor of the Immediate Acceptance mechanism: a student may benefit by not bidding on courses that she likes but that are too competitive.

The course bidding mechanism is often advertised as a competitive market mechanism: students have a budget constraint and through their bids express their demand for the courses. In a competitive market, equilibrium prices reflect how courses are demanded: a low price means that a course is not demanded, and a high price means that a course is strongly demanded. In the context of course bidding, a market equilibrium is given by an assignment (a list of courses for each student), a list of bids, and *market-clearing prices*, such that:

- Each student chooses her bids for the different courses that maximize her expected payoff.
- For each course c, the market-clearing price is
  - 0 if the course does not fill its capacity;
  - the bid of the last student admitted in course c (i.e., the lowest bid for course c among all the bids for that course by the students assigned to course c).

In example 15.1, the market-clearing prices for courses X, Y, and Z are respectively 47, 35, and 0.

- If a student can afford a set of courses $\{c_1, c_2, \ldots, c_k\}$ that is different from the one she gets (i.e., if the bids for the courses $c_1, c_2, \ldots, c_k$ are higher than the market-clearing prices for these courses), then she prefers her assignment to the set of courses $\{c_1, c_2, \ldots, c_k\}$.

The idea of having an auction that serves as a proxy for a competitive market makes sense because achieving an equilibrium in a competitive market can take substantial time (when there is no central clearinghouse computing the equilibrium), and expressing demand over schedules can be cognitively difficult. In a competitive market, *competitive prices* (or equilibrium prices) are obtained by equating demand and supply. In principle, if an auction is efficient, it should produce competitive prices. The problem is that the course bidding mechanism is not equivalent to a competitive market.

To see this, note that in competitive equilibrium any student whose bid for a course is higher than the market-clearing price would be assigned to that course. As an illustration, consider the case of Liza, whose favorite course is A. She knows that this course is underdemanded. That is, Liza knows that the market-clearing price in the bidding mechanism will be 0. Liza will then allocate her budget on bidding for other courses, say courses B and C. Suppose that her budget is 101 tokens and that her bids are:

- 1 token for course A. We assume here that in order for a student to be enrolled in a course, a positive bid is necessary. Although Alice predicts that she will be enrolled for sure in course A, she still has to bid a nonzero amount to signal that she is willing to enroll in course A.
- 50 tokens each for courses B and C.

Suppose that Liza can take at most two courses and that the bids of the other students are such that she is enrolled in courses B and C. In that case, the market-clearing price for course A is 0, but since Liza has been enrolled in courses B and C, she cannot be enrolled in course A. Her bid for course A is higher than the market-clearing price, so if the assignment is a competitive equilibrium, she should be enrolled in that course.

To sum up, the course bidding mechanism does not elicit students' true preferences over courses, and the market-clearing prices that are obtained are not equivalent to competitive prices. Consequently, the final allocation is likely to be inefficient.

15.2.3 Deferred Acceptance with Bids

As Sönmez and Ünver explain in their article, one of the main reasons for the inefficiency of the course bidding mechanism outlined in section 15.2.1 is that it uses students' bids to

- infer students' preferences among courses; and
- determine which student has a bigger claim (i.e., priority) over each course. A student with a higher bid for a course has a higher priority.

Course Allocation

The problem is that because the bidding process is strategic (i.e., it is not strategyproof), bids do not reveal students' preferences over courses.

Sönmez and Ünver's solution to this problem is to separate the two problems by using a modified version of the bidding mechanism. In this algorithm, we will denote by k the number of courses each student has to take.

Algorithm 15.1 Gale-Shapley Pareto-Dominant Market Mechanism

Step 1 Students are randomly ordered. That ordering will be used to break ties between students.

Step 2 Each student submits her preferences over courses (and not schedules).

Step 3 Each student submits a bid for each course.

Step 4 The Deferred Acceptance algorithm is run with students proposing using the preference lists submitted in step 2. In that algorithm, each student proposes to her most preferred k courses in the first step of the Deferred Acceptance algorithm. In the subsequent step of the Deferred Acceptance algorithm, if a student has been rejected in the previous step by p courses ($p > 0$), then she proposes to the most preferred p courses among the courses she did not propose to at a previous step.

For the courses, each course, at each step, accepts the students one at a time, starting with the students with the highest priority, up to its capacity and rejects the other students, where the priorities are given by the bids: the student with the highest priority is the student with the highest bid, the student with the second-highest priority is the student with the second-highest bid, and so on.

This algorithm enjoys several properties. First, revealing one's true preferences over courses (in step 2 of algorithm 15.1) is a dominant strategy for the students. Choosing how much to bid for each course is still a strategic decision, however. So this means that we need to consider *equilibria* in the bidding game.

Result 15.2 (Sönmez and Ünver) If a mechanism uses the Gale-Shapley Pareto-dominant algorithm, where the students choose the bids for each course that maximize their expected payoff, then the course assignment that is obtained corresponds to a market equilibrium.

To obtain their result, Sönmez and Ünver make several assumptions. First, unlike the assignment models we saw in the previous chapters, they assume that students have utilities, not simply preferences, over courses. Preferences (i.e., orderings over courses) can be easily derived when students have utility: a student prefers course X to course Y if, and only if, the utility for course X is higher than the utility for course Y. Second, the authors assume that students have beliefs about the market-clearing prices of each course (which could be obtained easily by looking at the market-clearing prices of previous years) and use those beliefs to maximize their expected payoff (i.e., the bidding game is a Bayesian game; see section A.4 in appendix A). Finally, the authors also assume that, when choosing their bids, students behave as *price takers*. That is, the authors suppose that, when a student chooses her bid, the student assumes that it will not affect the market-clearing bids (this assumption is usually made in microeconomics when analyzing competitive markets).

The solution proposed by Sönmez and Ünver is also attractive with respect to the students' overall welfare.

Result 15.3 (Sönmez and Ünver) The course assignment obtained with a mechanism using the Gale-Shapley Pareto-dominant algorithm Pareto dominates the assignment of any competitive equilibrium whenever the bids used constitute an equilibrium. Also, it cannot be Pareto dominated by the assignment obtained with the course bidding mechanism (using the same bids).

15.3 The Harvard Business School Method

Course assignment at the Harvard Business School takes a different approach than the one we saw in section 15.2. The procedure is relatively simple and has been studied in detail by Eric Budish and Estelle Cantillon.[3] As we will see, the Harvard mechanism is interesting in several aspects: the mechanism is not strategyproof or efficient, yet it produces outcomes that can be more efficient than a strategyproof mechanism!

15.3.1 The Harvard Draft Mechanism

The procedure used at Harvard is a modification of the Random Serial Dictatorship (see section 12.2).

Algorithm 15.2 Harvard Draft

Step 1 Students submit a preference list over courses.

Step 2 Students are assigned a random number (so that no two students have the same number).

Step $k \geq 3$, k odd Each student who still needs to add a course to her schedule is assigned her most preferred course among the courses that are still available (i.e., those with remaining capacity) and that she did not obtain at a previous step, starting with the student with the *highest* random number.

Step $k \geq 3$, k even Each student who still needs to add a course to her schedule is assigned her most preferred course among the courses that are still available (i.e., those with remaining capacity) and for which she is not yet assigned, starting with the student with the *lowest* random number.

End The algorithm ends when each student is assigned to as many courses as she needs or when all courses have enrolled students up to their capacities.

In words, the Harvard Draft algorithm consists of running several rounds of the Serial Dictatorship algorithm, where

- for each round each student picks one course; and
- the order of dictators is reversed between each round.

For instance, if there are three students and their order is, say, Alice, Bob, and Carol, then the Harvard Draft algorithm proceeds as follows:

1. Run the Serial Dictatorship with Alice choosing first, then Bob, and then Carol.
2. Run the Serial Dictatorship with Carol choosing first, then Bob, and then Alice.
3. Run the Serial Dictatorship with Alice choosing first, then Bob, and then Carol.
4. And so on.

15.3.2 Strategic Behavior

The Harvard Draft mechanism is not strategyproof. The reason is that there is a conflict between the preferences of a student and the popularity of a course. Suppose that a student likes course X, which is very popular (i.e., most students will put it on top of their list of submitted preferences), but that is not her most preferred course. When it is that student's turn to pick a course, course X is likely to be filled and thus unavailable. Consequently, that student will have to pick a course that is ranked lower (in the submitted preferences) than course X. If instead she says that course X is the most preferred, she is more likely to take it. The following example illustrates this point.

Example 15.2 There are three students (Alice, Bob, and Carol) and four courses (c_1, c_2, c_3, and c_4). Each student has to pick two courses, and each course can enroll two students at most. The preferences of these students are in the following table.

P_{Alice}	P_{Bob}	P_{Carol}
c_1	c_2	c_1
c_2	c_1	c_3
c_3	c_3	c_4
c_4	c_4	c_2

Suppose that all students submit their true preferences. Note that, for any ordering, Bob is certain to obtain course c_2. Whether he is first, second, or third in the random queue, he is always the first student to ask for course c_2.

Note also that, for any ordering of the students, at the end of the first round, course c_1 is no longer available (it is taken by Alice and Carol).

So, at the beginning of the second round, there is one seat left for course c_2 and two seats left for courses c_3 and c_4. The top choices for Alice, Bob, and Carol among those courses are c_2, c_3, and c_3, respectively. (Bob's most preferred course is c_2, but he already picked it, so we take the most preferred available course that he has not yet taken, c_3). So, for any ordering of the students, Bob is certain to obtain course c_3. Hence, when Bob submits his true preferences, he obtains the schedule $\{c_2, c_3\}$.

Suppose that Bob submits instead the preferences $P'_{Bob} = c_1, c_2, c_3, c_4$. In that case, he gets course c_1 only if he is either the first or the second in the queue, which occurs with probability $\frac{2}{3}$.

Suppose this is the case, so at the end of the first round, course c_1 is no longer available and there is one seat left for course c_2 (if Alice is last in the queue) or one seat left for course c_3 (if Carol is last in the queue). If Alice is last in the queue, the top choices for Alice, Bob, and Carol among the available courses at the beginning of the second round are c_3, c_2, and c_3. So Bob is the only one asking for course c_2 in the second round and thus he is certain to obtain it. If Carol is last in the queue, the top choices for Alice, Bob, and Carol among the available courses at the beginning of the second round are c_2, c_2, and c_4. Again, Bob is certain to obtain course c_2, because there are still two seats available for that course.

It remains to consider the case when Bob is last in the order. In this case, course c_1 is no longer available when it is his turn, and thus he gets course c_2. Being last in the order means that he is the first deciding in the second round, and thus he picks course c_3.

To sum up, we have the following outcomes for Bob:

- If Bob is truthful, he obtains the schedule $\{c_2, c_3\}$.
- If Bob submits the preference list P'_{Bob}, he obtains the schedules
  - $\{c_2, c_3\}$ with probability $\frac{1}{3}$ (he is last in the order).
  - $\{c_1, c_2\}$ with probability $\frac{2}{3}$ (he is not last in the order).

Obviously, Bob prefers submitting the list P'_{Bob}.

The intuition behind the manipulations that students can make in the Harvard Draft mechanism is similar to what we observed with the Immediate Acceptance algorithm in chapter 13. A student should not "waste time" by asking for courses that are not popular and should instead first try to get the popular courses. This intuition can be precisely characterized. Before we present the result, it is important to remind ourselves that because the order of students is random, the assignments are random (e.g., in example 15.2, Bob obtains with the submitted preference P'_{Bob} course c_1 with probability $\frac{2}{3}$).

Result 15.4 (*Budish and Cantillon*) (a) Students should not reverse the relative rankings of two courses in their submitted preference lists (with respect to their true preferences) if by doing so they do not obtain the preferred course for sure.

(b) Students should reverse the relative rankings of two courses unless it comes at the cost of not obtaining the more preferred course.

Result 15.4 reads as follows. Suppose that a student truly prefers course c to course c', and suppose that when the student submits a preference list ranking c' above c she is never assigned to course c. Statement (a) says that in this case the student should always rank c above c' in the submitted preference list. However, statement (b) says that if c' is a popular course, then declaring that c' ranks above c is a good strategy if it does not alter the probability of getting c. In other words, students should move up in their rankings popular courses that are not the most preferred, but should do so only if it does not alter the probability of getting the truly preferred courses.

15.3.3 Welfare

The Harvard Draft mechanism is relatively easy to game as soon as one knows which courses are popular, where popularity means that the course fills quickly its capacity. In example 15.2, course c_1 is popular, while course c_4 is not popular. Course c_1 is also more popular than course c_2, as it fills capacity before c_2, for any order of play (when all students submit their true preferences). Rearranging one's ranking of courses by putting them in order of popularity is an easy task that pays off.

But the strategic manipulations highlighted in the previous sections may also have some negative consequences. If in their submitted preferences students give high ranks to popular courses, it artificially increases congestion and thus may hurt those students who value such courses highly.

Since the Harvard Draft mechanism is a random mechanism (the ordering of students is random), there are two possible ways to analyze students' welfare under this mechanism: *ex-ante* and *ex-post*. Budish and Cantillon show that for both approaches the Harvard Draft mechanism performs poorly.

Result 15.5 (*Budish and Cantillon*) The outcome of the Harvard Draft mechanism under equilibrium play can be ex-post inefficient.

It is possible that all students strictly prefer the distribution of outcomes they receive from truthful play to that which they receive under equilibrium play.

Budish and Cantillon analyzed a series of data from students who had to submit preference lists over courses in July 2005. The data includes:

- a poll conducted in May asking students their *true* preferences over courses;
- preference lists submitted for a *trial* run in May 2005; and
- preference lists submitted for the *real* run.

Analysis of the data shows that the preferences submitted in the trial run in May or the preferences submitted for the assignment in July differ significantly from the true preferences. Perhaps more interesting is that the changes between the true preferences and the preferences submitted change according to result 15.4. That is, students move up in their rankings popular courses.

Budish and Cantillon also find evidence of errors in the students' submitted preference lists. Some errors result from incorrect beliefs about the degree of popularity of some courses. Between the trial and the real run, some courses gained popularity while others lost popularity, and some students in the authors' data behaved in July according to the popularity degrees observed in the trial run. Another type of mistake results from the complexity of the task at hand. A student may identify a course that she has to move up in her submitted preference list but how much the ranking should change can be a difficult question to answer. For instance, a popular course may rank, say, tenth in a student's true preferences, but determining whether it should rank second, fourth, or fifth in the submitted preference list is a task prone to errors.

One interesting exercise that can be conducted is to measure the welfare loss (if any) of the Harvard Draft mechanism. This is possible with the data because it contains students' true preferences. Also, comparisons with other mechanisms can be done. The results obtained by Budish and Cantillon are surprising. To make comparisons between various assignments, we consider the average ranks (in the true preferences) of the courses assigned to the students. So a low average rank corresponds to a good outcome for a student: she is assigned, on average, to courses that rank high in her preference list.

Result 15.6 (Budish and Cantillon)

(a) On average, about 64% of the students would benefit by trading at least one of the courses they obtain when playing the Harvard Draft mechanism.

(b) When students are strategic, the average rank of their assignments with the Harvard Draft mechanism is *higher* than when they submit their true preferences.

(c) When students are strategic, the average rank of their assignments with the Harvard Draft mechanism is *lower* than when using a Random Serial Dictatorship mechanism.

Result 15.6 contains several messages that might be surprising. First, the assignment that is obtained is inefficient. We have seen that students' strategic behavior leads them to rush for popular courses (by putting them in a high position in their submitted preferences). The first counterproductive effect is that it makes such courses look even more popular than they really are. Second, some students may be overlooking (and thus not obtaining) courses that they truly prefer. If an assignment is inefficient, some students can exchange between themselves some of the courses they are assigned to. Statement (a) says that when running several simulations (so as to obtain different random orders), the number of students who would be willing to trade their assignments is relatively high. That is, the Harvard Draft mechanism is not efficient.

If all students were truthful (i.e., running the Harvard Draft algorithm with the true preferences), Budish and Cantillon observe that students would be better off (statement (b) of result 15.6). So there are two sources of inefficiency: the mechanism itself and the strategic behavior of the students.

What if we eliminate the strategic aspect by using instead a strategyproof mechanism? Budish and Cantillon did this exercise using the Random Serial Dictatorship, where each student, when it is her turn, chooses her entire assignment. As it turns out, such a mechanism would result in a welfare loss: on average, students would be assigned to less preferred courses. So, in spite of being a nonstrategyproof mechanism that generates a welfare loss, the Harvard Draft mechanism turns out to perform better than a strategyproof mechanism. One key feature that can explain the popularity of this mechanism is that strategic manipulation of one's reported preferences is relatively easy to undertake.

15.4 The Wharton Method

One issue with the previous approaches we have studied is that they implicitly assume that students' preferences over courses are *responsive*. Students are asked to submit a preference list over courses instead of schedules. As we discussed in section 15.1, one of the key features of a course assignment problem is that students may view some courses as complements and other courses as substitutes. For instance, a student may want to take a course on security pricing only if she also takes a course on probability, and she may want to take either a course on marketing or a course on human resources, but not both. The problem of course assignment is part of the family of **combinatorial assignment problems**, where each individual can be assigned several objects (and/or each object can be assigned to several individuals) and we should, in principle, consider all possible combinations. The VCG auction studied in chapter 4 is an example of combinatorial assignment.

One aspect of mechanism design that economists need to care about when implementing solutions in real-life settings is to have mechanisms that are practical and relatively easy to use. In the case of course assignments, this aspect is challenging because there are typically a large number of possible assignments, as many as there are combinations of courses. Expressing one's preferences over a large number of combinations of courses can be cognitively difficult. It is easy to compare two courses or two schedules but very difficult to rank several schedules, because they typically have multidimensional attributes (e.g., which days and at what times the courses are taught, the list of courses). The difficulty of reporting preferences over combinations also increases the risk of making errors.

To sum up, the course assignment problem suffers from two problems:

(a) It is in general not possible to obtain a nice and robust mechanism unless we adopt a Serial Dictatorship mechanism, which is ex-post unfair (see result 15.1).

(b) Asking students to express preferences over schedules is, in practice, not feasible.

The solution we will present in the following sections aims at solving these two problems.

15.4.1 Approximate Competitive Equilibrium from Equal Incomes

As result 15.1 suggests, combinatorial assignment is a nearly impossible problem to solve once we try to combine at the same time efficiency and strategyproofness. The tension between these properties becomes too strong for the case of many-to-many assignments, unless we restrict ourselves to the use of Serial Dictatorship mechanisms. While such mechanisms are widely used in practice, they are inevitably unfair: the first student in the random order gets to choose all her most preferred courses, and as we go down along the ordering of students, their choices are more and more constrained.

To escape this conundrum, we need to make some compromises. The objective is to make them as small as possible. A solution proposed by Budish does exactly this and, surprisingly in light of all the previous chapters, it uses "traditional" competitive analysis.[4] Auctions or matching/assignment mechanisms are used to find allocations in situations where the traditional supply/demand analysis is not suitable. This is the case for auctions, where the problem of price discovery is paramount, or assignment problems, where the objects to be allocated are typically indivisible. Budish's contribution consists of showing that if we accept an *approximate* competitive equilibrium instead of an "exact" equilibrium, then, under certain conditions, we can show that such an equilibrium exists.

Let us start with a quick reminder of what a competitive equilibrium is. Consider a situation where we have individuals, each with a budget and having to decide which goods to consume and in what quantities. Given a price for each good, using the individual's preferences over the bundles of goods, we can derive the individual's demand for each good. For instance, I can have a budget of, say, $10, and if the price of a cup of coffee is $2 and a cup of soup is $3, then my best choice is, say, to buy two cups of coffee and two cups of soup. If the prices of the coffee and the soup change, then I may want to buy different amounts. If we look at what I want to buy for all possible combinations of prices, we have my demand for coffee and soup. This demand is obtained by looking at the most preferred bundle that I can afford given the prices and my budget.

A competitive equilibrium consists of prices for coffee and soup such that the total quantities of coffee and soup that all buyers want to purchase are no more than the total quantities of coffee and soup that the sellers offer. If coffee and soup are both $0.01 per cup, then perhaps the total demand will be a large amount of coffee and soup but the total supply will be much less. So those prices cannot constitute an equilibrium.

The approach used by Budish for course allocation is similar. First, each student is given some budget, as we did in section 15.2. Next, we consider students' preferences over schedules and from them we derive their demand for courses. For instance, suppose that my budget is 1000 tokens and I consider the three courses Marketing, Corporate Finance, and Accounting. Let p_M, p_C, and p_A be the respective prices of these courses. Suppose I need to take two courses. My demand could be, for instance:

- If $p_M = 10$, $p_C = 800$, and $p_A = 200$, then I want to take the Corporate Finance and Accounting courses.
- If $p_M = 100$, $p_C = 800$, and $p_A = 500$, then I want to take the Marketing and Corporate Finance courses.
- If $p_M = 200$, $p_C = 900$, and $p_A = 500$, then I want to take the Marketing and Accounting courses.

A competitive equilibrium in such a situation will be a price for each course such that:

- Given the prices and the budget, the choice of courses of each student is her most preferred option.
- For each course, the total number of students requesting that course is no more than the capacity of that course (we say that the prices in this case *clear the market*).

Budish's approximation method consists first of allowing for prices that do not exactly clear the market. That is, it allows having some courses that are *slightly* oversubscribed. Second, instead of giving each student the exact same budget, Budish gives each student a *slightly* different budget (but not too different). An equilibrium will be *approximate* in that context if the course prices do not exactly clear the market (e.g., some courses are oversubscribed). With this approach, it is possible to show that an approximate equilibrium always exists.

The approximation could be worrisome, as it results in having some courses that are oversubscribed. But Budish shows that the market-clearing errors can be relatively small. He calculates, for instance, that at Harvard Business School, with about 900 students having to choose 5 courses among 50 possible courses, the number of errors is around 11. That is, there are only 11 instances of oversubscription. Also, since the approximate equilibrium also requires that students not have the same budget, the final allocation may not be fair. But the budget differences can be relatively small. For the implementation of that mechanism at Wharton (see section 15.4.2), students are first given an initial budget of 5000 tokens and then some random numbers are drawn to make the budgets slightly different. The maximum difference between students' budgets is 80, which is just 1.6% of the base budget.

15.4.2 The Wharton Experiment

The approximate competitive equilibrium concept developed by Budish is attractive because, provided we accept some minor "errors" in the assignment (i.e., a small amount of overenrollment), we can obtain an assignment that is relatively fair and efficient. Yet, a major obstacle is that it relies on having access to students' complete preferences over schedules.

Around 2011–2012, the Wharton School at the University of Pennsylvania started to consider overhauling its course assignment mechanism. The mechanism in place at that time was a bidding mechanism like the one we saw in section 15.2. Budish's approximate competitive equilibrium was an attractive solution, except that it is virtually impossible to put in practice: to compute an equilibrium assignment, students need to report their preferences over all possible course schedules. For a school like Wharton, the number of possible schedules is likely to be over a hundred million.

To test the suitability of the approximate competitive equilibrium, Eric Budish and Judd Kessler ran an experiment that led them to design a method to elicit from the students in a simple way their preferences over schedules.[5] Their experiment brings several novelties and also constitutes a lesson about how to design mechanisms for real-life settings using theoretical complex mechanisms.

15.4.2.1 The design The experiment they set up departs from the traditional experiments found in the economics literature. The usual way consists of endowing subjects with some preferences and incentivizing them by offering some monetary reward based on the outcomes (which depend on the subjects' decisions). In the case of course assignment, this methodology cannot be used, because the objective is precisely to elicit students' preferences over schedules. Another possibility would be to endow subjects with some preferences in a format different from the one used by the interface (where they would report their preferences). But this does help test whether subjects can report their *real* preferences over schedules. The objective is to design a procedure to be used in real life.

Also, participants in the experiment were Wharton students, and they had to report preferences over real courses (i.e., courses taught at Wharton) instead of abstract objects or courses. The familiarity with the object of the experiment allowed subjects to focus more on the preference reporting task but also simplified the comparison with the bidding mechanism that was being used at that time. One of the objectives of the experiment was indeed to convince the Wharton School of the advantage of this new assignment mechanism. Running the experiment with Wharton students and course names constituted a more realistic demonstration. It also helped to identify what Budish and Kessler call "side effects" of the mechanism. In real-life settings, there are many aspects that may influence individuals' behavior that may be difficult (if not impossible) to identify in a theoretical setup. For instance, the experiments on school assignments discussed in chapter 13 showed that individuals can be sensitive to the capacity of the school in spite of the theory predicting that such parameters should not influence individuals' decisions.

The mechanism proposed by Budish and Kessler for preference reporting works as follows:

Step 1 Students give a score between 1 and 100 for any course they are interested in (1 for the least desired courses, 100 for the most desired courses). For courses they are not interested in, a score of 0 is assigned by default.

Step 2 Students can report "adjustment" values for any pair of courses. These values can be negative or positive. For instance, a negative adjustment for courses A and B means that the student sees these two courses as substitutable: she would prefer to have one or the other but not both. A positive adjustment for two courses signals complementarity: the student equally values both courses.

It is clear that with this procedure students cannot report extremely complex preferences over schedules. For instance, complementarities or substitutabilities can be expressed only among pairs of courses. However, the procedure is relatively simple.

Preferences over schedules were obtained by simply adding all the reported values of the courses listed in a schedule. For instance, suppose that a student reports the following values:

- Course A: 90.
- Course B: 50.
- Course C: 40.
- Course D: 20.

This means that the student prefers getting courses A and D ($90 + 20 = 110$) to getting courses B and C ($50 + 40 = 90$). If a student puts an adjustment of $+25$ to courses B and D together, then it means that the student prefers getting courses B and D ($50 + 20 + 25 = 95$) to getting courses B and C. Looking at the total scores of all possible schedules allows the computer to construct a complete preference ordering over schedules.[6]

Subjects were also able at any time to see the list of what the computer deduced to be the ten most preferred schedules given the reported scores. While relatively simple, this design allowed students to report their preferences with greater accuracy. Once the subjects submitted their preferences, a heavy computation was made to calculate the approximate competitive equilibrium.

15.4.2.2 Results The results of the experiment were largely positive. For various measures of efficiency and fairness, the approximate competitive equilibrium outperformed the bidding mechanism used at Wharton. In light of those results, Wharton implemented this new mechanism in fall 2013. Since then, a majority of the students have agreed that the new system is fairer and more effective than the previous bidding mechanism.

Besides using the new mechanism, students were also asked in the experiment to construct their schedules using the bidding mechanism that was used at Wharton. At the end of the auction, they were asked to compare the schedules obtained by the two mechanisms. Reporting one's preferences over a combination of courses is, as we said, an extremely difficult task, but comparing two different schedules is a relatively easy one. Budish and Kessler ran eight sessions, each comprising between 14 and 19 students. For six sessions, the majority of students preferred the schedule obtained with the new mechanism, and there was a tie for the two other sessions. In other words, the new mechanism increases efficiency, as it gives students their preferred schedules.

Subjects were also asked to compare their schedules with the schedules of other students. Such comparison permits seeing whether a mechanism generates envyness, which is one of the most common ways to assess the fairness of a mechanism. In a problem like course allocation, envy is almost impossible to eliminate.[7] In the experiment, the percentage of students who envy the schedule of another student dropped by about 30% after the switch to having the schedule assignment computed by the new mechanism.

Making binary comparisons of schedules can also indicate to the designers of the experiment whether the reporting language is accurate. If the scores assigned to the courses are such that the computer considers that a subject prefers, say, schedule I to schedule II (e.g., *A* and *D* preferred to *B* and *C*) but when making a binary comparison of these two schedules the subject says she prefers schedule II, that suggests that the reporting method (giving scores to courses) does not permit subjects to report their preferences over schedules accurately. In total, 1662 binary comparisons were made by the subjects. Nearly 85% of the comparisons were consistent with the scoring method. That is, for 85% of the comparisons, the total score computed for a schedule that is said to be preferred to another schedule was higher for the former. So the proposed scoring methodology (with the adjustments) allows students to report their preferences accurately. Budish and Kessler note, however, that in some cases students did experience difficulties in reporting their preferences. One task that turns out to be particularly difficult is to assign a numerical value (the score) to a course. While students did not indicate any particular difficulty in saying which course they preferred, assigning a number that captures the intensity of those preferences is a more difficult task.

16   Kidney Exchange

There exist various applications of matching or assignment mechanisms, but perhaps one of the most spectacular is the design of a kidney exchange platform.

Kidney transplantation is one of the most common types of organ transplantation. But kidneys are not like any other organ. Since each person has two kidneys and most of us can live with only one, grafts come not only from deceased donors but also from living ones. Traditional economic analysis with supply and demand has no bite here; commerce in human organs is prohibited in almost all countries in the world.[1] There is a black market for organs, where people sell and buy them, but in the "official market," monetary transactions between buyers and sellers are not legal. It is easy to see that organ transplantation can be an assignment or matching problem: there is on one side a set of available kidneys and on the other side a set of patients seeking a kidney. In now famous research published in 2004, Alvin Roth, Tayfun Sönmez, and Utku Ünver showed that the exchange procedure set by the Top Trading Cycle algorithm can be adapted to manage kidney assignment.[2]

16.1   Background

As of January 2016, there were a bit more than 100,000 people in the United States waiting for a kidney transplant in the United States.[3] Each year, almost 5000 patients die while waiting for a kidney, and nearly 4000 patients become too sick to receive a transplant. So, the problem of maximizing the number of transplants is an important one, as it can save a number of lives (and money: one year of dialysis can cost $80,000, if not more).

When a patient needs a kidney transplant, there are two possibilities:

- Obtain a kidney from a deceased donor. This is managed using a waiting list that takes into account the waiting time, the state of the patient, and other factors.
- Obtain a kidney from a living donor. The human body has two kidneys but, fortunately for the patients, most of us can live with only one.

In 2014, 17,107 kidney transplants took place in the United States. Donations from the deceased accounted for 11,570 transplants, and there were 5,537 transplants from living donors. Donations from the deceased are relatively "easy" to manage: once the doctors have an available kidney, they just have to identify on the waiting list the patient with the highest priority who is compatible with the kidney. Since there are, unfortunately, many people on the waiting list, it is never a problem to find an adequate patient.

The main problem is when doctors have at their disposal a kidney from a living donor. In most cases, the living donor is a person who offers one of his or her kidneys to a relative or close friend who needs a kidney. The donor is willing to have a kidney removed if the relative or friend is receiving one. A problem arises when the kidney of the donor is not compatible with the recipient. Two factors determine whether a kidney is compatible for a patient:

- **Tissue type compatibility.** This relates to our immunological system. Each cell in our body has some markers so that our immunological system can recognize them as being "theirs." When performing a transplant, there is the risk that the immunological system will view the graft as a foreign element and start attacking it. This is when the graft is rejected.

  The markers on our cells come from a combination of proteins, and some people may have similar combinations. If the difference is not too big, the doctors can administer *immunosuppressants*, which are drugs that will diminish the body's ability to reject the graft. This comes at the expense of taking immunosuppressants forever, but it has a great benefit in that it can considerably expand the number of kidneys compatible for a patient (following the earlier sections in this chapter, we can say that it increases the number of kidneys that would be *acceptable* for a patient).

- **Blood type compatibility.** The blood comes in different types. The main types are A, B, AB, and O, and each human has one type of blood. Blood types are very well understood in modern medicine, and they act a little bit like the tissue type compatibility. That is, some people can receive blood of a certain type but not other types. For instance, people with blood type AB can receive blood of type A or type B. However, a person of blood type A cannot receive type B blood.

Unlike tissue type incompatibility, blood type incompatibility constitutes a major obstacle when considering a transplant and is almost impossible to overcome. So, blood type compatibility is the main factor that will determine whether a kidney is compatible for a patient.

The question of tissue type and blood type compatibility is not confined to the case of living donors. Before transplanting a kidney from a deceased donor to a patient, doctors have to check compatibility. Since there is no "intended" recipient in the case of a deceased donor, a kidney from a deceased donor is almost never wasted. This constrasts with the case of living donor, where the kidney may be wasted in that a donor may opt not to donate if her kidney is not compatible for her partner.

16.2 Trading Kidneys

The idea of a living transplant is not new. Because the emotional cost attached to giving one's organ is relatively high, donors are usually relatives of patients. A positive aspect of this is that it generally increases the likelihood of tissue and blood type compatibility. But it is sometimes not enough, so it often occurs that the patient and the donor are sent back home and the patient is added to the waiting list. It is in this sense that the kidney is wasted: we lose the opportunity of performing a transplant.

Trading kidneys can overcome the loss of opportunities to transplant kidneys, and the idea is very simple. Suppose there are two patients, Alice and Bob, and each of them has a donor: Alistair is Alice's donor, and Bernard is Bob's donor. Suppose that, as is too often the case, Alistair's kidney is not compatible with Alice, and Bernard's kidney is not compatible with Bob. But what if Bernard's kidney is good for Alice and Alistair's kidney is good for Bob? Without an organized exchange market, both Alice and Bob would have to remain on the waiting list. With an organized exchange, it is possible to organize a trade between these two couples, and the example makes it clear that such exchanges can be performed without putting a price on kidneys.

Remark 16.1 Instead of allowing patient-donor pairs to exchange kidneys, another possibility is to implement *indirect exchanges* (also called *list exchanges*). In such an exchange, the donor's kidney is given to some patient on the waiting list, and the patient (in the patient-donor couple) obtains in return a high-priority position on the waiting list.

Roth, Sönmez, and Ünver note that most people in the transplantation community agree that this type of exchange is harmful for type O patients. Because the kidney of a type O donor is compatible with any other blood type (once we control for the tissue type compatibility), very few type O kidneys (from living donors) will be offered to the waiting list. Whenever there is a patient-donor pair where the donor is of type O, the donor will be able to give her kidney directly to her partner (again, as long as the tissue type incompatibility can be overcome). However, since type O patients can only receive a type O kidney, it increases the likelihood that the patients who are added to the waiting list with a high priority are type O patients.

To sum up, a type O patient who is already on the waiting list is doubly harmed by indirect exchanges, once because they lose their priority on the waiting list (to the patient involved in the exchange) and a second time because the type O kidneys available to patients on the waiting list become more scarce.

16.2.1 Trades versus Waiting List

The Top Trading Cycle algorithm is a perfect algorithm to identify potential trades between patient-donor pairs (as long as the medical information of each donor and patient is put in a common database). There are, however, some aspects that need to be taken into account before running the Top Trading Cycle algorithm right away.

The problem of assigning kidneys to patients is slightly more complex than the problem of exchanging objects, as it mixes a problem with private endowments and a problem with public endowments. In a kidney exchange problem, we have two "types" of kidneys and two "types" of patients.

- Some patients come with a donor (and thus a kidney). In the context of object allocations, such patients have a private endowment (the kidney of their donor). But there are also some patients who do not have a living donor. Those patients are theoretically only eligible for cadaveric kidneys, as they cannot participate (theoretically) in an *exchange*.
- Some kidneys are "held" by some patients, namely the patients who have a living donor. The other kidneys are those coming from cadaveric donors. The kidneys that come from cadaveric donors differ from those coming from living donors in that they cannot be considered kidneys that are available (but not yet assigned). In other words, kidneys from deceased donors are not part of the *public endowment* (see chapter 11). So, for a patient, there are two options:

1. Obtain a kidney from a living donor.
2. Join the waiting list (or remain on it).

To sum up, the problem of designing an exchange procedure for kidneys is to combine, at the same time,

- trades between donors and patients, and
- the management of the waiting list for cadaveric donors.

16.2.2 The Kidney Exchange Algorithm

The solution proposed by Roth, Sönmez, and Ünver consists of allowing a patient-donor pair to trade a kidney for a priority in the cadaveric kidney waiting list: the donor gives the kidney to some patient, and the patient gets a higher priority on the waiting list. The following example illustrates this idea. Suppose that Alice needs a kidney transplant and has a relative (or friend), Alistair, who is willing to give one of his kidneys. Unfortunately, there is no possible chain of exchange that Alice and Alistair can be part of (like Bob and Bernard in the previous example). However, there is a patient, say Carol, who could benefit from one of Alistair's kidneys but who has no donor. The idea is then to let Alistair give one of his kidneys to Carol and in exchange Alice is put at the top of the waiting list.

Before presenting the kidney exchange algorithm, let us give a more formal description of the model we will analyze. We have a set of **patients**. Some patients come with a donor, and other patients come alone. There is a set of **kidneys**, which are all proposed by living donors (and each donor is paired to a patient).

In a kidney exchange problem, each patient has a **preference relation** over the available kidneys (proposed by the living donors) and the option of entering the waiting list. Note that these preference relations do not have to be seen as genuine *preferences* (e.g., in terms of *taste*) but rather can be seen as fitness relations. In practice, these "preferences" are usually deduced by the doctors and depend on the patient and the donor's kidney's characteristics. So, in the assignment problems we will study, when we say that a patient, say Alice, "prefers" Bob's kidney to Carol's kidney, we have to understand that, medically speaking, Bob's kidney is a better match for Alice than Carol's kidney is.

For a patient, a kidney is **acceptable** if it is compatible for the patient's body (i.e., the doctors believe the kidney can be transplanted into the patient's body). Otherwise the kidney is **unacceptable** for the patient.

Remark 16.2 In real life, a patient may be indifferent between two or more kidneys. For the sake of simplicity, we will assume here that this is not the case: each patient has a **strict** preference ordering over the kidneys brought by the donors. In section 16.3, we will study the impact of indifference.

The algorithm we will present is very similar to the Top Trading Cycle algorithm in the sense that each individual and each object will have to "point" (to an object or to an individual). However, since we have at the same time an allocation problem (to whom should we assign the kidneys proposed by the living donors?) and a queuing problem (patients, with or without a donor, are added to the waiting list), the pointing phase can yield two different types of situations:

- **Cycle**. This is the same concept as the one we saw for the assignment problems. A cycle is a sequence of patients and kidneys (with patients and kidneys denoted by the letters p and k, respectively),

$k_1, p_1, k_2, \ldots, k_m, p_m,$

where kidney k_1 points to patient p_1, patient p_1 points to kidney $k_2, \ldots$, and kidney k_m points to patient p_m, who points to kidney k_1.

- **Chain**. A chain is also a sequence of patients and kidneys, except that now the "last" patient in the sequence does point to the waiting list. So a chain is of the form

$$k_1, p_1, k_2, \ldots, k_m, p_m, w,$$

where kidney, k points to patient p_1, patient p_1 points to kidney $k_2, \ldots$, and kidney k_m points to patient p_m, who points to the waiting list (denoted by w).

Algorithm 16.1   Top Trading Cycles and Chains

The algorithm is a multistep procedure, where all steps are identical. For each $h \geq 1$, proceed as follows.

Step h.1 Each patient points to his most preferred acceptable kidney among the available kidneys. If, for a patient, none of the kidneys is acceptable, then that patient points to the waiting list option.

Each kidney points to its paired patient (e.g., if Alice is a patient and Alistair is Alice's donor, then the kidney of Alistair points to Alice).

Step h.2 If there is one or more cycles, proceed to the exchange as follows. For each cycle, each patient in the cycle is allocated the kidney she is pointing to. All the patients and kidneys involved in a cycle are removed from the problem. Then go to step (h + 1).

If there is no cycle, go to step (h).3.

Step h.3 Select *one* chain, and allocate the kidneys in the following way:

- The last patient in the chain is added to the waiting list.
- The other patients in the chain (if any) are assigned the kidney they are pointing to.

For all patients involved in the selected chain, the assignment is *final*. A *chain selection rule* determines whether the selected chain is removed from the problem. Then go to step (h + 1).

End The algorithm stops when all patients have been either assigned to a kidney or added to the waiting list.

Two remarks about the algorithm are in order. First, one may argue that it is possible for the algorithm to never stop. This would occur when there are still some unassigned patients but there are no cycles and no chain. Fortunately, Roth, Sönmez, and Ünver show that, as long as the number of patients is finite, there must always exist either a cycle (involving only patients that have a donor) or a chain. Whether there is a chain or a cycle, at each step there is at least one patient removed from the problem. Therefore, at each step, the pool of remaining patients shrinks, and we must eventually end up with a situation where all the patients are removed. This is when the algorithm stops.

Second, algorithm 16.1 does not define one algorithm but rather a *family* of algorithms, for it does not fully define how, in step h.3, the chain is selected and whether the selected chain is removed from the problem. In section 16.2.3, we discuss the impacts of various selection rules.

In September 2004 (roughly the same time Roth, Sönmez, and Ünver published their research on kidney exchange), the Renal Transplant Oversight Committee of New England approved the creation of a clearinghouse for kidney exchange that was proposed by Roth, Sönmez, and Ünver and Drs. Francis Delmonico and Susan Saidman. Since then, the size and scope of kidney exchange programs have increased, significantly increasing the number of transplants that could be realized (and opening new directions for academic research on kidney exchange).

16.2.3 Chain Selection Rules

As we have commented in the previous section, one aspect about which the Top Trading Cycles and Chains algorithm is relatively vague is how we should select the chain (when there is no cycle).

The first issue with chains is that, unlike with cycles, an individual or a kidney can be part of different chains. To see this, suppose, for instance, that the patient-donor pairs are (Alice, Albert), (Beth, Bob), and (Carol, Calvin), and that the preferences over the donor's kidneys are:

- Alice: None of the available kidneys is acceptable for her, so in the algorithm she points to the waiting list.
- Beth: She prefers Albert's kidney to the waiting list (but finds Calvin's kidney unacceptable).
- Carol: She prefers Albert's kidney to the waiting list (but finds Bob's kidney unacceptable).

In the algorithm, we would then have the graph depicted in figure 16.1, where the patient-donor couples are indicated with the dashed shape and the arrows indicate to whom or what each patient and donor is pointing to. We can see that there are two chains:

Bob, Beth, Albert, Alice, waiting list

Calvin, Carol, Albert, Alice, waiting list

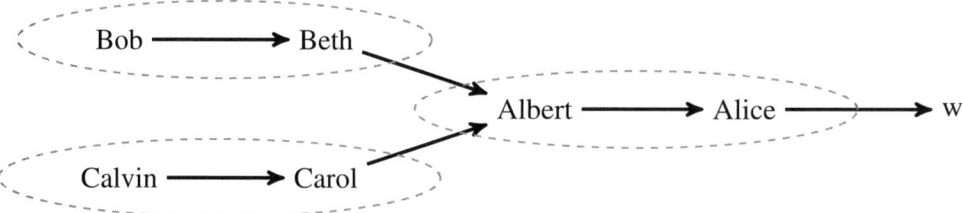

Figure 16.1.
Two overlapping chains

Here we clearly have Albert and Alice being part of two different chains. Therefore, if such a configuration arises, the algorithm will have to select only one of those two chains.

Another question we have to address when treating with chains is whether we should remove from the problem the patients and the kidneys involved in a selected chain. As an illustration, consider the chain starting with Bob in figure 16.1, and suppose that the algorithm selected it. The algorithm states that once this chain is selected, Beth will obtain Albert's kidney and Alice will join the waiting list, independently of what happens at the following steps (i.e., the algorithm states that this assignment is *final* for Beth and Alice). However, note that in this assignment we do not specify what we do with Bob's kidney. There are two solutions:

- It is assigned to someone on the waiting list. In this case, since Bob's kidney is no longer available (i.e., for any patient that is still in the pool of patients without an assigned kidney), we can remove Bob and all the other donors and the patients involved in the chain.
- We keep Bob's kidney available. In this case, we do not remove the chain from the problem.

To see the difference between these two solutions, let us alter the example a little bit by changing Carol's preferences. For Carol, Albert's and Bob's kidneys are both acceptable, but she prefers Albert's kidney. For the other patients, their preferences are unchanged. In this case, when Bob, Beth, Albert, Alice, Calvin, and Carol are still present in the problem, the pointing phase yields the graph depicted in figure 16.1. Suppose that the chain selection rule selects the chain that starts with Bob. So Beth will receive Albert's kidney and Alice will join the waiting list.

If we remove all the patients and donors of the chain that has been selected, then only Calvin and Carol remain in the problem and there is no possible transplant for Carol. If, however, we do not remove that chain, then in the next step Carol

Kidney Exchange

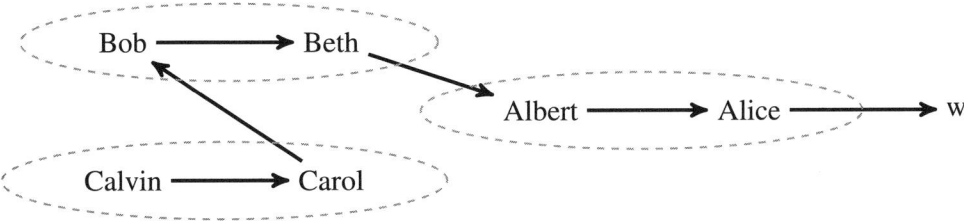

Figure 16.2.
Two overlapping chains

will no longer point to Albert (his kidney has been assigned) but instead she will point to Bob. We obtain the graph depicted in figure 16.2.

Step 2 of the algorithm will now identify a new chain: Calvin, Carol, Bob, Beth, Albert, and Alice. Roth, Sönmez, and Ünver propose and analyze different selection rules:

Rule a Choose the smallest chains, and remove from the problem the patients and kidneys in those chains once the assignment is determined.

Rule b Choose the longest chain, and remove from the problem the patients and kidneys in that chain (pick one chain at random if it is not unique).

Rule c Choose the longest chain, and keep in the problem the patients and kidneys in that chain (pick one at random if it is not unique).

Rule d Choose the chain that starts with the highest-priority patient, and remove from the problem the patients and kidneys in that chain.

Rule e Choose the chain that starts with the highest-priority patient, and keep in the problem the patients and kidneys in that chain.

Rule f Prioritize patient-donor pairs so that pairs with a type O donor have a higher priority (than the pairs whose donor is not of type O). Then choose the chain that starts with the highest-priority pair. If the starting pair in the chain has a type O donor, then remove from the problem the patients and kidneys in that chain. Otherwise keep in the problem all the patients and donors that are in the chain.

16.2.4 Efficiency and Incentives

The first question that Roth, Sönmez, and Ünver addressed is whether we can assign kidneys in an efficient way. The notion of efficiency here is similar to the one we saw in the previous chapters: an assignment is (Pareto) efficient if there is no other assignment that is weakly preferred by all patients and strictly preferred by at least one patient. Given this, a kidney exchange mechanism is **efficient** if, for any problem (sets of patients and for donors their preferences), it always selects an efficient assignment.

Result 16.1 (*Roth, Sönmez, and Ünver*) Consider a chain selection rule such that any chain selected at a nonterminal step remains in the procedure (and thus the kidney of the first donor in the chain remains available for the following steps). The Top Trading Cycle and Chains mechanism, implemented with any such chain selection rule, is efficient.

Result 16.1 is not surprising in light of the discussion we had when we compared the two possible solutions for selected chains (removing or not removing the chain from the problem). In our example, Carol clearly benefits if we do not remove the chain that starts with Bob.

The second question raised by the use of an assignment mechanism is whether the people involved in a kidney exchange program have an incentive to report their preferences truthfully. We have already stressed that, for kidney exchange, patients' preferences are not to be understood as reflecting "tastes" over kidneys. But this difference in the way one has to interpret preferences over kidneys does not eliminate the question of whether a kidney exchange mechanism can be gamed.

A strategic manipulation in a kidney exchange program consists of a physician altering a patient's data so as to alter that patient's preferences over kidneys, which in turn yields a different assignment (for that patient). The manipulation is successful if the assignment obtained with the misrepresentation yields a more preferred outcome than the one obtained when being truthful. For the "classic" assignment problem, we have seen that the Top Trading Cycle mechanism is strategyproof; that is, it is a dominant strategy to submit one's true preferences. For kidney exchange, the proposed algorithm is not exactly the same, as the presence of cycles is not a prerequisite to assign kidneys. The selection rule thus has an impact on the final assignment, and therefore it is possible that some rules may give incentives to misrepresent a patient's preferences over kidneys.

Result 16.2 (*Roth, Sönmez, and Ünver*) An assignment mechanism that uses the Top Trading Cycles and Chains algorithm with the chain selection rule a, d, e, or f is strategyproof.

Roth, Sönmez, and Ünver note that chain selection rules e and f are particularly appealing. Under rule e, both result 16.1 and result 16.2 are valid. Rule f, on the contrary, does not yield efficient assignments but has the merit of increasing the inflow of type O kidneys.

16.3 On the Number of Exchanges

The use of a kidney exchange mechanism that uses an algorithm like the Top Trading Cycles and Chains algorithm seems to be a good way to increase the number of transplants. However, when considering real-life settings, such an algorithm cannot be used as is. The reason is that the trading cycles identified by the algorithm may involve too large a number of patient-donor pairs.

When there is a kidney exchange between two or more patient-donor pairs, it is desirable, if not necessary, that the surgeries with the donors and patients involved in an exchange occur *all at the same time* (the surgeries to extract the organs from the donors and the surgeries to transplant the organs to the patients). Having the surgeries made at the same time eliminates the risk that some donor retracts once her partner (the patient) has received another donor's kidney. If there is an exchange between two patient-donor pairs, we then need four surgeries (two extractions and two transplantations). An exchange involving three patient-donor pairs involves six operating rooms (each with a different team). This creates a problem because the transplantation of an organ to a patient (and the extraction of an organ from a living donor) is a relatively complex operation that requires a large team of surgeons and nurses, and heavy equipment. Since hospitals are constrained regarding the number of operating rooms and surgical teams, there is thus a natural limit on the *length* of the exchange cycles we can reasonably consider.

The algorithm we saw in section 16.2 does not have any constraint on the length of the cycles. If we want to implement such an algorithm, we need to consider a constrained version. The most extreme version is to allow only cycles involving two patient-donor pairs, also called **two-way** kidney exchanges. If we consider exchanges involving three pairs, we add also the possibility to have **three-way** kidney exchanges. Obviously, increasing the maximum number of pairs in an exchange will increase the number of transplants that can be *theoretically* performed. But how much is lost if we consider only two-way or up to three-way exchanges?

In a standard assignment problem (see chapter 11), such a question is extremely difficult to answer because there are potentially many objects, and individuals may have very different preferences over the objects. But the case of kidney exchange is special for several reasons. First, as we explained in section 16.1, the tissue type compatibility and the blood type are the two main parameters that determine whether a donor's kidney is acceptable for a patient. Without entering into too much (biological) detail, the possible values these parameters can take imply that for most patients there are not too many different types of kidneys. Also, blood type compatibility is not patient-specific: two persons of the same

Table 16.1
Average number of exchanges

♯ pairs	2-way	Up to 3-way	Up to 4-way	Unrestricted
25	8.86	11.272	11.824	11.992
50	21.792	27.266	27.986	28.09
100	49.708	59.714	60.354	60.39

blood type can receive the same types of blood. It follows that many patients will have the same preferences over kidneys (if they have the same blood type and similar tissue type). Another simplification added by the kidney exchange model is that in most cases patients have very simple preferences over kidneys: acceptable or unacceptable. All these simplifications allow a precise analysis of the impact of constraints on the number of patient-donor pairs involved in exchange cycles.

Roth, Ünver, and Sönmez showed that if the population of patients and donors is sufficiently large, then the maximum length of an exchange cycle that is needed if we want to maximize the number of transplants is equal to the number of blood types.[4] This result is obtained thanks to a series of assumptions. First, the authors assume that blood type compatibility is sufficient to determine whether a donor's kidney is compatible for a patient (i.e., tissue type compatibility is absent in their model). A second assumption they make is related to the distribution of blood types across the patient-donor pairs. For instance, since a person of type O can only receive type O blood, pairs where the patient is of type O and the donor is of type A, B, or AB are more frequent than pairs where the patient is of type A, B, or AB and the donor is of type O.[5]

To refine their results further, they also ran a numerical analysis using simulated data (aimed at reproducing the characteristics of real-life data). The benefit of such simulations is that we do not need to make any assumptions like the ones made to derive theoretical results. Table 16.1 shows the results obtained by Roth, Ünver, and Sönmez for different population sizes. The first column gives the number of patient-donor pairs in the pool. The second column (two-way) reports the average number of patient-donor pairs that can be matched when we restrict exchange cycles to involve at most two pairs. The third and fourth columns report the average number of pairs when restricting cycles to have at most three and four pairs, respectively. The last column shows the results when there is no restriction on the length of the cycles. For instance, when there are 50 pairs in the population, restricting to only two-way exchange yields, on average, a match for 21.792 pairs. This means that with 50 patients we can find a kidney for nearly 22 of them.

Table 16.2
Gains from k- to $(k+1)$-way exchanges

♯ pairs	2 → 3-way	3 → 4-way	4-way → unrestricted
25	27.2%	4.9%	1.4%
50	25.1%	2.6%	0.4%
100	20.1%	1.1%	0.1%

The simulations first show the obvious: weakening the constraint on the length of the exchange cycles cannot have a negative impact on the number of transplants. A less intuitive result is that the gains quickly diminish. Table 16.2 calculates the percentage gains (from table 16.1) when weakening the constraint.

Table 16.2 shows that allowing up to three-way exchanges significantly increases the average number of patients that can be matched to a donor, but allowing exchanges involving five, six, or more pairs has almost no impact at all (insofar as such exchanges are logistically possible).

There is here an interesting lesson for market design. At the outset, the problem of kidney exchange looks very similar to a "standard" assignment problem. There is a set of kidneys that can be assigned and a set of patients waiting for a transplant. However, unlike most assignment problems, the logistics of kidney exchange drastically limits the number of patient-donor pairs that can be involved in an exchange. A surprising (and fortunate) result obtained by Roth, Ünver, and Sönmez is that the specificities of kidney transplantation almost eliminate the need to worry about logistic constraints.

Appendix A: Game Theory

Game theory provides a structured language that helps in analyzing interactive situations; that is, when several individuals have to make decisions, and the outcome (and the individuals' welfare) depends on each person's choice. This chapter is aimed at briefly presenting the basic game theory concepts that are related to the analysis of auctions and matching markets.

A.1 Strategic Form Games

A.1.1 Definition

A strategic form game captures situations in which agents make decisions without knowing the decisions of the other players.

Definition A.1 A **strategic form game** (also called a **normal form game**) is composed of three elements:

- a set of individuals, called **players**, denoted $N = \{i_1, i_2, \ldots, i_n\}$;
- for each player $i \in N$, a **set of pure strategies**, denoted S_i; and
- for each player $i \in N$, a payoff function u_i that gives the payoff player i will receive for all possible combinations of actions chosen by all the players. So u_i is a function from $S = S_1 \times S_2 \times \cdots \times S_n$ to $\mathbb{R}$.

A game is denoted $G = \langle N, (S_i)_{i \in N}, (u_i)_{i \in N} \rangle$. A **strategy profile** $s = (s_1, \ldots, s_n)$, also called an **outcome**, is a collection of strategies (one for each player). When the action set of every player is *finite*, then we say that the game is *finite*.

Example A.1 A finite game can be, in general, easily represented by a table. Consider the game $G = \langle N, (A_i)_{i \in N}, (u_i)_{i \in N} \rangle$, where

- $N = \{\text{Alice}, \text{Bob}\}$; and
- $S_{\text{Alice}} = \{U, D\}$ and $S_{\text{Bob}} = \{L, R\}$. So in this game Alice has two pure strategies, U and D, and Bob also has two pure strategies, L and R.

It remains to describe the payoff function of Alice and Bob in order to complete the description of the game. That is, we need to describe Alice's and Bob's payoffs for each possible strategy profile.

There are four possible combinations of play, (U, L), (U, R), (D, L), and (D, R), where (U, L) means that Alice plays the strategy U and Bob plays the strategy L, and similarly for the other combinations of strategies.

Table A.1
A simple game with two players, each with two strategies

	L	R
U	3, 2	0, 0
D	1, 0	2, 9

In such a simple example, it is easier to represent the payoffs as a table like table A.1.

In each cell, we have two numbers. The first number is Alice's payoff, and the second number is Bob's payoff. For instance, when Alice plays D and Bob plays L, Alice gets a payoff of 1 and Bob's payoff is 0.

Table A.1 is thus a complete description of a game. We have two players, one choosing the row and one choosing the column, and for each possible combination of strategies by Alice and Bob, we have a description of their payoffs.

When considering a strategy profile s, we will often consider the strategy profile of all but one player. For instance, we may be interested in the actions of all players except player i. In this case, the action profile is denoted s_{-i} and is given by the profile

$$s_{-i} = (s_1, s_2, \ldots, s_{i-1}, s_{i+1}, \ldots, s_n). \tag{A.1}$$

So the strategy profile s can be written as the profile (s_i, s_{-i}). It contains the strategy of player i, s_i, and the strategies of all the other players, s_{-i}.

A.1.2 Pure and Mixed Strategies

As we have explained, in game theory, the strategies U, D, L, and R that are given in the game in table A.1 are called pure strategies. They are the strategies, or actions, that the players will eventually play. It is sometimes useful to extend the notion of strategies to also include nondeterministic play, allowing players to pick a pure strategy with some probability. A strategy that consists of a randomization over pure strategies is called a **mixed strategy**.

It is important to understand that a mixed strategy is a plan of play *before* the game is played. That is, when a player uses a mixed strategy, it does *not* mean that the player is playing a little bit of each of the pure strategies constituting the mixed strategy. It just means that each pure strategy has some probability of being played.

Definition A.2 A **mixed strategy** for a player i is a probability distribution over her set of pure strategies.

To avoid any confusion, we will always use Latin letters to describe pure strategies and Greek letters to describe mixed strategies. So, for a player i, with S_i being her set of pure strategies, a mixed strategy is thus given by a function $\sigma_i : S_i \to [0,1]$ that assigns to each pure strategy $s_i \in S_i$ a probability $\sigma_i(s_i) \geq 0$. The other way to describe a mixed strategy is the following. Let $\Delta(S_i)$ be the set of all probability distributions over the set of pure strategies S_i. Then a mixed strategy is simply an element of $\Delta(S_i)$. Note that since σ_i is a probability distribution, we have $\sum_{s_i \in S_i} \sigma_i(s_i) = 1$.

For instance, in the game of table A.1, a mixed strategy is

- playing U with a probability of $\frac{2}{3}$;
- playing D with a probability of $\frac{1}{3}$.

Appendix A: Game Theory

This mixed strategy, which we denote σ_i, can be rewritten in a more compact way as

- $\sigma_{Alice}(U) = \frac{2}{3}$;
- $\sigma_{Alice}(D) = \frac{1}{3}$.

Remark A.1 Note that since playing a pure strategy is equivalent to playing that pure strategy with probability 1 and all the other pure strategies with probability 0, a pure strategy is also a mixed strategy.

A **mixed strategy profile** $\sigma = (\sigma_1, \ldots, \sigma_n)$ is a collection of mixed strategies (one for each player). Note that if σ is a mixed strategy profile, the probability that the pure strategy profile s is played, $\sigma(s)$, is given by the product of each player's pure strategy in the profile $s = (s_1, \ldots, s_n)$,
$$\text{Prob}(s \text{ is played}) = \sigma(s) = \sigma_1(s_1) \times \sigma_2(s_2) \times \cdots \times \sigma_n(s_n).$$

It follows that a mixed strategy profile is also a probability distribution over pure strategy profiles. To see this, again we use the game of table A.1, and let us consider the mixed strategy profile

- $\sigma_{Alice}(U) = \frac{2}{3}$ and $\sigma_{Alice}(D) = \frac{1}{3}$;
- $\sigma_{Bob}(L) = \frac{1}{4}$ and $\sigma_{Bob}(R) = \frac{3}{4}$.

Then the probability that the pure strategy profile (U, R) is played is

$$\sigma_{Alice}(U) \times \sigma_{Bob}(R) = \frac{2}{3} \times \frac{3}{4} = \frac{1}{2}.$$

We do similarly for the other pure strategy profiles, and for (U,L), (D,L), and (D,R), we find respectively $\frac{1}{6}$, $\frac{1}{12}$, and $\frac{1}{4}$. Adding these four probabilities, we have $\frac{1}{2} + \frac{1}{6} + \frac{1}{12} + \frac{1}{4} = 1$.

Players' payoffs when considering mixed strategies are easily derived from the payoff functions defined over (pure) strategy profiles using **expected payoffs**, where the expectation is taken over the probability distribution over the pure strategy profiles. To define payoffs over mixed strategies, we simply extend the payoff functions $(u_i)_{i \in N}$ as follows. Let $G = \langle N, (S_i)_{i \in N}, (u_i)_{i \in N} \rangle$ be any game. The payoff of a player i under the mixed strategy profile $\sigma = (\sigma_1, \ldots, \sigma_n)$ is

$$u_i(\sigma) = \sum_{s \in S} \sigma(s) u_i(s). \tag{A.2}$$

If we take the mixed strategy profile σ we described earlier, Alice's payoff is

$$u_{Alice}(\sigma) = \sigma(U,L) u_{Alice}(U,L) + \sigma(U,R) u_{Alice}(U,R)$$
$$+ \sigma(D,L) u_{Alice}(D,L) + \sigma(D,R) u_{Alice}(D,R)$$
$$= \frac{1}{6} \times 3 + \frac{1}{2} \times 0 + \frac{1}{12} \times 1 + \frac{1}{4} \times 2 = \frac{13}{12}.$$

A.2 Extensive Form Games

Unlike strategic form games, extensive form games allow capture of situations where the players may observe the play of other players before they have to play. In other words, extensive form games allow us to consider situations where some players play before other players.

A.2.1 Definition

An extensive form game is a more complex object than a normal form game, because we not only have to specify the set of players and their payoffs for each possible choice made by the players but also have to describe the order of play and what the players can observe or not observe as the game is played. The key concept is that of **history**, which simply captures the sequence of actions taken by the players. Together with the history, we also have to specify after each nonterminal history the player who has to play. Before giving a formal definition, we first present an example of a sequential game.

Example A.2 The graph in figure A.1 describes an extensive form game.

In this game, there are two players, Alice and Bob. The game starts with Alice, who has to choose between two **actions**, *A* and *B*.

Once Alice has played, it is Bob's turn. If Alice has played *A*, then Bob can play either *C* or *D*, and if she has played *B*, then Bob can play *E* or *F*.

We can see that representing strategic interactions this way enables us to capture situations where some players play *before* others, and their play can be observed by the other players. In this game, the possible actions that Bob can pick depend on Alice's play. In other words, if Bob is told that he has to play either *C* or *D*, then he can deduce that Alice has played *A*.

After Bob's play, we observe the players' payoffs. For instance, if Alice plays *A* and Bob plays *C*, then Alice's payoff is 1 and Bob's payoff is 0. Similarly, if Alice plays *B* and Bob plays *E*, then Alice's payoff is 4 and Bob's payoff is -1.

In example A.2, one key aspect is that when it is Bob's turn to play, he *observes* Alice's play. That is, Bob knows *perfectly* where he is. Games where all players always perfectly observe the history of play are called games with **perfect information**. We will see in section A.2.3 how to deal with imperfect information. We now turn to the formal definition of an extensive form game.

Definition A.3 An **extensive form game of perfect information** is composed of the following elements:

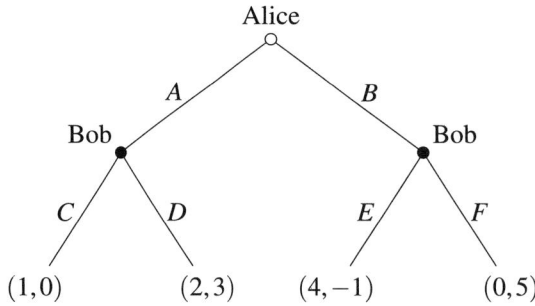

Figure A.1.
An extensive form game

Appendix A: Game Theory

- A set of **players**, denoted $N = \{i_1, i_2, \ldots, i_n\}$.
- A set of **histories**, denoted H. A typical history is a sequence of **actions** taken by the players. The set H satisfies the following properties:
  - The empty history, $\emptyset$, is a member of H (no action has been taken yet).
  - If $(a^1, a^2, \ldots, a^k)$ is a history in H (i.e., first action a^1 has been taken, then action a^2, etc.), then the **subhistory** $(a^1, a^2, \ldots, a^h)$ where $h < k$ is also a history in H.

  A history $(a^1, \ldots, a^k) \in H$ is **terminal** if there is no a^{k+1} such that the history $(a^1, \ldots, a^k, a^{k+1}) \in H$. The set of terminal history is denoted Z.
- A function P that assigns to each nonterminal history h a player in N.
- For each player $i \in N$, a payoff function u_i that gives the payoff player i will receive at each terminal history.

In the game of example A.2, the actions are A, B, C, D, E, and F. The empty history is at the first vertex, where Alice has to play. At this stage, no action has been taken. The collection of all histories is

$$h_0 = (\emptyset), \quad h_1 = (\emptyset, A), \quad h_2 = (\emptyset, B),$$

$$h_3 = (\emptyset, A, C), \quad h_4 = (\emptyset, A, D),$$

$$h_5 = (\emptyset, B, E), \quad h_6 = (\emptyset, B, F).$$

History h_3 is a terminal history, because there is no action that can be played once we reach the end of that history. In contrast, h_2 is not a terminal history, because at the end of history h_1, actions E or F can be played. So the terminal histories are h_3, h_4, h_5, and h_6.

The function P in definition A.3 indicates the player who has to play at the end of each nonterminal history. The nonterminal histories are h_0, h_1, and h_2. At the end of history h_0, it is Alice's turn to play. That is, she is the player who will decide whether history h_0 will become h_1 or h_2. So we have $P(h_0) = $ Alice. Similarly, we have $P(h_1) = P(h_2) = $ Bob.

A.2.2 Strategies

So far, when defining or describing extensive form games, we referred to *actions* and avoided the term *strategies*. In extensive form games, players also have strategies, but they have a precise definition.

Before giving the formal definition of a strategy, we need one additional piece of notation. Let h be any nonterminal history. We will denote by $A(h)$ the actions that are available for the player who has to play at history h,

$$A(h) = \{a : (h, a) \in H\}.$$

So $A(h)$ is the set of all actions "a" such that history h followed by action a is a well-defined history. For instance, in example A.2, we have $A(h_1) = \{C, D\}$. That is, at history h_1, it is Bob's turn, and he can play either C or D. However, $E \notin A(h_1)$. This is because the history $(h_1, E) = (\emptyset, A, E)$ is not a history of our game.

Definition A.4 A **strategy** of a player i in an extensive form game is a function that assigns, for each nonterminal history h such that $P(h) = i$, an action in $A(h)$.

In words, a player's strategy **must** specify an action for each history where it is that player's turn to play.

Consider the game in example A.2. The case of Alice is simple because there is only one history for which it is her turn to play, h_0. So Alice's strategies are A and B. Bob's case is

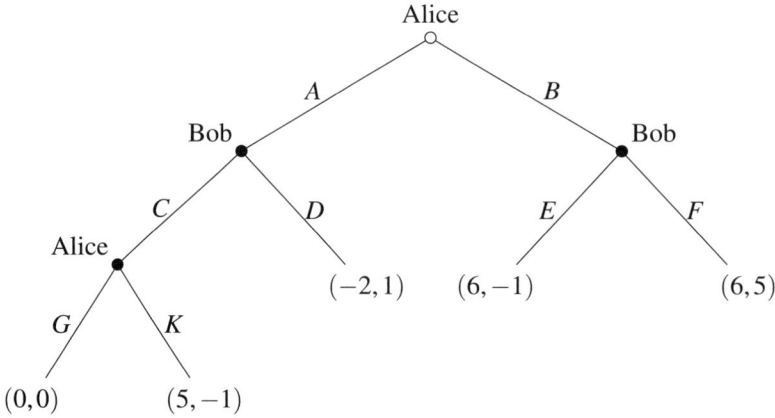

Figure A.2.
An extensive form game

more delicate. Observe that for him there are two histories for which it is his turn to play, h_1 and h_2. So any strategy for Bob must specify what he would do at h_1 and what he would do at h_2. For instance, "play C at h_1 and F at h_2" is a strategy. But C is not a strategy for Bob because we do not specify what he would do.

Example A.3 Consider the extensive form game depicted in figure A.2.

Observe that in this game Alice can play twice, if she initially chose the action A and Bob chose the action C. So there are two histories at which Alice has to play, (∅) and (∅, A, C). So any strategy for Alice must specify an action for these two histories. For instance, (A, G) is a strategy for Alice (adopting the convention that the first entry is the action at the empty history and the second entry is the action at the history (∅, A, C)). So Alice's strategy set in this game is

$\{(A, G), (A, K), (B, G), (B, K)\}$.

One may argue that Alice's strategy (B, K) in example A.3 is not very consistent, because if Alice plays B, she will never have the opportunity to play K. There are some reasons to define strategies this way.

Game theory is aimed at analyzing strategic *interactions*, how players' choices are affected by other players' choices. In the game of example A.3, Alice may want to choose to play A, because Bob is planning to play D if Alice plays A. Why would Bob play D? Simply because Alice is planning to play K if Bob plays C. In other words, Alice's plan to play K if Bob plays C influences Bob's choice, which in turn influences Alice's choice between A and B.

A.2.3 Imperfect Information

In the extensive form games we defined in section A.2.1, each player, when it is her turn to choose an action, knows *perfectly* which actions have been played so far. This is why they are called games with *perfect information*. Extensive form games can accommodate broader situations, where players may only have an *imperfect* knowledge of the history of play.

Appendix A: Game Theory

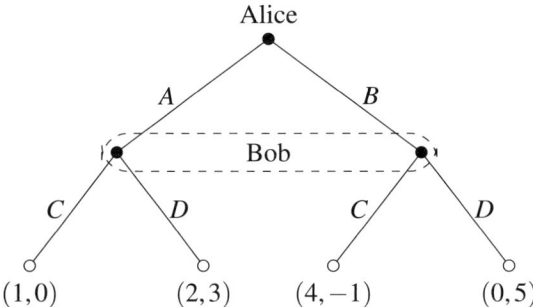

Figure A.3.
An extensive form game

Example A.4 The graph in figure A.3 describes an extensive form game with *imperfect information*. It is very similar to the game in example A.1.

For Alice, nothing has changed: she plays at the empty history, and she has to choose between playing A or B.

Bob plays after Alice, but now, when it is his turn to play, he does not know which action Alice played. To capture the fact that the two histories that correspond to Bob's turn are wrapped together in the same set, figure A.3 depicts them with dashed lines. That set is called an **information set**.

Observe that now Bob has only two possible actions, C and D. The set of actions available to Bob cannot depend on what Alice played (for otherwise he would be able to deduce Alice's choice).

We are now ready to give the formal definition of an extensive game with imperfect information. Most of the elements defining such games are identical to those found in definition A.3, so we list them without details (the new element is indicated with the symbol •, the other elements with the symbol ○).

Definition A.5 An **extensive form game** is composed of the following elements:
○ A set of players, denoted N.
○ A set of histories.
○ A function P that assigns a player to each nonterminal history.
• For each player $i \in N$, a partition of all the histories h such that $P(h) = i$. If two histories h and h' are in the same element of the partition, then $A(h) = A(h')$. An element of the partition is called an **information set**.
○ For each player, a payoff function that gives the payoff the player will receive at each terminal history.

In example A.4, Bob's partition has only one element, $\{h_1, h_2\}$, where $h_1 = (\emptyset, A)$ and $h_2 = (\emptyset, B)$.

In the game in figure A.4, Bob can play after the following three histories: $h_1 = (\emptyset, A)$, $h_2 = (\emptyset, B)$, and $h_3 = (\emptyset, C)$. The partition of these three histories described in definition A.5 is $\{\{h_1, h_2\}, \{h_3\}\}$. It contains two elements.

After histories h_1 and h_2, Bob does not know what Alice played, so these two histories belong to the same information set, $\{h_1, h_2\}$. Therefore, the actions at his disposal for both

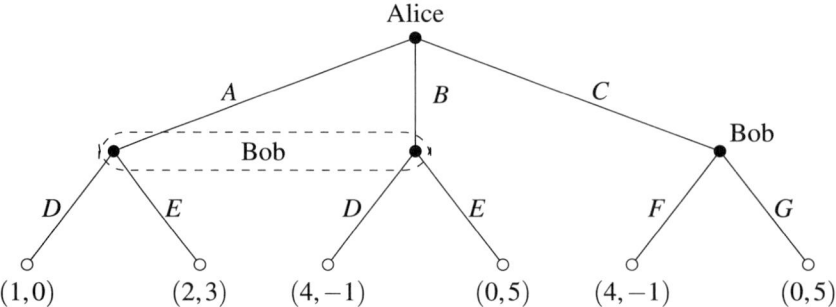

Figure A.4.
An extensive form game

Table A.2
A game with a dominated strategy

	L	R
U	3, 1	9, 5
M	5, 3	7, 6
D	4, 10	4, 1

histories are the same, D and E. After Alice has played, say, A, Bob does not know whether she has played A or B, but he knows she did not play C, because the history h_3 is in a different information set. Similarly, when Alice plays C, Bob knows it because the element of the partition that contains h_3 only contains h_3. In other words, there is no doubt about which history yields Bob playing either F or G.

The definition of a strategy extends naturally to the case of an extensive game of imperfect information. A strategy will simply consist of a plan of action *for each information set*. So, in the game of figure A.4, Bob has the following four strategies (one action per information set):

(D, F), (D, G), (E, F), (E, G).

Remark A.2 Note that a game with perfect information is necessarily also a game with imperfect information. In a game with perfect information, no two histories belong to the same information set.

A.3 Solving Games

A.3.1 Dominated and Dominant Strategies

Once we have defined a game, one of the first questions we can ask is whether we can make some prediction about which strategies the players will choose. For some games, such predictions can be obvious. To see this, consider the game depicted in table A.2, where Alice is the player choosing the row (so her pure strategy set is $\{U, M, D\}$) and Bob is the player choosing the column (so his pure strategy set is $\{L, R\}$).

Observe that in this game, no matter how Bob is playing, playing M gives Alice a strictly higher payoff than playing D. Indeed, if Bob plays L, then Alice's payoff is 5 if she plays

Table A.3
Domination by a mixed strategy

	L	R
U	2,2	7,4
M	3,5	3,2
D	6,1	2,2

M and 4 if she plays D. If Bob plays R, then Alice's payoff is 7 if she plays M and 4 if she plays D. If Bob uses a mixed strategy, Alice's payoff playing M is also higher than the payoff she obtains if she plays D. Let σ_{Bob} be any mixed strategy for Bob. Note that since σ_{Bob} is a probability distribution over $\{L, R\}$, we must have $\sigma_{Bob}(R) = 1 - \sigma_{Bob}(L)$. So we have

$u_{Alice}(M, \sigma_{Bob}) = \sigma_{Bob}(R) \times 5(1 - \sigma_{Bob}(R)) \times 7 = 7 - 2\sigma_{Bob}(R)$

and

$u_{Alice}(D, \sigma_{Bob}) = \sigma_{Bob}(R) \times 4(1 - \sigma_{Bob}(R)) \times 4 = 4.$

Since $0 \leq \sigma_{Bob}(R) \leq 1$, we have $5 \leq u_{Alice}(M, \sigma_{Bob}) \leq 7$ and thus $u_{Alice}(M, \sigma_{Bob}) > u_{Alice}(D, \sigma_{Bob})$.

In the language of game theory, we say that the strategy D is *strictly dominated*.

Definition A.6 A pure strategy s_i is **strictly dominated** for player i if there exists a mixed strategy σ_i for player i such that

$$u_i(\sigma_i, s_{-i}) > u_i(s_i, s_{-i}) \quad \text{for all } s_i \in S_{-i}. \tag{A.3}$$

A pure strategy s_i is **weakly dominated** for player i if there exists a mixed strategy σ_i for player i such that

$$u_i(\sigma_i, s_{-i}) \geq u_i(s_i, s_{-i}) \quad \text{for all } s_i \in S_{-i} \tag{A.4}$$

and there exists at least one pure strategy profile s_{-i} such that $u_i(\sigma_i, s_{-i}) > u_i(s_i, s_{-i})$.

In definition A.6, we only considered the pure strategies of player i's opponents. This is without loss of generality, because the inequalities in equations (A.3) and (A.4) are satisfied for all pure strategy profiles s_{-i} of player i's opponents if, and only if, they are satisfied for all mixed strategy profiles σ_{-i}.

When considering whether a strategy is dominated, it is important to not only consider not only pure strategies but also mixed strategies. The game depicted in table A.3 illustrates this. Here again, the players are Alice, who chooses the row, and Bob, who chooses the column.

In this game, no pure strategy is dominated by any other pure strategy. However, the pure strategy M is dominated by the mixed strategy $\sigma_{Alice}(U) = \frac{1}{2}, \sigma_{Alice}(D) = \frac{1}{2}$.

A more extreme case than the one we saw in the previous section is when there is a strategy that always gives the highest payoff. Consider, for instance, the game depicted in table A.4, with Alice and Bob the players, where $S_{Alice} = \{U, M, D\}$ and $S_{Bob} = \{L, R\}$.

In this game, it is not difficult to check that for Alice both the strategy M and the strategy D are strictly dominated. If Alice is a *rational player* (i.e., if she always chooses the strategy that maximizes her payoff), then she will never choose the strategies M or D, for the strategy U always yields a higher payoff.

Table A.4
A game with a dominant strategy

	L	R
U	7,2	5,4
M	3,1	2,2
D	4,1	1,2

Table A.5
Dominant and dominating strategies

	L	R
U	1,2	7,4
M	3,2	3,1
D	0,1	2,2

Definition A.7 A pure strategy s_i is **strictly dominant** for player i if, for any mixed strategy σ_i for player i,

$$u_i(s_i, s_{-i}) > u_i(\sigma_i, s_{-i}) \quad \text{for all } s_i \in S_{-i}. \tag{A.5}$$

A pure strategy s_i is **weakly dominant** for player i if, for any mixed strategy σ_i for player i,

$$u_i(s_i, s_{-i}) \geq u_i(\sigma_i, s_{-i}) \quad \text{for all } s_i \in S_{-i} \tag{A.6}$$

and there exists at least one pure strategy s_i such that $u_i(\sigma_i, s_{-i}) > u_i(s_i, s_{-i})$.

Note that if a player's strategy is strictly dominant, then all the other strategies of that player are strictly dominated. Similarly, if a player's strategy is weakly dominant, then all the other strategies of that player are weakly dominated.

Note that if a strategy, say s_i, is dominated by a strategy, say s'_i, then it does not mean that s'_i is a dominant strategy. The game in table A.5 illustrates the difference. For the player choosing the rows, the strategy D is dominated (by the strategy M): whether the column player chooses L or R, the payoff from playing M is always higher than the payoff from playing D. However, strategy M is not a dominant strategy.

A.3.2 Elimination of Dominated Strategies

When a strategy is strictly dominated, we can predict that it will never be played by a rational player. In the game in table A.2, we can thus predict that Alice will never play the strategy D. But we can push the argument further. If Bob knows that Alice is rational, he knows that she will never play D. So Bob can ignore that strategy from Alice, anticipating that she will only play the strategy U, D or a mixed strategy putting a positive probability only on U and D.

So for Bob it is as if the game becomes the game depicted in table A.6.

In this "restricted" game, the pure strategy L for Bob is now strictly dominated. If Alice knows that

(i) Bob is rational, and

(ii) Bob knows that Alice is rational,

Appendix A: Game Theory

Table A.6
The game in table A.2 without strategy D

	L	R
U	3, 1	9, 5
M	5, 3	7, 6

Table A.7
The game of table A.6 without strategy L

	R
U	9, 5
M	7, 6

Table A.8
The order of elimination of weakly dominant strategies matters

	L	R
U	3, 4	4, 3
M	5, 3	3, 5
D	5, 3	4, 3

then Alice can anticipate that Bob will play the strategy R (because the other strategy, L, is dominated). In this case, Alice will consider the game even more restricted, depicted in table A.7.

In this game strategy, M is now strictly dominated, and thus we end up with a unique possible outcome, the strategy profile (U, R). The procedure we just followed is called the **iterated deletion of strictly dominated strategies**.

Definition A.8 A game is **strict dominance solvable** if the iterated deletion of strictly dominated strategies gives a unique strategy profile.

The iterated deletion of strictly dominated strategies can be sustained if players' rationality is **common knowledge**. By this we mean of course first that each player in the game is rational. But it also means that:

(i) each player knows that each player is rational;

(ii) each player knows *(i)*;

(iii) each player knows *(ii)*;

(iv) each player knows *(iii)*;

... and so on.

How long we have to repeat the statement "each player knows..." depends on the number of strategies in the game. For finite games, a finite number of steps is enough.

The concept of iterated deletion can also be applied using weakly dominated strategies. However, while the order by which we delete the strategies has no impact when considering strictly dominated strategies, it does matter (and will have an impact) when considering weakly dominated strategies. To see this, consider the game in table A.8.

Table A.9
There is no Nash equilibrium in pure strategies

	L	R
U	0,1	4,0
D	2,0	0,3

In this game we can see that both M and U are weakly dominated (by D). If we eliminate M then we can eliminate R, and then we can eliminate U and end up with the profile (D, L). However, if we start by eliminating U then it is now L that is weakly dominated, and now we can delete M to end up with the profile (D, R). Because of this problem, there is unfortunately no unique or standard definition of **weak dominance solvability**. The following approaches are all acceptable definitions:

• A game is weak dominance solvable if *there is at least one order* of deletion of weakly dominated strategies that leads to a unique strategy profile.

• A game is weak dominance solvable if *any order* of deletion of weakly dominated strategies leads to a unique strategy profile.

• A game is weak dominance solvable if deleting at each iteration *all the weakly dominated strategies* leads to a unique strategy profile.

A.3.3 Nash Equilibrium

Not all games are dominance solvable. Can we still make some predictions in that case? The most common concept in game theory is that of a **Nash equilibrium**.

Definition A.9 A Nash equilibrium of a game $\langle N, (S_i)_{i \in N}, (u_i)_{i \in N} \rangle$ is a profile of strategies σ such that for each player $i \in N$ we have

$$u_i(\sigma_i, \sigma_{-i}) \geq u_i(\sigma'_i, \sigma_{-i}) \quad \text{for each } \sigma'_i \in \Delta(S_i). \tag{A.7}$$

In definition A.9, we invoke mixed strategies (recall that any pure strategy is also a mixed strategy). Mixed strategies are especially useful when considering Nash equilibria. To see this, consider the game depicted in table A.9, where Alice chooses the row and Bob the column.

In this game, if we consider only pure strategies, there is no Nash equilibrium. Consider the profile (U, L). For this profile to be a Nash equilibrium, it must be that for Alice we have

$$u_{\text{Alice}}(U, L) \geq u_{\text{Alice}}(D, L) \tag{A.8}$$

and for Bob we have

$$u_{\text{Bob}}(U, L) \geq u_{\text{Bob}}(U, R). \tag{A.9}$$

Equation (A.8) clearly is not satisfied, so (U, L) cannot be a Nash equilibrium. However, if we consider mixed strategies, we can find a Nash equilibrium.

Example A.5 Consider the game in table A.9, with $S_{\text{Alice}} = \{U, D\}$ and $S_{\text{Bob}} = \{L, R\}$. We have seen that there is no pure strategy equilibrium. Let σ_{Alice} be a mixed strategy of Alice, *i*. Since there are only two strategies,

$$\sigma_{\text{Alice}}(D) = 1 - \sigma_{\text{Alice}}(U).$$

Similarly, let σ_{Bob} be Bob's mixed strategy.

Appendix A: Game Theory

We look for a Nash equilibrium. We already know that there is no Nash equilibrium in pure strategies. Hence, without loss of generality, we can assume that $0 < \sigma_{\text{Alice}}(U) < 1$ and $0 < \sigma_{\text{Bob}}(L) < 1$.

To find a Nash equilibrium, we need to find σ_{Alice} and σ_{Bob} that satisfy equation (A.7). First consider Alice. If she plays the pure strategy U, her payoff is

- $u_i(U, L) = 0$ with probability $\sigma_{\text{Bob}}(L)$ (i.e., the probability that Bob plays L) and
- $u_i(U, R) = 4$ with probability $\sigma_{\text{Bob}}(R)$.

So Alice's expected payoff if she plays the pure strategy U is

$$u_{\text{Alice}}(U, \sigma_{\text{Bob}}) = 0 \times \sigma_{\text{Bob}}(L) + 4 \times \sigma_{\text{Bob}}(R) = 4\sigma_{\text{Bob}}(R).$$

Similarly, if she plays the pure strategy D, then her expected payoff is $u_{\text{Alice}}(D, L) = 2$ with probability $\sigma_{\text{Bob}}(L)$ and $u_{\text{Alice}}(D, R) = 0$ with probability $\sigma_{\text{Bob}}(R)$. So her expected payoff is

$$u_{\text{Alice}}(D, \sigma_{\text{Bob}}) = 2 \times \sigma_{\text{Bob}}(L) + 0 \times \sigma_{\text{Bob}}(R) = 2\sigma_{\text{Bob}}(L).$$

If Alice plays a mixed strategy σ_{Alice}, she gets $4\sigma_{\text{Bob}}(R)$ with probability $\sigma_{\text{Alice}}(U)$ and $2\sigma_{\text{Bob}}(L)$ with probability $\sigma_{\text{Alice}}(D)$. So we have

$$u_{\text{Alice}}(\sigma_{\text{Alice}}, \sigma_{\text{Bob}}) = 4\sigma_{\text{Alice}}(U)\sigma_{\text{Bob}}(R) + 2\sigma_{\text{Alice}}(D)\sigma_{\text{Bob}}(L).$$

If $(\sigma_{\text{Alice}}, \sigma_{\text{Bob}})$ is an equilibrium, it satisfies equation (A.7), and thus

$$u_{\text{Alice}}(\sigma_{\text{Alice}}, \sigma_{\text{Bob}}) > u_{\text{Alice}}(U, \sigma_{\text{Bob}})$$

and

$$u_{\text{Alice}}(\sigma_{\text{Alice}}, \sigma_{\text{Bob}}) > u_{\text{Alice}}(D, \sigma_{\text{Bob}}).$$

The key step is to compare $u_{\text{Alice}}(U, \sigma_{\text{Bob}})$ and $u_{\text{Alice}}(D, \sigma_{\text{Bob}})$.

Suppose first that $u_{\text{Alice}}(U, \sigma_{\text{Bob}}) > u_{\text{Alice}}(D, \sigma_{\text{Bob}})$ (i.e., playing U is better than playing D). So $4\sigma_{\text{Bob}}(R) > 2\sigma_{\text{Bob}}(L)$.

$$u_i(\sigma_i, \sigma_j) = 4\sigma_{\text{Alice}}(U)\sigma_{\text{Bob}}(R) + 2\sigma_{\text{Alice}}(D)\sigma_{\text{Bob}}(L)$$
$$< 4\sigma_{\text{Alice}}(U)\sigma_{\text{Bob}}(R) + 4\sigma_{\text{Alice}}(D)\sigma_{\text{Bob}}(R)$$
$$= 4\sigma_{\text{Alice}}(U)\sigma_{\text{Bob}}(R) + 4(1 - \sigma_{\text{Alice}}(U))\sigma_{\text{Bob}}(R)$$
$$= 4\sigma_{\text{Bob}}(R) = u_{\text{Alice}}(U, \sigma_{\text{Bob}}).$$

That is, if $u_{\text{Alice}}(U, \sigma_{\text{Bob}}) > u_{\text{Alice}}(D, \sigma_{\text{Bob}})$, then σ_{Alice} does not satisfy equation (A.7). A similar calculus shows that if $u_{\text{Alice}}(U, \sigma_{\text{Bob}}) < u_{\text{Alice}}(D, \sigma_{\text{Bob}})$, then σ_{Alice} does not satisfy equation (A.7) either.

Therefore, the only possibility for σ_{Alice} to satisfy equation (A.7) is for

$$u_{\text{Alice}}(U, \sigma_{\text{Bob}}) = u_{\text{Alice}}(D, \sigma_{\text{Bob}})$$
$$\Leftrightarrow \quad 4\sigma_{\text{Bob}}(R) = 2\sigma_{\text{Bob}}(L).$$

Since $\sigma_{\text{Bob}}(L) = 1 - \sigma_{\text{Bob}}(R)$, we have

$$4\sigma_{\text{Bob}}(R) = 2\sigma_{\text{Bob}}(L) \quad \Leftrightarrow \quad 4\sigma_{\text{Bob}}(R) = 2(1 - \sigma_{\text{Bob}}(R)) \quad \Leftrightarrow \quad \sigma_{\text{Bob}}(R) = \frac{1}{3}.$$

We can run a similar analysis for Bob, and we find $\sigma_{\text{Alice}}(U) = \frac{3}{4}$. The Nash equilibrium in mixed strategies is then

Alice plays U with probability $\dfrac{3}{4}$ and D with probability $\dfrac{1}{4}$,

Bob plays L with probability $\frac{2}{3}$ and R with probability $\frac{1}{3}$.

When finding the mixed strategy equilibrium in example A.5, we used an important property: *if a player uses a mixed strategy in a Nash equilibrium, then she obtains the same (expected) payoff with the mixed strategy as with any pure strategy that is played with positive probability.*

Mixed strategies are also important, as they guarantee the existence of a Nash equilibrium. The following theorem was established by John Nash himself.[1]

Result A.1 (*Nash*) Any game with a finite set of players and a finite number of pure strategies always admits a Nash equilibrium (possibly in mixed strategies).

A.4 Bayesian Games: Games with Incomplete Information

The games we studied in the previous sections are called **games with complete information**. In such games, it is assumed that each player knows the strategy sets and the payoff functions of the other players. This is clearly restrictive, as in most real-life situations individuals do not know everything about the persons they interact with. **Games with incomplete information**, also called **Bayesian games**, are designed to capture such situations.

A.4.1 Introductory Example

When a player, say Alice, interacts with other players, she may not know the payoffs of her opponents. However, it is reasonable to assume that there is a collection of possible payoffs of Alice's opponents. These players know what their payoffs are, but Alice does not. For her, there are various possibilities for her opponents' payoffs.

The "trick" that we use in this case is to invoke Nature, who chooses for each player a payoff function from a set of possible payoff functions and communicates to each player her payoff function. If there are two players, Alice and Bob, this would go as follows. Alice has two possible payoff functions, u_A and $\hat{u}_A$, and Bob also has two possible payoffs functions, u_B and $\hat{u}_B$. Before the game starts, Nature chooses a payoff function for Alice and another one for Bob. Alice will know her payoff function but not Bob's, and similarly Bob will know his payoff function but not Alice's.

In a Bayesian game, it is generally assumed that Nature chooses the payoff functions of each player with some probability (like a mixed strategy). That is, Nature will choose u_A with some probability and $\hat{u}_A$ with the complementary probability (1− the probability that Nature chooses u_A). For Bob, it is the same (but not necessarily with the same probabilities). These probabilities are known to all players. Finally, in an incomplete information game, Nature has a special status in that it does not have a payoff function.

Example A.6 We have two players, Alice and Bob. Alice has two strategies, U and D, and Bob also has two strategies, L and R. Alice has two possible payoff functions, u_A and $\hat{u}_A$, and Bob has two as well, u_B and $\hat{u}_B$. Table A.10 describes the game for each possible combination of payoff functions for Alice and Bob.

When Nature chooses Alice's and Bob's payoff functions, we then have four possibilities:

- u_A and u_B. In that case, Alice and Bob play the game that is in the top-left corner of table A.10 (the game where Alice's and Bob's payoffs are 5 and 10, respectively, if they play U and R, respectively).

Appendix A: Game Theory

Table A.10
An incomplete information game

		u_B					$\hat{u}_B$	
		L	R				L	R
u_A	U	3, 1	5, 10		u_A	U	0, 1	1, 6
	D	1, 0	3, 3			D	8, 0	4, 3

		u_B					$\hat{u}_B$	
		L	R				L	R
$\hat{u}_A$	U	0, 4	4, 1		$\hat{u}_A$	U	5, 1	3, 0
	D	2, 3	6, 2			D	1, 0	1, 3

- $\hat{u}_A$ and u_B. In that case, Alice and Bob play the game that is in the bottom-left corner of table A.10 (the game where Alice's and Bob's payoffs are 0 and 4, respectively, if they play U and L, respectively).
- u_A and $\hat{u}_B$. In that case, Alice and Bob play the game that is in the top-right corner of table A.10 (the game where Alice's and Bob's payoffs are 8 and 0, respectively, if they play D and L, respectively).
- $\hat{u}_A$ and $\hat{u}_B$. In that case, Alice and Bob play the game that is in the bottom-right corner of table A.10 (the game where Alice's and Bob's payoffs are 5 and 1, respectively, if they play U and L, respectively).

If Nature chooses the payoffs u_A and $\hat{u}_B$, then we have that:

- Alice knows that her payoff is u_A, but she does not know Bob's payoff. So she does not know whether she is playing the top-left (Bob's payoff is u_B) or the top-right (Bob's payoff is $\hat{u}_B$) game.
- Bob knows that his payoff is $\hat{u}_B$, but he does not know Alice's payoff. So he does not know whether he is playing the top-right (Alice's payoff is u_A) or the bottom-right (Alice's payoff is $\hat{u}_A$) game.

A.4.2 Definition

We first give the formal definition of a Bayesian game and then comment in detail on each element of the definition.

Definition A.10 A **Bayesian game** consists of:

- a finite set of individuals, called **players**, denoted $N = \{i_1, i_2, \ldots, i_n\}$;
- for each player $i \in N$, a set of **actions**, denoted A_i;
- for each player i, a set T_i of possible **signals**;
- a probability distribution f over the set of signal profiles $T = T_1 \times T_2 \times \cdots \times T_n$;
- for each player $i \in N$, a payoff function u_i that gives the payoff player i will receive for each possible combination of actions chosen by all the players and each possible profile of signals received by the players. So u_i is a function from $A \times T$ to $\mathbb{R}$, where $A = A_1 \times A_2 \times \cdots \times A_n$.

In a Bayesian game, a **pure strategy** for a player i is a function s_i that gives for each signal t_i an action in A_i. That is, $s_i(t_i)$ is the action player i plays according to the strategy s_i when she receives the signal t_i.

Thus, in a Bayesian game, players first receive a signal, drawn according to the probability distribution f. In the standard auction model presented in chapter 2, the signal of a player (bidder) is the valuation of the bidder, and in the calculations the distribution f is the uniform distribution (i.e., signals/valuations are drawn at random within some interval). In example A.6, $T_{\text{Alice}} = \{u_A, \hat{u}_A\}$.

Once each player learns her signal, she plays the action prescribed by her strategy. The notion of a strategy in a Bayesian game is similar to the one we have for extensive form games. Recall that, in an extensive form game, a player's strategy must specify an action for each possible history at which the player has to play. Here we have the same, a strategy specifies an action for each possible signal the player has received.

Now consider a strategy profile for the players, s. If $t = (t_1, t_2, \ldots, t_n)$ is the signal profile, then the players play the action profile $(s_1(t_1), s_2(t_2), \ldots, s_n(t_n))$. Players' payoffs depend on the action profile *and* the signal profile. So, if the signal profile t is realized, player i obtains the payoff

$$u_i(s(t), t).$$

Note that each player observes only her signal. So, from a player's perspective, the payoff when she is considering a strategy is in the expected term, where the expectation is with respect to the signals (and thus the actions) of the other players. Note, however, that this expectation is conditional on her own signal, which is known to her.

We are now ready to define an equilibrium concept for Bayesian games, which is an adaptation of the Nash equilibrium to the case of Bayesian games in order to obtain the Bayesian-Nash equilibrium.

Definition A.11 A profile of (pure) strategies s is a **Bayesian-Nash equilibrium** if for each player i and each signal $t_i \in T_i$, and for each action $a_i \in A_i$,

$$E\left[u_i\big((s_i(t_i), s_{-i}(t_{-i})), (t_i, t_{-i})\big) \mid t_i\right] \geq E\left[u_i\big((a_i, s_{-i}(t_{-i})), (t_i, t_{-i})\big) \mid t_i\right]. \tag{A.10}$$

On the left-hand side of equation (A.10), $s_i(t_i), s_{-i}(t_{-i})$ is the action profile when each player j receives the signal t_j. Player i's payoff u_i depends on both the action profile and the signal profile, (t_i, t_{-i}). The expectation is conditional on player i receiving signal t_i because she knows this.

Equation (A.10) says that the expected payoff of player i must be higher when playing the equilibrium strategy than when playing a different strategy. Note that, in such a deviation, the player only changes the action prescribed by her strategy: she plays a_i instead of $s_i(t_i)$. However, player i's payoff function still depends on his or her true signal, t_i, and the expectation is still taken conditional on i's signal being t_i. So that is why the only differences between the two sides of the inequality in equation (A.10) are in $s_i(t_i)$ (on the right-hand side) and a_i (on the left-hand side).

Example A.7 For the sake of simplicity, we consider an incomplete information game with two players, Alice and Bob, where Alice can have only one signal and Bob can have two signals. The story is the following. Alice has a bank and has to decide whether to give a loan to Bob, who wants to launch a new business. Bob's signal can be either "lazy" or "hard working." The probability that Bob receives the "lazy" signal is p (and thus the probability that he received the "hard working" signal is $1 - p$).

Appendix A: Game Theory

Table A.11
Alice and Bob's incomplete information game

	Bob is NOT lazy			Bob is lazy	
	work	fun		work	fun
Lend	9, 10	3, 4	Lend	6, 3	1, 6
Not lend	5, 5	5, 1	Not lend	5, 0	5, 3

Bob has two possible actions. He can *work* for his business or he can just *have fun*, spending the money on leisure. The game is depicted in table A.11.

When Bob is not lazy, we can see that working dominates having fun: he gets a payoff of 10 or 5 (depending on whether Alice lends him money) if he works, and only 4 or 1 if he has fun. However, if Bob is lazy, then having fun dominates working.

Since Bob always knows his signal (and thus which payoff table matters), and Alice can have only one signal, the strategy of Bob in any equilibrium must be the following:

- Work if not lazy.
- Have fun if lazy.

We call s_{Bob} Bob's strategy.

Alice does not know Bob's signal; that is, she does not know whether Bob is lazy or not. However, she knows that if Bob is not lazy then working dominates having fun, and in that case she is better off lending him some money. If Bob is lazy, then the situation is reversed. Since having fun dominates working when Bob is lazy, Alice knows that if this is Bob's signal, he will have fun. Therefore, Alice is better off not lending (her payoff is 4) than lending (her payoff is 1).

Alice does not know Bob's signal, but she knows the probability of each signal. Alice has only one possible signal, so she has to choose only one action. Recall that a strategy consists of a choice of an action for *each possible* signal. In other words, Alice must choose between lending and not lending.

- *If she lends*:
  – Bob is lazy with a probability p. In that case, he has fun, and thus with probability p Alice's payoff is 1 (and Bob's payoff is 6).

  – Bob is not lazy with a probability $1-p$. In that case, he works, and thus with probability $1-p$ Alice's payoff is 9 (and Bob's payoff is 10).

So Alice's expected payoff if she lends money is

$$u_{\text{Alice}}(\text{lending}, s_{\text{Bob}}) = 1 \times p + 9 \times (1-p).$$

- *If she does not lend*:
  – Bob is lazy with a probability p. In that case, he has fun, and thus with probability p Alice's payoff is 5 (and Bob's payoff is 3).

  – Bob is not lazy with a probability $1-p$. In that case, he works, and thus with probability $1-p$ Alice's payoff is 5 (and Bob's payoff is 5).

So Alice's expected payoff if she lends money is

$$u_{\text{Alice}}(\text{not lending}, s_{\text{Bob}}) = 5 \times p + 5 \times (1-p).$$

Now we have to find the best strategy for Alice. Alice prefers to lend if

$u_{Alice}(\text{lending}, s_{Bob}) > u_{Alice}(\text{not lending}, s_{Bob})$

$\Leftrightarrow \quad 1 \times p + 9 \times (1-p) > 5 \times p + 5 \times (1-p)$

$\Leftrightarrow \quad p < \dfrac{1}{2}.$

So if the probability that Bob is lazy is less than 1/2, Alice's best strategy is to lend. Her expected payoff is higher than if she does not lend. So Alice's equilibrium strategy is:

- Lend if $p \leq 1/2$.
- Do not lend if $p > 1/2$.

What happens if $p = 1/2$? In that case, Alice is indifferent between lending and not lending. Her expected payoff is the same for both actions. So another possible equilibrium strategy for Alice is to lend if $p < 1/2$ and not lend if $p \geq 1/2$.

For simplicity, here we have defined the Bayesian equilibrium using pure strategies. As for the Nash equilibrium, it is possible also to consider mixed strategies. In this case, we have the following.

Result A.2 Any game with a finite set of players, a finite number of pure strategies, and a finite number of signals always admits a Bayesian equilibrium (possibly in mixed strategies).

Appendix B: Mechanism Design

In Appendix A, we outlined the basic game theory model and presented some tools to analyze games. This chapter can be seen as a follow-up to Appendix A, where we ask (and somehow answer) the following question: when we study a game, where does the game come from? The area devoted to this question is known as *mechanism design*. Auctions and matching or assignment problems are in fact special cases of mechanism design. The general purpose of mechanism design is to *design* games in order to achieve specific outcomes. As we will see, mechanism design is particularly interesting in situations with incomplete information.

B.1 Preliminaries

Mechanism design aims at understanding, in a very general or abstract manner, how outcomes are affected by the rules that govern how individuals interact. The general approach in mechanism design theory consists of designing a set of rules, the **mechanism**, to achieve certain objectives. That design is made taking into account that individuals may be strategic and also hold private information that can affect the final objective.

For instance, consider an auction. We studied in chapter 2 how bidding rules can affect the way bidders bid and ultimately the final allocation (who gets the item at what price). An auction is thus a mechanism where the individuals are the buyers and the private information they have is their valuations (or the signal they received about the value of the object). An objective could be, for instance, to maximize the sellers' revenue and/or allocate the object to the bidder with the highest valuation. Similarly, if we consider instead a matching or assignment problem (see chapters 9 and 11, respectively), the algorithm used to match or assign individuals can have an impact on individuals' behavior and thus affect the final outcome. A matching procedure is thus an example of a mechanism where the individuals' private information is their preferences over potential partners. A possible objective is to obtain a stable matching.

Mechanism design theory often takes a very broad approach. For instance, when studying buyer-seller situations, mechanism design theory can address the problem of having an efficient trade between buyers and sellers without specifying the protocol used by the individuals. For that specific example, each auction format defines a protocol (and thus each auction format is one example of a mechanism), but there can be other protocols, such as a bargaining procedure that buyers and sellers have to follow to determine the final allocation. A typical question for mechanism design theory would then be whether there exists a mechanism (without necessarily identifying it) that satisfies some properties (e.g., that a trade occur whenever the seller's valuation is less than or equal to the buyer's valuation). Such a situation is studied in section B.4.2.

The situations we consider are those where individuals have private information. For instance, the individuals could be car dealers, and the private information they have is the quality of the cars they sell. Those individuals are the ones who will play the game we will design. In mechanical design theory, they are usually called the **agents**. There is an additional individual, called the **principal**. The role of the principal is to design the game played by the agents. The objective of the principal is to obtain an outcome that satisfies some property, which may necessitate that the agents act in a certain way.

As we have already mentioned, the starting point of a mechanism design problem is a situation where agents have some private information and the principal has to design a game that will be played by the agents. Since the principal has complete freedom in the design of the game, she can well design a game where each agent has only one strategy. If, for some reason, the set of strategies cannot be modified by the principal, then she can still choose the payoffs such that all the strategies that each agent is supposed to play are dominant strategies. So, where is the difficulty?

First, in some situations, the principal may not have complete freedom in designing payoffs. (A classic way to affect agents' payoffs is to introduce monetary transfers.) But the crux of the problem comes from the fact that the principal's objective may depend on the information held by the agents, and the principal's and the agents' objectives may not be aligned. A mechanism is defined as a set of possible messages that the agents can send to the principal and a function that specifies which outcome is realized for each possible set of messages sent by the agents. The following (very simple) example illustrates what a mechanism can look like.

Example B.1 Alice, Bob, and Carol are three kids. Their father, Denis, wants to prepare dinner but hesitates between cooking green beans and lentils. Here Alice, Bob, and Carol are the agents and Denis is the principal. Each child's favorite vegetable is his or her private information.

Denis has to prepare dinner for his children. He wants the kids to eat vegetables, so he needs to know the children's favorite vegetable.

We can have, for instance, the following two mechanisms.

Mechanism 1

• Each kid tells Denis whether he or she prefers the lentils or the green beans. So "lentils" and "green beans" are the possible messages for each agent.
• Upon receiving the messages, Denis cooks the dish that has received the most votes. The outcome function is thus cooking lentils if two or more messages are "lentils" and cooking green beans otherwise.

Mechanism 2

• Each kid gives Denis a number between 1 and 10, so each kid has a choice between ten different messages.
• Upon receiving each kid's numbers, Denis computes the sum of the numbers. If the sum is an odd number, he cooks lentils; otherwise he cooks green beans.

Appendix B: Mechanism Design

B.2 The Model

Formally, denote the set of agents by $N = \{i_1, i_2, \ldots, i_n\}$. There is a collection of possible decisions that may be made, which we denote D. In an auction, a decision would be to deciding whom the object is allocated, and in a matching problem a decision would be simply a matching (who is matched with whom). Note that a decision does not specify a price (which is needed when considering an auction). The notion of monetary transfer will come later.

Each individual $i \in N$ holds some private information θ_i, and we denote by Θ_i the set of possible pieces of information that i may hold.[1] In mechanism design theory, the information held by an individual is linked to her payoff function or preferences over decisions, which are usually represented by a utility function, $u_i : D \times \Theta_i \to \mathbb{R}$. The information t_i of an individual is also called the **type** of the individual and is similar to the notion of a signal that we saw when defining Bayesian games in Appendix A. In this setting, the payoff that individual i enjoys when the decision is d and her type is θ_i is thus $u_i(d, \theta_i)$.

Example B.2   In an auction setting, a bidder's type is her valuation. The payoff function is

$$u_i(d, \theta_i) = \begin{cases} \theta_i & \text{if } d \text{ is "bidder } i \text{ wins the object"} \\ 0 & \text{if } d \text{ is "bidder } i \text{ does not win the object"} \end{cases}$$

(In the chapters on auctions, the valuation of a bidder i was usually denoted v_i.) Again, note that the description of the model does not yet include side payments (such as the price paid by the bidders).

The description we give here corresponds to the case of *private values*: the utility that an individual enjoys from a decision d depends only on her type and not the types of the other individuals.

Remark B.1   In settings such as matching or assignment problems, the standard approach consists of endowing individuals with a preference relation that only gives an ordering over the outcomes (as opposed to a utility function that gives a numerical representation of the individuals' welfare). In such environments, the only relevant information about individuals' welfare is *ordinal*, so we can only compare two alternatives. In such cases, the type of an individual is simply the preference relation itself. For instance, if there are three objects (a, b, and c), then the preference relations $P_i : a, b, c$ and $P'_i : c, a, b$ correspond to two different types. In spite of this difference in the way we model agents' preferences, the intuition is the same: "how much" individuals "enjoy" an outcome depends on their type.

B.2.1 Mechanism

A mechanism is defined by two elements: a **message space** and an **outcome function**. In a mechanism (set by the principal), the agents are required to send a message to the principal. Upon receiving the messages, the principal implements a decision (using a outcome function).

Before giving a more formal definition of a mechanism, we need to add a third element, the **transfer function**. In some situations, such as a buyer-seller relation (or an auction), decisions are accompanied by a transfer (e.g., a price, a tax, etc.). Some problems thus come with a transfer function that maps each vector of messages (i.e., a message from each individual) to a number for each agent. The number corresponding to each agent describes the payment that the agent receives (if positive) or makes (if negative) from or to the principal. So the outcome function is a combination of two elements: a decision function and a transfer function.

Definition B.1 A **mechanism** is a pair (M, g) where

- $M = M_1 \times M_2 \times \cdots \times M_n$ is the message space, where for each individual $i \in N$, M_i is the set of messages that individual i can send and
- g is an outcome function that maps each vector of messages to a decision and a transfer, $g : M \to D \times \mathbb{R}^n$.

In many cases, it proves useful to distinguish the decision from the transfer. In this case, we define the outcome function g as being the combination of a decision function f and a transfer function t, both of which depend on the messages sent by the agents. Note that the transfer function t is in fact a vector of transfer functions, one for each agent, $t = (t_1, t_2, \ldots, t_n)$.

A mechanism (M, g), where the outcome function specifies a decision and a transfer, $g = (f, t)$, works as follows. First, each agent $i \in N$ selects a message m_i. The principal thus receives the vector of messages $m = (m_1, \ldots, m_n)$. Next, the principal implements the decision prescribed by the decision function and calculates the transfers. So the decision is $d = f(m)$ and the transfers are $(t_1(m), t_2(m), \ldots, t_n(m))$. The final payoff of an agent i whose type is θ_i is a function v_i,

$$v_i(\theta_i, m, d, t) = u_i(d, \theta_i) + t_i(m). \tag{B.1}$$

Example B.3 An auction is a typical example of a mechanism: the principal is the seller, and the agents are the buyers. A buyer's type is the maximum price she is willing to pay, which is private information.

So the problem for the seller is to design an auction so that the buyers tell the seller the maximum price they are willing to pay. The messages sent by the buyers can be bids, and the outcome function is simply a function that says who wins the auction (usually the highest bidder) and what price was paid by the bidder.

But mechanism design theory allows us to approach the problem of a seller facing several buyers from a much broader perspective. The outcome function does not need to describe an auction. The benefit of this more general (albeit abstract) approach is that we can check whether an auction is the best way for a seller to, for instance, maximize her revenue.

It is tempting (and not useless) to consider a mechanism in a more "compact" way:

- Each agent selects a message she wants to send.
- The payoff of each agent depends on the combination of messages sent by the agents— given by equation (B.1).

This compact description looks very similar to that of a Bayesian game (see Appendix A) where the messages play the role of strategies (and the agent's types are the signals), and thus we would like to say that a mechanism is nothing but a (Bayesian) game. There is a crucial difference, though. In a game, we have the agents, their strategies, and a *payoff function* (if we change the payoffs, we have a different game). In contrast, in a mechanism, we do not have payoffs. According to definition B.1, in a mechanism we only have messages and *outcomes* (the decision and the transfers). Once the payoff functions are defined

Appendix B: Mechanism Design

(a function $u_i(d, t)$, for each $i \in N$), we have a game. But notice that we can change the payoff functions without changing the mechanism. A mechanism thus describes a (very large) family of games. For this reason, a mechanism is often referred to as a **game form**.

B.2.2 Implementing Social Choice Functions

To sum up, we have the following situation. The principal designs an outcome function, selects a set of messages the agents can send, and the agents then have to choose which message they send, having in mind that the outcome depends on the message they send. In that context, what is the objective of the principal? As we said earlier, the outcome the principal would like to obtain generally depends on the types of the agents. That is, for each possible combination of agents' types, there is an "ideal" outcome for the principal. Formally, we say that the principal would like to **implement** a function

$$\gamma : \Theta \to D \times \mathbb{R}^n.$$

That function γ is called a **social choice function** and specifies for each type a decision and a transfer.[2]

Example B.4 In example B.1, an example of a social choice function would be:

- Cook lentils if all three kids prefer lentils.
- Cook green beans otherwise.

So the social choice function is defined over the types of the agents (what the kids like) and not what they say.

The problem for the principal is then the following. On the one side, there is the social choice function that constitutes the objective of the principal. It describes the decision the principal would like to see for each possible combination of the agents' types. Put differently, it is the decision the principal would choose if she knew the agents' types. On the other side, there is the mechanism and the agents' behavior. Given their types, the messages available to them, and the decision function, the agents will select a particular message and thus a decision. The issue for the principal is then whether

$$g(m_1, m_2, \ldots, m_m) = \gamma(\theta_1, \theta_2, \ldots, \theta_n); \tag{B.2}$$

that is, whether the decision and transfer selected by the outcome function g the same as the objective given by the social choice function γ. In the language of mechanism design, if equation (B.2) holds, we say that the **mechanism implements the social choice function** γ.

Example B.5 A seller (the principal) would like to sell an object, he is facing n potential buyers, and each buyer (agent) has a maximum price she is willing to pay (which is her type). Here an outcome is who the buyer is (the decision) and what price is paid by each agent. (In general, only the buyer pays a nonzero amount, but we may think of situations where all individuals pay something, such as a raffle.)

If $\theta_1, \theta_2, \ldots, \theta_n$ are the agents' types, an example of a seller's objective is the following social choice function:

$$\gamma(\theta_1, \theta_2, \ldots, \theta_n) = \begin{cases} \text{agent } i \text{ buys the good and pays } \theta_i & \text{if } \theta_i = \max\{\theta_1, \ldots, \theta_n\} \\ \text{agent } i \text{ does not buy the good and pays } 0 & \text{if } \theta_i < \max\{\theta_1, \ldots, \theta_n\}. \end{cases}$$

In other words, the function γ says that the agent with the highest willingness to pay (i.e., the highest type) is the one buying the good, and the price the buyer pays is her type (i.e., her willingness to pay). If an agent does not have the highest type, then she pays nothing (and does not get the good).

Now the objective for the principal is to design a mechanism such that the agents behave in such a way that it is always the agent with the highest type that gets the good.

An obvious set of messages are prices (i.e., bids). That is, the mechanism consists of each agent announcing a price, and then the good is awarded to the agent announcing the highest price (and she is asked to pay that price).

So here the mechanism and the function γ are identical. But would it work? Certainly not. To see this, suppose that there are only two agents, Alice and Bob. Alice's willingness to pay is $5 and Bob's is $10. Will Alice and Bob send the message $5 and $10? It is easy to see that they will not. If they send these messages, then the seller is happy, but it is not in Bob's interest to do so. If he instead sends the message $6, then he still has the good (Alice's message is lower) but he now pays less.

This very simple example shows that the problem of designing the right mechanism is trickier than it seems. If the mechanism is identical to the social choice function, then the agents may behave in such a way that the outcome that is implemented does not correspond to the principal's objective.

If Alice's and Bob's willingnesses to pay are $5 and $10, respectively, the seller wants to sell to Bob at a price of $10. This is the outcome of the social choice function. But in this case Alice and Bob may announce different prices, for instance $3 and $6, which will give the outcome "sell to Bob at a price of $6."

The strategic behavior of the agents in example B.5 suggests that the optimal mechanism for a principal is, in general, not identical to the social choice function the principal wants to implement. If the agents are not strategic, then our problem is solved: it is enough for the principal to use as a mechanism the social choice function she wants to implement. However, if the agents are strategic, then the principal needs to somehow anticipate the agents' strategic choice so that the outcome that is obtained from their messages corresponds to the outcome targeted by the principal.

As for games, there are various concepts we can use to "predict" agents' behavior in a mechanism. In fact, since once we have a payoff function for each agent a mechanism describes a Bayesian game, we can use the various solution concepts to analyze such games.

B.2.3 Direct versus Indirect Mechanism

So far, we have not been very precise about the nature of the messages. The way we have defined a mechanism design problem, the messages could be anything; that is, they may not be directly related to the types of the agents. This is the case, for instance, in mechanism 2 in example B.1.

In the mechanism design literature, we distinguish between two classes of mechanisms: direct and indirect. A **direct mechanism** is a mechanism where the messages are the types of the agents. That is, the principal simply asks the agents to reveal their types. It follows that any social choice function γ, which depends on the agents' types, is a direct mechanism, where the message space is given by the type space, $\Theta = \Theta_1 \times \Theta_2 \times \cdots \Theta_n$, and the outcome function is the social choice function itself, γ.

Of course, as example B.5 shows, the use of a direct mechanism does not necessarily imply that the agents will be truthful and reveal their types. In contrast, in an **indirect mechanism**, the messages do not consist of types; they could be anything. In example B.1, mechanism 2 is an indirect mechanism, while mechanism 1 is a direct mechanism.

Appendix B: Mechanism Design

B.3 Dominant Strategy Implementation

When we design a mechanism, we would like to have some notion that can help us predict how the agents will behave. The most cogent notion is that of a dominant strategy. Let (M, g) be a mechanism where $g = (f, t)$, and for each agent $i \in N$, let u_i be the utility function of agent i (which depends on the decision on i's type).

A strategy m_i (i.e., a message) is a **dominant strategy** for a type θ_i if for any message profile $m_{-i} = (m_1, \ldots, m_{i-1}, m_{i+1}, \ldots, m_n)$, and any strategy m',

$$u_i(f(m_i, m_{-i}), \theta_i) + t_i(m_i, m_{-i}) \geq u_i(f(m'_i, m_{-i}), \theta_i) + t_i(m'_i, m_{-i}). \tag{B.3}$$

Not surprisingly, this concept of a dominant strategy is the same as the one we saw for games (see definition A.7 in Appendix A).

Definition B.2 A social choice function γ is implemented in dominant strategies by a mechanism (M, g) if there exists for each i a function $h_i : \Theta_i \to M_i$ such that $h_i(\theta_i)$ is a dominant strategy for each θ_i and we have

$$g(h(\theta)) = \gamma(\theta) \quad \text{for all } \theta \in \Theta.$$

Definition B.2 reads as follows. We want to implement the social choice function γ using a mechanism (M, g). In this mechanism, the agents will send a message, and we want there to be, for each type, a particular message that is the dominant strategy. If we have for each type a dominant strategy, then we can say that there exists a function, called h_i in definition B.2, that simply names which strategy is dominant (the message) for each possible type. Finally, we want the outcome obtained when the agents play the dominant strategy, $g(h(\theta))$, to be the same as the one the principal would have implemented had she known the type of each agent, $\gamma(\theta)$.

So far, we have built a very general approach, but it starts to be a bit cumbersome to manipulate: we have on the one one side the agents' types and on the other side the messages that they have to send. If we want to know whether there exists a mechanism that implements a particular social choice function, we may have to consider complex message spaces, which may make the search of the h functions described in definition B.2 tedious. When we consider direct mechanisms, we do not have to worry about the existence of the functions h_i for each agent $i \in N$. So definition B.2 becomes simpler.

Definition B.3 A social choice function γ (i.e., a direct mechanism (Θ, γ)) is **dominant strategy incentive compatible** if for each agent $i \in N$, for each $\theta_i \in \Theta_i$, θ_i is a dominant strategy at θ_i.

In the mechanism design literature, a social choice function that is dominant strategy incentive compatible is often called *strategyproof*. In a direct mechanism, agents reveal a type, so it is easy to see that if we impose that, for each $i \in N$, $h_i(\theta_i) = \theta_i$ for each $\theta_i \in \Theta_i$, then definitions B.2 and B.3 are identical.

B.3.1 The Revelation Principle

Analyzing mechanisms can be a daunting task. If we want to find a mechanism that implements a specific social choice function, we first have to decide whether we will use a direct or an indirect mechanism. Clearly, a direct mechanism seems preferable, because it exempts us from defining the set of messages. But maybe there are situations where we would need to consider indirect mechanisms. Fortunately for us, a celebrated result in the literature simplifies our task considerably.[3]

Result B.1 (Revelation Principle) Let (M, g) be a mechanism that implements a social choice function γ in dominant strategies. Then the direct mechanism (Θ, γ) is dominant strategy incentive compatible.

The result says the following. Suppose you want to see whether a particular social choice function f can be implemented. To do this, you would have to consider *all possible* direct or indirect mechanisms. If there exists such a mechanism, then you can invoke the Revelation Principle, which says that if there exists a mechanism that implements the social choice function f, then

- you can use the social choice function f itself as the mechanism, and
- the mechanism f is such that in equilibrium the agents truthfully reveal their types (i.e., their equilibrium strategies consist of revealing their types).

The intuition behind the Revelation Principle is relatively straightforward. Let (M, g) be a mechanism where M is the message space for all agents and g is the outcome function, and suppose that (M, g) implements the social choice function γ in dominant strategies. So there exists for each agent i a function h_i such that $h_i(\theta_i)$ is a dominant strategy for agent i when her type is θ_i. Hence, when the agents, types are $\theta = (\theta_1, \ldots \theta_n)$, they send the messages $(h_1(\theta_1), \ldots, h_n(\theta_n))$, and the outcome of the function g with those messages is the same as that of the social choice function γ (because (M, g) implements γ),

$$g(h_1(\theta_1), \ldots, h_n(\theta_n)) = \gamma(\theta_1, \ldots \theta_n).$$

The "trick" of the Revelation Principle is that instead of asking the agents to play the mechanism (M, g), she asks them to reveal their types, and she will produce, *on their behalf*, the messages prescribed by the functions $(h_1, \ldots, h_n)$ and feed those messages into the outcome function g. It is easy to see that if sending the message $h_i(\theta_i)$ is a dominant strategy for agent i, then it is also a dominant strategy to send the type θ_i. This is because the principal, upon receiving the "message" θ_i, will map it to the message $h_i(\theta_i)$. To see that the incentives are identical in the indirect and direct mechanisms note that if in the direct mechanism an agent i can be better off pretending to be of type θ_i' (when she is in reality of type θ_i), then she should also be better off sending the message $h_i(\theta_i')$ in the indirect mechanism because the principal will run the outcome function g with the message $h_i(\theta_i')$.

B.3.2 The Gibbard-Satterthwaite Theorem

The model we have presented and briefly analyzed so far is very much inspired by the problem of trading with prices in that it includes, besides the decision, a monetary transfer between the agents and the principal. There are many situations, however, where such transfers may not be possible (or should be avoided). The chapters on matching and assignment deal with such situations.

To begin with, note that the model we have presented with transfers can be almost straightforwardly rewritten for the case when transfers are not feasible. It suffices to set the transfer to be equal to 0 for any collection of messages received by the principal.

When transfers are not an option and there are a finite number of possible decisions, it is often the case that agents' preferences are represented not by a utility function but simply by an *ordering*. That is, each agent has a "personal ranking" over the possible decisions (such a case is discussed in chapter 9). In this context, what does mechanism design theory have to say? The answer is rather negative. Before showing what we mean

Appendix B: Mechanism Design

by this, we need one more definition. A decision function f is *dictatorial* if there exists an agent i such that, for any type profile (i.e., whatever the types of agents are), the decision selected by the function f is always i's most preferred decision. In other words, when the decision function is dictatorial, the choice of the decision that will be implemented only depends on one agent, and it is always the same (that agent is called the *dictator*). A fundamental result about the implementation of a social choice function is the following.[4]

Result B.2 (Gibbard; Satterthwaite)   Suppose that D contains a finite number of decisions, that the decision function f is such that at least three different decisions can be selected (depending on agents' types), and agents' types can consist of any possible preference ordering over the decisions. Then the decision function f is dominant strategy incentive compatible if, and only if, it is dictatorial.

Result B.2 does not seem like good news; it suggests that the only possible dominant strategy incentive compatible decision functions are dictatorial. We say "suggests" because it comes from a superficial reading of result B.2. As for any result, there are some assumptions. Here, the main assumption is that agents' preferences are unrestricted. Any preference ordering is possible. This means the following. Suppose that there are, say, four different decisions that can be made: d_1, d_2, d_3, and d_4. There are $4! = 24$ different ways to rank those decisions, and any ranking corresponds to a possible type. There are, fortunately, some situations where some preferences, orderings, or types are irrelevant. For instance, there could be situations such that if it is possible that an agent prefers d_1 to d_2 and d_2 to d_3 or prefers d_3 to d_2 and d_2 to d_1, then it is not possible to prefer, for instance, d_1 to d_3 and d_3 to d_2.[5]

A second assumption that is restrictive in result B.2 is the absence of transfers. Hence, this result does suggest that in some situations nonzero transfers may be needed to implement some social choice functions in dominant strategies. Section B.3.3 addresses the existence of such mechanisms when there are transfers.

B.3.3 The Vickrey-Clarke-Groves Mechanism

When transfers between the agents and the principal are feasible, there exists a mechanism that is not only dominant strategy incentive compatible but also *efficient*. Here, efficiency means that for each vector of type θ, the decision that is chosen is such that the *sum* of individuals' payoffs is the highest (but ignoring the transfers). Formally, a decision function f is efficient if

$$\sum_{i \in N} u_i(f(\theta), \theta_i) \geq \sum_{i \in N} u_i(d', \theta_i) \quad \text{for all } d' \in D \text{ and all } \theta \in \Theta. \tag{B.4}$$

That is, for any type profile of the agents θ, the decision made, $f(\theta)$, is such that the sum of each agent's payoff for that decision is greater than or equal to the sum of agents' payoffs for any other decision d'. Given a decision d, the sum $\sum_{i \in N} u_i(d, \theta_i)$ is the social welfare at the profile θ. Thus, a decision function is efficient if, for any type profile θ, it always maximizes the social welfare.

We now consider the transer function defined as follows. Let $d \in D$ be any decision.

$$t_i(\theta) = \sum_{j \neq i} u_j(d, \theta_j) - \max_{\widehat{d} \in D} \sum_{j \neq i} u_j(\widehat{d}, \theta_j). \tag{B.5}$$

The function described in equation (B.5) is the following. Make a decision d, and compute the sum of each agent's payoff for that decision but ignoring agent i's payoff. This is the term

$$\sum_{j \neq i} u_j(d, \theta_j).$$

The second part of equation (B.5) describes the maximum social welfare when we ignore agent i; it makes the decision d that yields the largest sum of the payoff of all agents (except agent i). Hence, the transfer function described in equation (B.5) is the difference between the social welfare of all agents (except agent i) when a decision d is made and the maximum social welfare of all agents (except agent i).

Definition B.4 The Vickrey-Clarke-Groves (VCG) mechanism is the direct mechanism (f, t), where f is an efficient decision function and the transfer t is given by equation (B.5), where the decision d is $f(\theta)$.

By construction, the VCG mechanism is efficient. But it also enjoys another desirable property.

Result B.3 The Vickrey-Clarke-Groves mechanism is dominant strategy incentive compatible.

B.4 Bayesian Mechanism Design

In some settings, we may not be able to identify a dominant strategy incentive compatible mechanism even if we allow transfer. This is the case, for instance, when we want the transfers to be feasible or budget balanced.

A transfer function t is feasible if, for any type profile θ, the sum of each individual's transfers is less than or equal to zero,

$$\sum_{i \in N} t_i(\theta) \leq 0 \quad \text{for all } \theta \in \Theta.$$

Recall that transfers are added to an agent's utility function—see equation (B.1). So if the sum of all transfers is strictly positive, then it means that some money must be brought from outside.

A more stringent requirement is that the transfers be balanced. This means not only that the transfer be feasible but also that the agents do not pay "too much,"

$$\sum_{i \in N} t_i(\theta) = 0 \quad \text{for all } \theta \in \Theta.$$

B.4.1 Bayesian Incentive Compatibility

We presented in the previous section a dominant strategy incentive compatible mechanism, the VCG mechanism. The problem with this mechanism is that the transfer function may not be balanced. One way out consists of relaxing the requirement that reporting one's true type is a dominant strategy and instead require that reporting one's true type be an equilibrium strategy. Since a mechanism is in fact a Bayesian game (once we have specified each individual's payoff function), we can use the Bayesian-Nash equilibrium concept. We first define incentive compatibility for direct mechanisms; that is, we consider mechanisms (Θ, γ) with $\gamma = (f, t)$, where f is the decision function and t the transfer function.

Appendix B: Mechanism Design

Definition B.5   A direct mechanism (Θ, γ) is **Bayesian incentive compatible** if, for each type profile θ, the strategy profile that consists of each agent announcing her type is a Bayesian-Nash equilibrium,

$$E\Big[u_i(f(\theta_i, \theta_{-i}), \theta_i) + t_i(\theta_i, \theta_{-i}) \mid \theta_i\Big] \geq E\Big[u_i(f(\theta'_i, \theta_{-i}), \theta_i) + t_i(\theta'_i, \theta_{-i}) \mid \theta_i\Big], \quad (B.6)$$

for each $i \in N$, for each $\theta_i \in \Theta_i$, and for $\theta'_i \in \Theta_i$.

Equation (B.6) is identical to the condition for a Bayes-Nash equilibrium (equation (A.10) in Appendix A) but rewritten with the notation and the framework of a mechanism. For the general case (indirect and direct mechanisms), we then have that a mechanism (M, g) implements a social choice function in Bayesian-Nash equilibrium if there exists a Bayesian equilibrium m of (M, g) such that, for each θ, $g(m(\theta)) = f(\theta)$.

For dominant strategy incentive compatible mechanisms, we have seen that the Revelation Principle assures us that it is enough to consider only direct mechanisms. The same principle also holds for Bayesian equilibrium implementation.[6]

Result B.4 (*Revelation Principle*)   Let (M, g) be a mechanism that implements a social choice function γ in Bayesian equilibrium. Then the direct mechanism (Θ, γ) is Bayesian incentive compatible.

B.4.2   Trading: The Myerson-Satterthwaite Theorem

The generality of mechanism design theory allows us to study a wide range of situations, and since the setup is relatively abstract, we can study situations by abstracting from particular details. To see this, consider the situation of two parties engaged in trading. There is a seller, Sarah, and a buyer, Bob. Sarah has an object that is valuable for her and for Bob, but they may value the object differently. Of course, Sarah is willing to sell the object only if the price she gets is at least as high as her valuation, and Bob is willing to buy the object only if the price is at most as high as his valuation.

The situation we have just described is very general, and we have not given any details about how Sarah and Bob will perform a transaction. One way would be that Sarah announces a price at which she is willing to sell the object, and Bob either accepts and purchases the object or simply refuses to buy it. Another possibility is that Bob announces a price he is willing to pay. A third possibility is that Sarah and Bob engage in a repeated bargaining process in which one party makes an offer and the other party responds by making a counteroffer, and a transaction is performed as soon as the offers coincide. A fourth possibility is that they pick a random number between the minimum price Alice asks and the maximum price Bob is willing to pay, and that random number is the price Bob will pay. Those are only a few examples of how Sarah and Bob can proceed, and they are not restricted to proceeding in one of these ways.

The problem becomes more interesting when we assume that Sarah and Bob have private information: each knows his or her valuation but does not know the valuation of the other party. To see how mechanism design can be useful, notice that any procedure that Sarah and Bob adopt to perform a transaction can be seen as a mechanism.

The question we address now is the following: does there exist a (transaction) mechanism that satisfies some properties? Our question is thus very simple, for we do not want to identify a specific mechanism but simply know whether a mechanism exists. The properties we want to impose are relatively easy to justify.

- **Ex-ante individual rationality.** The expected payoff for both Sarah and Bob must be nonnegative.

Sarah and Alice will play a mechanism that gives some outcome (a transaction with a price). Since neither Sarah nor Bob know the valuation of their opponent, the final payoff of Sarah or Bob is random: the probability of each outcome is the probability of each type.

Suppose, for instance, that Sarah's valuation is $10, that the mechanism consists of announcing a price, and the outcome function is:
- A transaction occurs if Bob's price is higher than Sarah's price; otherwise there is no transaction.
- If a transaction occurs, the price is the midpoint between Sarah's and Bob's prices.

Suppose that Bob's valuation is $14 with probability 0.6 and $8 with probability 0.4, and assume that both Sarah and Bob are truthful. So, for Sarah, there is a transaction with probability 0.6, and the price is ($14 + $10)/2 = $12. With probability 0.4, Bob announces a price lower than Sarah's, meaning there is no transaction. So Sarah's expected gain is

Expected revenue − expected loss = (0.6 × $12 + 0.4 × $0) − (0.6 × $10 + 0.4 × $0) = $1.2.

Her expected payoff is positive, so this mechanism is individually rational. Now suppose instead that the mechanism is such that there is always a transaction and the price is the smallest difference between Alice's and Bob's valuations. In this case, Alice's expected gain is (0.6 × $10 + 0.4 × $8) − (0.6 × $10 + 0.4 × $10) = −$0.8. Her expected payoff is negative, so that mechanism is not individually rational.

- **Weak budget balanced.** The principal does not need to add money to the mechanism in order to subsidize the trade.

Suppose, for instance, that the price announced by Sarah is $10 and Bob announces $8, yet the mechanism decides that there is a trade. In this case, there is no price that is higher than Sarah's price and lower than Bob's price. If there is a transaction, for any price one of the parties always has a negative payoff. To guarantee that the mechanism is ex-ante individually rational, we may then want to ask the principal to pour some money into the game so that neither Sarah nor Bob ends up with a negative payoff. A mechanism that has a balanced budget is such that the principal does not need to add money to the game. The "weak" term in the condition refers to the fact that the price paid by the buyer may exceed the price received by the seller. This allows, for instance, consideration of the presence of a sales tax.

- **Bayesian equilibrium incentive compatibility.** For both Sarah and Bob, telling the truth (i.e., reporting one's true valuation) is a best reply if the other party does so.

- **Ex-post efficiency.** The object must go to the agent who values it the most. If Sarah values the object more than Bob, then she should keep the object; otherwise it must be Bob who gets it.

Roger Myerson and Mark Satterthwaite showed the following, surprising result.[7]

Result B.5 (Myerson and Satterthwaite) In a bilateral trading environment with private information, there is no mechanism that is ex-ante individually rational, budget balanced, Bayesian equilibrium incentive compatible, and ex-post efficient.

In other words, the Myerson-Satterthwaite theorem says that there is no efficient way for two agents to trade an object when valuations are private information without taking

Appendix B: Mechanism Design

the risk that one of the parties trades at a loss, that an outsider will have to put up some money, or that one of the parties will not be honest.

This result is rather surprising and counterintuitive. Indeed, it is not difficult to imagine situations where such trading mechanisms are possible. To see this, suppose that Sarah's valuation is always below $10 and that Bob's valuation is always above $10. A mechanism such that

- there is always a transaction, for any possible price announced by Sarah or Bob (i.e., a type, which is a valuation) and
- the transaction price is $10

does satisfy the four properties stated in result B.5. But the theorem considers all possible situations! In other words, to illustrate the theorem, it suffices to present a situation where we cannot design a transaction procedure that satisfies our requirements. In fact, the key property on which this result relies is the incentive compatibility condition: being truthful is an equilibrium.

Suppose that Sarah's valuation for the object is either $0 or $0.9 and that Bob's valuation is either $0.1 or $1. For both Sarah and Bob, the probability of either valuation is 0.5.

The budget balance condition says that the price paid by the buyer should be at least as high as the price received by the seller. For simplicity, let us assume that the price paid by Bob is equal to the price received by Sarah (if there is a transaction). So we abstract from the presence of a sales tax.

We then have a price function $p(v_{Sarah}, v_{Bob})$, where v_{Sarah} and v_{Bob} are Sarah's and Bob's valuations reported to the mechanism, respectively. (Recall that, thanks to the Revelation Principle, we can assume that the mechanism in place is a direct revelation mechanism, that Sarah and Bob both announce a valuation.)

Our purpose here is to analyze how the price function $p(v_{Sarah}, v_{Bob})$ should look for each possible combination of types (valuations). There are four possible combinations of Sarah's and Bob's valuations.

Case 1 $v_{Sarah} = \$1$ and $v_{Bob} = \$0.9$
So here we look at the value of $p(\$0.9, \$0.1)$. Individual rationality says that the price must be at least $0.9, for otherwise Sarah has a negative payoff. So we have
$p(\$0.9, \$1) \geq \$0.9$.

Case 2 $v_{Sarah} = \$0$ and $v_{Bob} = \$0.1$
Individual rationality says that the price must be at most $0.1, for otherwise Bob has a negative payoff,
$p(\$0, \$0.1) \leq \$0.1$.

Case 3 $v_{Sarah} = \$0.9$ and $v_{Bob} = \$0.1$
There is no transaction possible that is individually rational for both Sarah and Bob. So we set the price to $0:
$p(\$0.9, \$0.1) = 0$.

Case 4 $v_{Sarah} = \$0$ and $v_{Bob} = \$1$
The characterization of $p(\$0, \$1)$ is where the incentive compatibility condition kicks in. When Sarah's valuation is $0, we want her to be truthful. She does not know Bob's valuation. If Bob's valuation is $0.1 (which occurs with probability 0.5), we know that the price must be at most $0.1 (this is case 2). But with probability 0.5, Bob's valuation is $1. In this

case, Sarah would like the price to be very high. Since she does not know Bob's valuation, she can only compute an expected payoff, and since we look for an equilibrium in which she is truthful, the expected payoff she obtains when announcing $0 must be at least as high as when announcing $0.9:

$$\underbrace{\frac{1}{2} \times p(\$0, \$1) + \frac{1}{2} \times p(\$0, \$0.1)}_{\text{expected payoff when Sarah announces \$0}} \geq \underbrace{\frac{1}{2} \times p(\$0.9, \$1) + \frac{1}{2} \times p(\$0.9, \$0.1)}_{\text{expected payoff when Sarah announces \$0.9}}. \quad (B.7)$$

In equation (B.7), the expected revenue is simply the expected price, because Sarah's valuation is 0, and the probability of each price is the probability of Bob's valuation being $1 or $0.1.

Using what we know about the price function $p(\cdot, \cdot)$ from cases 1, 2, and 3, we can simplify equation (B.7) as

$$p(\$0, \$1) \geq p(\$0.9, 1) - p(\$0, \$0.1) \quad \Rightarrow \quad p(\$0, \$1) \geq \$0.8. \quad (B.8)$$

We now consider Bob's perspective. When his valuation is $1, we want him to announce $1 and not $0.1. As earlier, Bob does not know Sarah's valuation; he can only compute an expected payoff. So we need his expected payoff announcing $1 to be at least as high as his expected payoff announcing $0.1. Bob's payoff is his gain (the value of the object if he gets it, 0 if he does not) minus the price decided by the mechanism:

$$\text{Bob's payoff} = \begin{cases} \text{his valuation} - \text{the price} & \text{if he gets the object} \\ 0 & \text{if he does not get the object} \end{cases}$$

So we have

$$\underbrace{\frac{1}{2} \times (1 - p(\$0, \$1)) + \frac{1}{2} \times (1 - p(\$0.9, \$1))}_{\text{expected payoff when Bob announces \$1}} \geq \underbrace{\frac{1}{2} \times (1 - p(\$0, \$0.1)) + \frac{1}{2} \times (0 - p(\$0.9, \$0.1))}_{\text{expected payoff when Bob announces \$0.1}}.$$

(B.9)

Using what we know about the price function $p(\cdot, \cdot)$ from cases 1, 2, and 3, we can simplify equation (B.8) as

$$1 - p(\$0, \$1) + 1 - p(\$0, 9, \$1) \geq p(\$0, \$0.1) \quad \Rightarrow \quad p(\$0, \$1) \leq \$0.2. \quad (B.10)$$

Combining equations (B.8) and (B.10), we obtain

$$p(\$0, \$1) \leq \$0.2 \quad \text{and} \quad p(\$0, \$1) \geq \$0.8. \quad (B.11)$$

Obviously, equation (B.11) is impossible to satisfy. The Myerson-Satterthwaite theorem is counterintuitive for most people because usually we forget about the incentive compatibility condition. Once we require that the agents be truthful, we have a conflict between obtaining an optimal trade and incentives (i.e., reporting one's true valuation).

Appendix C: Order Statistics

Let $v_1, v_2, \ldots, v_n$ be n variables. Let $w_1^{(n)}, w_2^{(n)}, \ldots w_n^{(n)}$ be a reordering of these variables such that

$$w_1^{(n)} \geq w_2^{(n)} \geq \cdots \geq w_n^{(n)}.$$

That is, $w_1^{(n)}$ corresponds to the highest variable among $\{v_1, \ldots, v_n\}$, $w_2^{(n)}$ is the second highest, and so on.

We will make the following assumptions.

Assumption 1 For each $h = 1, \ldots, n$, v_h is a random number between 0 and 100. This implies that the variables $v_1, \ldots, v_n$ are uniformly distributed between 0 and 100.

Assumption 2 The n draws are independent.

Since the variables $v_1, \ldots, v_n$ are random, so are the variables $w_1^{(n)}, \ldots, w_n^{(n)}$. The variables $w_k^{(n)}$, with $k = 1, 2, \ldots, n$, are called the *order statistics*. The variable $w_1^{(n)}$ is called the first-order statistic, $w_2^{(n)}$ is called the second-order statistic, and so on. Our objective is to obtain formulas for the expectations of the first- and second-order statistics and the conditional expectation of the first-order statistic. As we will see, the formulas will depend on the number of variables, n (which is why we keep track of that number in the notation).

C.1 Expected Highest Valuation

The expected highest value among n random values $v_1, \ldots, v_n$ is the *expectation of the first-order statistic*. So we want to calculate $E(w_1^{(n)})$. To calculate it, we first need to know the probability distribution of the first-order statistic, $w_1^{(n)}$. The strategy we will use is the following:

1. Obtain the *cumulative density function*.
2. Derive the *probability density function*.
3. Calculate the expectation.

C.1.1 Obtaining the Cumulative Density Function

Our objective here is to obtain the cumulative density function of $w_1^{(n)}$, which we denote by $F_1^{(n)}$. Let v be any number between 0 and 100. Hence, $F_1^{(n)}(v)$, from the definition of a cumulative density function, has probability $w_1^{(n)}$ less than v. Clearly, "$w_1^{(n)}$ being less than v" is the same as "all the order statistics are less than v," which is the same as "all the values $v_1, \ldots, v_n$ are less than v." So we have

$$F_1^{(n)}(v) = Prob(v_1 \leq v \text{ and } v_2 \leq v \text{ and } \ldots \text{ and } v_n \leq v).$$

Since bidders' valuations are independent, the probability that all their valuations are less than v is the same as the *product* probability of each bidder's probability that her valuation is less than v,

$$F_1^{(n)}(v) = Prob(v_1 \leq v) \times Prob(v_2 \leq v) \times \cdots \times Prob(v_n \leq v).$$

Since bidders are symmetric, $F_1^{(n)}(v)$ is equal to n times the same probability. So we just have to calculate the probability that *a* bidder's valuation is less than v. For the uniform distribution on $[0, 100]$, we have, for any bidder $i = 1, \ldots, n$,

$$Prob(v_i \leq v) = \frac{v}{100}.$$

So we obtain

$$F_1^{(n)}(v) = \underbrace{\frac{v}{100} \times \cdots \times \frac{v}{100}}_{n \text{ times}} = \left(\frac{v}{100}\right)^n. \tag{C.1}$$

C.1.2 Obtaining the Probability Density Function

The probability density function is simply the derivative of the cumulative density function. Denote by $f_1^{(n)}$ the probability density function of the first-order statistics.

So we have

$$f_1^{(n)}(v) = \left(F_1^{(n)}\right)'(v) = n \times \left(\frac{v}{100}\right)^{n-1} \times \frac{1}{100} = \frac{nv^{n-1}}{100^n}.$$

C.1.3 Calculate the Expectation

The expectation of $w_1^{(n)}$, denoted $E(w_1^{(n)})$, is given by applying the following formula:

$$E(w_1^{(n)}) = \int_0^{100} v f_1^{(n)} dv.$$

So we have

$$E(w_1^{(n)}) = \int_0^{100} v f_1^{(n)} dv$$

$$= \int_0^{100} v \frac{nv^{n-1}}{100^n} dv$$

$$= \frac{n}{100^n} \int_0^{100} v^n dv$$

Appendix C: Order Statistics

$$= \frac{n}{100^n} \int_0^{100} v^n dv$$

$$= \frac{n}{100^n} \left[\frac{1}{n+1} v^{n+1} \right]_0^{100}$$

$$= \frac{n}{100^n} \left(\frac{100^{n+1}}{n+1} \right) = 100 \times \frac{n}{n+1}. \tag{C.2}$$

C.2 Expected Second-Highest Valuation

We now want to have an expression for the expectation of the second-highest valuation; that is, we want to compute $E(w_2^{(n)})$. We will follow the same strategy as the one we used to calculate $E(w_1^{(n)})$: first calculate the cumulative density function, then the probability density function, and finally the expectation.

Let $F_2^{(n)}$ be the cumulative density function of the second-order statistic:

$$F_2^{(n)}(v) = Prob(w_2^{(n)} \leq v). \tag{C.3}$$

If $w_2^{(n)} \leq v$, there are two cases.

Case 1 All the numbers $v_1, v_2, \ldots, v_n$ are less than or equal to v.
In this case, we have $w_1^{(n)} \leq v$, and we know from section C.1.1 that

$$F_1^{(n)}(v) = Prob(w_1^{(n)} \leq v) = \left(\frac{v}{100} \right)^n.$$

Case 2 $n-1$ numbers are less than or equal to v, and one number is greater than v.
For this case, there are n possibilities:
- $v_1 > v$ and $v_h \leq v$ for $h = 2, \ldots, n$.
- $v_2 > v$ and $v_h \leq v$ for $h = 1, 3, \ldots, n$.
- $\ldots$
- $v_n > v$ and $v_h \leq v$ for $h = 1, 2, \ldots, n-1$.

Take the first of these cases, $v_1 > v$ and $v_h \leq v$ for $h = 2, \ldots, n$. We have

$Prob(v_1 > v \text{ and } v_h \leq v \text{ for } h = 2, \ldots n)$

$= Prob(v_1 > v) \times Prob(v_h \leq v \text{ for } h = 2, \ldots, n)$

$= Prob(v_1 > v) \times Prob(v_2 \leq v) \times Prob(v_3 \leq v) \times \cdots \times Prob(v_n \leq v)$

$= \dfrac{100-v}{100} \times \underbrace{\dfrac{v}{100} \times \cdots \times \dfrac{v}{100}}_{n-1 \text{ times}}$

$= \dfrac{100-v}{100} \times \left(\dfrac{v}{100} \right)^{n-1}.$

If we take the second case, $v_2 > v$ and $v_h \leq v$ for $h = 1, 3, \ldots, n$, then we will end up with the same probability. So we have n times the same probability, and thus the probability of

case 2 is
$$n \times \frac{100-v}{100} \times \left(\frac{v}{100}\right)^{n-1}.$$

Now, observe that the events described in cases 1 and 2 are *disjoint* (i.e., if one of the cases occurs, then the other case cannot occur), so $Prob(w_2^{(n)} \leq v)$ is the sum of the probabilities of cases 1 and 2,

$$F_2^{(n)}(v) = Prob(w_2^{(n)} \leq v) = \left(\frac{v}{100}\right)^n + n \times \frac{100-v}{100} \times \left(\frac{v}{100}\right)^{n-1}$$
$$= n \left(\frac{v}{100}\right)^{n-1} - (n-1)\left(\frac{v}{100}\right)^n. \tag{C.4}$$

Equation (C.1) says that $F_1^{(n)} = \left(\frac{v}{100}\right)^n$. So $F_1^{(n-1)} = \left(\frac{v}{100}\right)^{n-1}$, and thus we can rewrite equation (C.4) as

$$F_2^{(n)}(v) = nF_1^{(n-1)}(v) - (n-1)F_1^{(n)}(v). \tag{C.5}$$

Taking the derivative with respect to v gives

$$f_2^{(n)}(v) = nf_1^{(n-1)}(v) - (n-1)f_1^{(n)}(v). \tag{C.6}$$

So we have

$$E(w_2^{(n)}) = \int_0^{100} v f_2^{(n)}(v) dv$$
$$= \int_0^{100} v \left(nf_1^{(n-1)}(v) - (n-1)f_1^{(n)}(v)\right) dv$$
$$= n \int_0^{100} v f_1^{(n-1)}(v) dv - (n-1) \int_0^{100} v f_1^{(n)}(v) dv$$
$$= nE\left(w_1^{(n-1)}\right) - (n-1)E\left(w_1^{(n)}\right).$$

It now suffices to apply equation (C.2), and we get

$$E(w_2^{(n)}) = n \times \left(100 \times \frac{n-1}{n}\right) - (n-1) \times 100 \times \frac{n}{n+1}$$
$$= 100 \times (n-1) \times \left(1 - \frac{n}{n+1}\right)$$
$$= 100 \times \frac{n-1}{n+1}. \tag{C.7}$$

C.3 Conditional Expectation of the Highest Valuation

Let v_i be any value among the n random values. We want to obtain the expectation of the second-highest value conditional on v_i being the highest value, which can be written as

$$E(\max_{j \neq i} v_j \mid v_j \leq v \; \forall j \neq i). \tag{C.8}$$

Appendix C: Order Statistics

The expression "conditional on v_i being the highest value" simply means that all the other valuations are, with certainty, below v_i. If all the n values are randomly distributed between 0 and 100, then if we happen to know that $n-1$ of them are below some number v_i, those $n-1$ are also randomly distributed (but between 0 and v_i).

We want the second-highest value given that v_i is the highest. So we want the highest value among $n-1$ numbers, all of them randomly distributed between 0 and v_i.

So equation (C.8) can be rewritten as follows:

$$E(w_1^{(n-1)} | w_1^{(n-1)} \leq v). \tag{C.9}$$

Therefore, we want the first-order statistic among $n-1$ variables, but unlike in section C.1, we want them between 0 and v_i (instead of between 0 and 100). The expectation of the second-highest value conditional on v_i being the highest value is then the expectation of the first-order statistic among $n-1$ values uniformly distributed on $[0, v_i]$. We can thus use the formula in equation (C.2), but replacing 100 by v_i and n by $n-1$, and we obtain

$$E(\max_{j \neq i} v_j \mid v_j \leq v \, \forall j \neq i) = E\left(w_1^{(n-1)} \mid w_1^{(n-1)} \leq v\right) = v_i \times \frac{n-1}{n}. \tag{C.10}$$

C.4 Changing the Upper and Lower Bounds

It is not difficult to modify assumption 1 and assume instead that the variables $v_1, \ldots, v_n$ are distributed uniformly on $[\underline{v}, \overline{v}]$ (i.e., in the previous sections, we had $\underline{v} = 0$ and $\overline{v} = 100$). In this (slightly) more general case, the cumulative distribution $F(v)$ is

$$F(v) = \frac{v - \underline{v}}{\overline{v} - \underline{v}},$$

and thus we have the probability density function $f(v)$ given by

$$f(v) = \frac{1}{\overline{v} - \underline{v}}.$$

The order statistics can be easily derived if we use the following property of the expectation. For each $h = 1, \ldots, n$, define the variable $\hat{v}_h = v_h - \underline{v}$. So, the variables $\hat{v}_1, \ldots, \hat{v}_n$ are distributed uniformly on $[0, \overline{v} - \underline{v}]$.

Let $\hat{w}_1^{(n)}, \hat{w}_2^{(n)} \ldots, \hat{w}_n^{(n)}$ be the order statistics of the variables $\hat{v}_1, \ldots, \hat{v}_n$. Since $v_h = \hat{v}_h + \underline{v}$ for any $h = 1, \ldots, n$, we have $w_h^{(n)} = \hat{w}_h^{(n)} + \underline{v}$. Therefore, for any $h = 1, \ldots, n$,

$$E\left(w_h^{(n)}\right) = E\left(\hat{w}_h^{(n)} + \underline{v}\right) = \underline{v} + E\left(\hat{w}_h^{(n)}\right).$$

Now, it is not difficult to see that the number 100, which appeared in all the calculations in sections C.1, C.2, and C.3, was in fact the *length* of the interval from which the variables $v_1, \ldots, v_n$ were taken. Here the length is $\overline{v} - \underline{v}$, so we have for this more general case:

$$E\left(w_1^{(n)}\right) = \underline{v} + \frac{n}{n+1}(\overline{v} - \underline{v}), \tag{C.11}$$

$$E\left(w_2^{(n)}\right) = \underline{v} + \frac{n-1}{n+1}(\overline{v} - \underline{v}), \tag{C.12}$$

$$E\left(w_1^{(n-1)} \mid w_1^{(n-1)} \leq v\right) = \underline{v} + (v_i - \underline{v})\frac{n-1}{n}. \tag{C.13}$$

Notes

Chapter 1

1. Canice Prendergast, "The Allocation of Food to Food Banks" (unpublished manuscript, 2016). See also, "The Pickle Problem," *Planet Money*, NPR, Episode 665, November 25, 2015, podcast, http://www.npr.org/sections/money/2015/11/25/457408717/episode-665-the-pickle-problem, and, "Canice Prendergast on How Prices Can Improve a Food Fight (and Help the Poor)," *EconTalk*, December 7, 2015, podcast, http://www.econtalk.org/archives/2015/12/canice_prenderg.html.

2. Feeding America's former name was America's Second Harvest.

3. Except produce, but this exception was only with respect to the "penalty" imposed on a food bank should it refuse some food.

Chapter 2

1. For money, the "price" is an interest rate.

2. This is the method used, for instance, by The Kellogg School of Management at Northwestern University, the University of Michigan Business School, the Columbia Business School, the Haas School of Business at the University of California-Berkeley, and the Yale School of Management.

3. If a bidder's valuation also depends on the unknown valuations of the other bidders, then the bidder cannot know her valuation.

4. One could argue that different companies may not have exactly the same operating costs (e.g., they use different extraction technologies). In this case, we would be back to the case of interdependent values.

5. This may not always be the case, as we can have a bidder who cares about the identity of the winner if she loses. For instance, we can have Alice, Bob, and Carol participating in an auction for, say, a bike. Alice would like to have the bike, but if Bob wins the auction, she knows that Bob will let her use it, while Carol will not. In that case, Alice prefers to see Bob win (if she loses the auction).

6. If the last bidders all stop bidding at the same price, then we can flip a coin to determine the winner. For instance, suppose that at $100 there are four bidders, so the auction does not stop, and the price continues to increase, say, to $101. If at that price all bidders stop, then to determine the winner we go back to the last price at which there were still active bidders, and the winner is picked randomly from among these active bidders.

7. William Vickrey, "Counterspeculation, Auctions, and Competitive Sealed Tenders," *Journal of Finance* 16, no. 1 (1961): 8–37.

8. We consider here the standard case where bidders do know their valuations.

9. This is the same method we use, for instance, when maximizing a firm's profit in microeconomics.

10. To be rigorous, we should also check the *second-order condition*: that the solution found in equation (2.4) corresponds to a maximum and not a minimum. This is done by showing that the second derivative is negative, meaning the derivative of equation (2.3) with respect to b.

11. In the English auction, I can decide to continue to bid above $50 if Mr. Smith stopped bidding at $40, and if Mr. Smith stopped bidding at $45, then I stop bidding at $57. The result that truthful bidding is a dominant strategy in the English auction implies that such complicated strategies are irrelevant. However, whenever we perform a formal and rigorous analysis, we cannot ignore them.

12. The expectation of the highest bid is the expectation of "the expectation of the second-highest valuation conditional on having the highest valuation." That is, it is the expectation of a conditional expectation. When applying the expectation operator a second time, the conditional term disappears and we get the expectation of the second-highest valuation.

13. For instance, in the English auction, we know that each bidder should bid her valuation. What we are supposing now is that, if the auction is the English auction, bidder i does not bid v_i but $\tilde{v}$. If we consider instead the first-price auction and use example 2.6, we would have Alice bidding something different from 54.

14. Integrating a function f consists of finding a function g such that the derivative of g is the function f itself.

15. $b_1^{(n)}$ and $b_2^{(n)}$ are the first- and second-order statistics. See appendix C.

16. Roger B. Myerson, "Optimal Auction Design," *Mathematics of Operations Research* 6, no. 1 (1981): 58–73; John Riley and William Samuelson, "Optimal Auctions," *American Economic Review* 71, no. 3 (1981): 381–392.

17. The Greek letter ψ is called "psi" (it does not have an equivalent letter in the Latin alphabet).

18. Jeremy Bulow and Paul Klemperer, "Auctions versus Negotiations," *American Economic Review* 86, no. 1 (1996): 180–194.

19. René Kirkegaard, "A Short Proof of the Bulow-Klemperer Auctions vs. Negotiations Result," *Economic Theory* 28, no. 2 (2006): 449–452.

20. Like Bulow and Klemperer, we assume that all bidders' valuations are nonnegative.

Chapter 3

1. If Carol is not the second-highest bidder, then there is a third bid that is higher than $21, say, for instance, $25. But in this case the minimum bid would be $25 + $0.50 = $25.50 and not $20.50, so Carol is necessarily the second-highest bidder.

2. Alvin Roth and Axel Ockenfels, "Last-Minute Bidding and the Rules for Ending Second-Price Auctions: Evidence from eBay and Amazon Auctions on the Internet," *American Economic Review, Papers and Proceedings* 92, no. 4 (2002): 1093–1103.

3. This does not mean that all buyers have the same valuation for a given laptop. Buyers may have different needs and thus be willing to spend different amounts of money for the same laptop.

Chapter 4

1. Edward H. Clarke, "Multipart Pricing of Public Goods," *Public Choice* 11, no. 1 (1971): 17–33; Theodore Groves, "Incentives in Teams," *Econometrica* 41 (1973): 617–631.

2. We subtract 1 because the 1024 different combinations includes the case where the bidder does not get any item.

Chapter 5

1. The most expensive keywords in the United States are usually "insurance," "loans," "mortgage," "attorney," or "credit." The price per click for these keywords easily reaches $40 or even $50.

2. This is called "pay per print."

3. Search engines allow bidding made by software (like a trading algorithm), but the software used to bid must first be approved by the search engine.

4. Search engines like Google or Bing have several areas for sponsored links. For this chapter, we only consider the links that are on the main part of the page, just above the result of the search query as in figure 5.1.

5. Benjamin Edelman, Michael Ostrovsky, and Michael Schwarz, "Internet Advertising and the Generalized Second-Price Auction: Selling Billions of Dollars Worth of Keywords," *American Economic Review* 97, no. 1 (2007): 242–259.

6. Recall that a game is simultaneous if all players choose their strategies at the same time, and it is a one-shot game if each player only plays once.

7. We could also say that the fourth bidder gets the fourth position, but the clickthrough rate for that position is $\alpha_4 = 0$.

8. Those ad spaces work (roughly) as follows. A website dedicates some space on its page to display ads, but instead of transacting directly with the advertisers, the website asks Google to fill those spaces with ads. Technically, when a user loads a page, there is a small script that "fetches" ads from Google and puts them on the page the user is visiting.

Chapter 6

1. The winner is the bidder with the highest valuation as long as the auction is designed such that the winner is the bidder with the highest bid and bids are increasing with respect to the valuation.

2. In contrast, for small or medium-sized countries such as European countries, it is reasonable to consider the sale of nationwide licenses.

3. The fifth-largest operator in the United States is U.S. Cellular, which has a license for only 23 states.

4. Peter Cramton, "Simultaneous Ascending Auctions," in *Combinatorial Auctions,* ed. Peter Cramton, Yoav Shoham, and Richard Steinberg (Cambridge, MA: MIT Press, 2006).

5. See his excellent and detailed book: Paul Klemperer, *Auctions: Theory and Practice*, The Toulouse Lectures in Economics (Princeton, NJ: Princeton University Press, 2004).

Chapter 7

1. If the security is a bill, then the interest rate corresponds to the discount, and if it is a note or a bond (called "coupon-bearing securities"), it is the yield.

2. In some markets, it is prohibited for sellers to sell a higher quantity than they have. In other markets, it is permitted (under some condition). If a seller sells more units than she currently possesses, we say that the seller is "short-selling."

Chapter 8

1. See, for instance, NSADAQ's OUCH specifications for orders, which specifies all the data any order should contains, at http://www.nasdaqtrader.com/content/technicalsupport/specifications/tradingproducts/ouch4.2.pdf.

2. Take figure 7.1. If there is no point where demand and supply meet, the demand curve must always be below the supply curve. But if one buyer submits a market order, then there is a point (for the quantity demanded by that buyer) where the demand curve goes all the way up to an infinite price. In that case, the demand must necessarily cross the supply.

3. The official exchanges are often referred to as the "lit market."

4. None of the "New York exchanges" are located in New York City. NASDAQ is located in Carteret, NJ, and Bats and IEX are located in Weehawken, NJ.

5. Eric Budish, Peter Cramton, and John, "The High-Frequency Trading Arms Race: Frequent Batch Auctions as a Market Design Response," *Quarterly Journal of Economics* 130, no. 4 (2015): 1547–1621.

6. SPY is one of the most traded securities.

7. ES and SPY are not *exactly* identical in the way they are traded. For instance, SPY has a tick size of 0.10 index points, while ES has a tick size of 0.25 points. Also, in figures 8.1 and 8.2, the price of SPY has been multiplied by 10 to reflect that it tracks 10% of the S&P 500 index.

8. A typical processor needs *much less* than 1 microsecond to process one instruction.

Chapter 9

1. The original matching model we study in this chapter was introduced in David Gale and Lloyd S. Shapley, "College Admissions and the Stability of Marriage," *American Mathematical Monthly* 69, no. 1 (1962): 9–15. In their article, the authors called the two sets *men* and *women*, and named their model the *marriage problem*.

2. Many people in the literature simply refer to this problem as a one-to-one matching problem, because the expression "one-to-one" implicitly refers to a two-sided matching situation.

3. We could well have monetary transactions. For instance, the singer earns some money and has to pay some royalties to the musician (like a firm paying its employees). Such a case is perfectly compatible with our model as long as for each musician-singer pair there is only one possible contract. The opposite case (and thus not compatible with our matching model) is when singers and musicians must also negotiate the terms of the contract.

4. This contrasts, for instance, with the consumer's model in microeconomics, where the preferences over bundles of goods are not enough to determine the consumer's demand (or choice); we also need to know the budget and the prices of each good.

5. As we explained, Arthur only cares about which singer he is matched with, and Alice only cares about which musician she is matched with. Whether $\mu'(\text{Bob}) = \text{Barbara}$ or $\mu'(\text{Bob}) = \text{Bob}$, and thus $\mu'(\text{Barbara}) = \text{Barbara}$ makes no difference at all for Alice and Arthur.

6. It is a real recipe; you can try it!

7. Alvin E. Roth, "The Economics of Matching: Stability and Incentives," *Mathematics of Operations Research* 7, no. 4 (1982): 617–628; Lester E. Dubins and David A. Freedman, "Machiavelli and the Gale-Shapley Algorithm," *American Mathematical Monthly* 88, no. 7 (1981): 485–494.

8. Fuhito Kojima, and Parag Pathak, "Incentives and Stability in Large Two-Sided Matching Markets," *American Economic Review* 99, no. 3 (2009): 608–627.

Chapter 10

1. The NRMP is a U.S.-based, private, nonprofit organization sponsored by the American Board of Medical Specialties, the American Hospital Association, the American Medical Association, the Association of American Medical Colleges, and the Council of Medical Specialty Societies. In its early years, the matching procedure was for interns only, and thus was called the National Intern Matching Program (a medical school graduate undergoing on-the-job training is called a *resident*, and an *intern* is a first-year resident).

2. Alvin E. Roth, "The Evolution of the Labor Market for Medical Interns and Residents: A Case Study in Game Theory," *Journal of Political Economy* 92, no. 6 (1984): 991–1016.

3. Observe that since hospitals have preferences among *groups* of doctors, the definition of blocking may seem incomplete; that is, a hospital may want to block with several doctors at the same time. It turns out that under responsive preferences nothing is lost by considering only simple doctor-hospital pairs.

4. Alvin E. Roth, "A Natural Experiment in the Organization of Entry-Level Labor Markets: Regional Markets for New Physicians and Surgeons in the United Kingdom," *American Economic Review* 81, no. 3 (1991): 415–440.

5. In his 1991 paper, Roth documents the longevity of the London Hospital and Cambridge procedures and attributes this to their being relatively reduced markets, where some kind of social pressure gives little incentive for students to circumvent the matching procedure.

6. Jhon Kagel and Alvin E. Roth, "The Dynamics of Reorganization in Matching Markets: A Laboratory Experiment Motivated by a Natural Experiment," *Quarterly Journal of Economics* 115, no. 1 (2000): 201–235.

7. Ties are broken in favor of workers. Consider, for instance, two workers, Alice and Barbara, and one firm, John. Alice ranks John first, while Barbara ranks John second. John ranks Alice second and Barbara first. We can see that both pairs Alice-John and Barbara-John have a priority product equal to 2. Giving priority to workers means that the pair Alice-John comes first, and thus it is Alice who ends up being matched with John.

8. The first part of result 10.1 is from Roth (1984) and from David G. McVitie and Leslie B. Wilson, "Stable Marriage Assignments for Unequal Sets," *BIT* 10, no. 3 (1970): 295–309. The second part of the result is from Alvin E. Roth, "On the Allocation of Residents to Rural Hospitals: A General Property of Two-Sided Matching Markets," *Econometrica* 54, no. 2 (1986): 425–427.

9. The proof when hospitals can hire more than one doctor is slightly more involved (albeit not too much), but it goes beyond the scope of this book.

10. We present here a simple version of the lemma. The full version can be found in a paper by David Gale and Marilda Sotomayor, "Some Remarks on the Stable Matching Problem," *Discrete Applied Mathematics* 11, no. 3 (1985): 223–232.

11. For instance, even if stable matchings exist, it may not be possible to find an algorithm such that it is a dominant strategy (for one side of the market) to reveal one's true preferences. Other results, such as the rural hospital theorem or the optimality of some stable matchings (see result 9.2), are no longer guaranteed.

12. Alvin E. Roth and Elliott Peranson, "The Redesign of the Matching Market for American Physicians: Some Engineering Aspects of Economic Design," *American Economic Review* 89, no. 4 (1999): 748–780.

Chapter 11

1. Jinpeng Ma, "Strategy-Proofness and the Strict Core in a Market with Indivisibilities," *International Journal of Game Theory* 23, no. 1 (1994): 75–83.

2. Alvin E. Roth and Andrew Postlewaite, "Weak versus Strong Domination in a Market with Indivisible Goods," *Journal of Mathematical Economics* 4, no. 2 (1977): 131–137.

3. Atila Abdulkadiroğlu and Tayfun Sönmez, " House Allocation with Existing Tenants," *Journal of Economic Theory* 88, no. 2 (1999): 233–260.

4. The assignment at NH4 consists of allocating rooms, not houses. To streamline the section, we will keep talking about houses.

Chapter 12

1. Atila Abdulkadiroğlu and Tayfun Sönmez, "Random Serial Dictatorship and the Core from Random Endowments in House Allocation Problems," *Econometrica* 66, no. 3 (1999): 689–701.

2. Anna Bogomolnaia and Hervé Moulin, " A New Solution to the Random Assignment Problem," *Journal of Economic Theory* 100, no. 2 (2001): 295–328.

Chapter 13

1. Separability means that the "value" of a group is *exactly* the sum of the values of the individuals in a group. For instance, the "value" of two students, say Alice and Bob, is under separability the "value" of Alice + the "value" of Bob. So the value of Bob for a group does not depend on whether Alice is already part of that group.

2. For instance, if there are three students, i_1, i_2, and i_3 (i.e., $I = \{i_1, i_2, i_3\}$), then 2^I is the collection of all sets of students $\{i_1\}$, $\{i_2\}$, $\{i_3\}$, $\{i_1, i_2\}$, $\{i_1, i_3\}$, $\{i_2, i_3\}$, $\{i_1, i_2, i_3\}$ and the empty set, $\emptyset$. The function μ takes any element in the big set made of the set of students *and* the set of schools ($I \cup S$) and associates it to an element of 2^I or S.

3. Haluk I. Ergin, "Efficient Resource Allocation on the Basis of Priorities," *Econometrica* 70, no. 6 (2002): 2489–2497.

4. Onur Kesten, "School Choice with Consent," *Quarterly Journal of Economics* 125, no. 2 (2010): 1297–1348.

5. Atila Abdulkadiroğlu, and Tayfun Sönmez, "School Choice: A Mechanism Design Approach," *American Economic Review* 93, no. 3 (2003): 729–747.

6. Abdulkadiroğlu and Sönmez initially called this algorithm the "Boston algorithm," but they note that it was widely used in many different cities; however, many authors still use the term "Boston algorithm."

7. It should be clear, however, that we *can* describe a Deferred Acceptance or Immediate Acceptance algorithm in a school choice model where schools are on the proposing side. Such algorithms would be well defined. Because of this, we could well use such algorithms in real-life situations. This is the case, for instance; in France, where it is used to assign students to high schools. What we are saying here is that, in a context of school choice, it is difficult to justify the use of an algorithm where schools are the ones making the proposals.

8. At that time, both were faculty members at the Harris School of Public Policy of the University of Chicago.

9. http://archive.boston.com/news/local/articles/2003/09/12/school_assignment_flaws_detailed/.

10. Atila Abdulkadiroğlu, Parag Pathak, Alvin E. Roth, and Tayfun Sönmez, "The Boston Public Schools Match," *American Economic Review, Papers and Proceedings* 95, no. 2 (2005): 368–371.

11. Atila Abdulkadiroğlu, Parag Pathak, and Alvin E. Roth, "The New York City High School Match," *American Economic Review, Papers and Proceedings* 95, no. 2 (2005): 364–367.

Chapter 14

1. Aytek Erdil and Haluk Ergin, "What's the Matter with Tie-Breaking? Improving Efficiency in School Choice," *American Economic Review* 98, no. 3 (2008): 669–689.

2. We can also use instead the highest random number. All that matters is that the strict ordering is random.

3. Atila Abdulkadiroğlu, Parag Pathak, and Alvin E. Roth, "Strategy-proofness versus Efficiency in Matching with Indifferences: Redesigning the NYC High School Match," *American Economic Review* 99, no. 5 (2009): 1954–1978.

4. This method of comparing two columns of numbers using a sum follows the same principle that we used to define stochastic dominance in section 12.1.3.

5. Guillaume Haeringer and Flip Klijn, "Constrained School Choice," *Journal of Economic Theory* 144, no. 5 (2009): 1921–1947.

6. Caterina Calsamiglia, Guillaume Haeringer, and Flip Klijn, "Constrained School Choice: An Experimental Study," *American Economic Review* 100, no. 4 (2010): 1860–1874.

7. The experiment we ran is similar to an earlier experiment made in Yan Chen and Tayfun Sönmez, "School Choice: An Experimental Study," *Journal of Economic Theory* 127, no. 1 (2006): 202–231.

8. Rosalind Rossi, "8th-Graders' Shot at Elite High Schools Better," *Chicago Sun-Times*, November 12, 2009.

Chapter 15

1. Szilvia Pápai, "Strategyproof and Nonbossy Multiple Assignments," *Journal of Public Economic Theory* 3, no. 2 (2001): 257–271.

2. Tayfun Sönmez and M. Utku Ünver, "Course Bidding at Business Schools," *International Economic Review* 51, no. 1 (2010): 99–123.

3. Eric Budish and Estelle Cantillon, "The Multi-unit Assignment Problem: Theory and Evidence from Course Allocation at Harvard," *American Economic Review* 102, no. 5 (2012): 2237–2271.

4. Eric Budish, "The Combinatorial Assignment Problem: Approximate Competitive Equilibrium from Equal Incomes," *Journal of Political Economy* 119, no. 6 (2011): 1061–1103.

5. Eric Budish and Judd Kessler, "Bringing Real Market Participants' Real Preferences into the Lab: An Experiment That Changed the Course Allocation Mechanism at Wharton" (unpublished manuscript, 2016).

6. The courses listed consisted of a course name with a section name (i.e., the day of the week and the time that course is taught). So, for instance, course C could be Marketing on Wednesday at 10 a.m. and course D Marketing on Tuesday at 6 p.m.

7. Budish and Kessler give as an example the case of a star professor whose course all students want to take. The students not enrolled in her course will envy those who are.

Chapter 16

1. The Islamic Republic of Iran is the only country in the world where organs can be sold and bought legally (under some conditions).

2. See Alvin E. Roth, Tayfun Sönmez, and M. Utku Ünver, "Kidney Exchange," *Quarterly Journal of Economics* 119, no. 2 (2004): 457–488.

3. Kidney transplant is the most common type of transplant. About 80%–85% of patient waiting for a lifesaving organ transplant are waiting for a kidney.

4. Alvin E. Roth, Tayfun Sönmez, and M. Utku Ünver, "Efficient Kidney Exchange: Coincidence of Wants in Markets with Compatibility-Based Preferences," *American Economic Review* 97, no. 3 (2004): 828–851.

5. Note that if only blood type compatibility matters, then we should not observe in a kidney exchange pool a pair with a donor of type O. The kidney is acceptable for the patient independently of the patient's blood type (A, B, AB, or O).

Appendix A

1. John F. Nash, Jr., "Equilibrium Points in n-Person Games," *Proceedings of the National Academy of Sciences of the United States of America* 36 (1950): 48–49.

Appendix B

1. The letter θ is from the Greek alphabet (pronounced "theta"). The letter Θ is the uppercase version of θ.

2. The Greek letter γ is called "gamma" and corresponds to the Latin letter "g."

3. There are various versions of the Revelation Principle. The version presented in result B.1 was introduced in Allan Gibbard, "Manipulation of Voting Schemes: A General Result," *Econometrica* 41 (1973): 587–601.

4. The result was proved independently by Allan Gibbard and Mark Satterthwaite. See the Gibbard article cited in the preceding note and Mark Allen Satterthwaite, "Strategy-proofness and Arrow's Conditions: Existence and Correspondence Theorems for Voting Procedures and Social Welfare Functions," *Journal of Economic Theory* 10, no. 2 (1975): 187–217.

5. A classic illustration is the case of political preferences across the spectrum from left to right when agents' preferences are assumed to be *single peaked*. Each agent has an ideal, most preferred political platform, and the "further away" from the ideal a platform is, the less preferred it is. For instance, someone whose most preferred political platform is on the left will prefer a central/neutral platform to a platform further to the right on the political spectrum.

6. The Revelation Principle for Bayesian equilibrium was introduced in Partha Dasgupta, P. Hammond, and Eric Maskin, "The Implementaton of Social Choice Rules: Some General Results on Incentive Compatibility," *Review of Economic Studies* 46, no. 2 (1979): 185–216; and Bengt Holmstrom, "On Incentives and Control in Organizations" (PhD diss., Stanford University, 1977).

7. Roger B. Myerson and Mark A. Sattherthwaite, "Efficient Mechanisms for Bilateral Trading," *Journal of Economic Theory* 29, no. 2 (1983): 265–281.

References

Abdulkadiroğlu, Atila, Parag Pathak, and Alvin E. Roth. "The New York City High School Match." *American Economic Review, Papers and Proceedings* 95, no. 2 (2005): 364–367.

Abdulkadiroğlu, Atila, Parag Pathak, and Alvin E. Roth. "Strategy-proofness versus Efficiency in Matching with Indifferences: Redesigning the NYC High School Match." *American Economic Review* 99, no. 5 (2009): 1954–1978.

Abdulkadiroğlu, Atila, Parag Pathak, Alvin E. Roth, and Tayfun Sönmez. "The Boston Public Schools Match." *American Economic Review, Papers and Proceedings* 95, no. 2 (2005): 368–371.

Abdulkadiroğlu, Atila, and Tayfun Sönmez. "Random Serial Dictatorship and the Core from Random Endowments in House Allocation Problems." *Econometrica* 66, no. 3 (1998): 689–701. doi:10.2307/2998580. http://dx.doi.org/10.2307/2998580.

Abdulkadiroğlu, Atila, and Tayfun Sönmez. "House Allocation with Existing Tenants." *Journal of Economic Theory* 88, no. 2 (1999): 233–260.

Abdulkadiroğlu, Atila, and Tayfun Sönmez. "School Choice: A Mechanism Design Approach." *American Economic Review* 93, no. 3 (2003): 729–747.

Bogomolnaia, Anna, and Hervé Moulin. "A New Solution to the Random Assignment Problem." *Journal of Economic Theory* 100, no. 2 (2001): 295–328. doi:10.1006/jeth.2000.2710. http://dx.doi.org/10.1006/jeth.2000.2710.

Budish, Eric. "The Combinatorial Assignment Problem: Approximate Competitive Equilibrium from Equal Incomes." *Journal of Political Economy* 119, no. 6 (2011): 1061–1103.

Budish, Eric, and Estelle Cantillon. "The Multi-unit Assignment Problem: Theory and Evidence from Course Allocation at Harvard." *American Economic Review* 102, no. 5 (2012): 2237–2271.

Budish, Eric, Peter Cramton, and John Shim. "The High-frequency Trading Arms Race: Frequent Batch Auctions as a Market Design Response." *Quarterly Journal of Economics* 130, no. 4 (2015): 1547–1621.

Budish, Eric, and Judd Kessler. "Bringing Real Market Participants' Real Preferences into the Lab: An Experiment That Changed the Course Allocation Mechanism at Wharton." Unpublished manuscript, 2016.

Bulow, Jeremy, and Paul Klemperer. "Auctions versus Negotiations." *American Economic Review* 86, no. 1 (1996): 180–194.

Calsamiglia, Caterina, Guillaume Haeringer, and Flip Klijn. "Constrained School Choice: An Experimental Study." *American Economic Review* 100, no. 4 (2010): 1860–1874.

Chen, Yan, and Tayfun Sönmez. "School Choice: An Experimental Study." *Journal of Economic Theory* 127, no. 1 (2006): 202–231. doi:10.1016/j.jet.2004.10.006. http://dx.doi.org/10.1016/j.jet.2004.10.006.

Clarke, Edward H. "Multipart Pricing of Public Goods." *Public Choice* 11, no. 1 (1971): 17–33.

Cramton, Peter. "Simultaneous Ascending Auctions." In *Combinatorial Auctions*, edited by Y. Shoham, P. Cramton, and R. Steinberg. Cambridge, MA: MIT Press, 2006.

Dasgupta, Partha, Peter Hammond, and Eric Maskin. "The Implementation of Social Choice Rules: Some General Results on Incentive Compatibility." *Review of Economic Studies* 46, no. 2 (1979): 185–216. doi:10.2307/2297045. http://dx.doi.org/10.2307/2297045.

Dubins, Lester E., and David A. Freedman. "Machiavelli and the Gale-Shapley Algorithm." *American Mathematical Monthly* 88, no. 7 (1981): 485–494. doi:10.2307/2321753. http://dx.doi.org/10.2307/2321753.

EconTalk. "Canice Prendergast on How Prices Can Improve a Food Fight (and Help the Poor)," December 7, 2015, podcast. http://www.econtalk.org/archives/2015/12/canice_prenderg.html.

Edelman, Benjamin, Michael Ostrovsky, and Michael Schwarz. "Internet Advertising and the Generalized Second-Price Auction: Selling Billions of Dollars Worth of Keywords." *American Economic Review* 97, no. 1 (2007): 242–259.

Erdil, Aytek, and Haluk Ergin. "What's the Matter with Tie-Breaking? Improving Efficiency in School Choice." *American Economic Review* 98, no. 3 (2008): 669–689.

Ergin, Haluk I. "Efficient Resource Allocation on the Basis of Priorities." *Econometrica* 70, no. 6 (2002): 2489–2497. doi:10.1111/1468-0262.00383. http://dx.doi.org/10.1111/1468-0262.00383.

Gale, David, and Lloyd S. Shapley. "College Admissions and the Stability of Marriage." *American Mathematical Monthly* 69, no. 1 (1962): 9–15. doi:10.2307/2312726. http://dx.doi.org/10.2307/2312726.

Gale, David, and Marilda Sotomayor. "Some Remarks on the Stable Matching Problem." *Discrete Applied Mathematics* 11, no. 3 (1985): 223–232. doi:10.1016/0166-218X(85)90074-5. http://dx.doi.org/10.1016/0166-218X(85)90074-5.

Gibbard, Allan. "Manipulation of Voting Schemes: A General Result." *Econometrica* 41 (1973): 587–601. doi:10.2307/1914083. http://dx.doi.org/10.2307/1914083.

Groves, Theodore. "Incentives in Teams." *Econometrica* 41 (1973): 617–631. doi:10.2307/1914085. http://dx.doi.org/10.2307/1914085.

Haeringer, Guillaume, and Flip Klijn. "Constrained School Choice." *Journal of Economic Theory* 144, no. 5 (2009): 1921–1947. doi:10.1016/j.jet.2009.05.002. http://dx.doi.org/10.1016/j.jet.2009.05.002.

Holmstrom, Bengt. "On Incentives and Control in Organizations." PhD diss. Stanford University, 1977.

Kagel, John, and Alvin E. Roth. "The Dynamics of Reorganization in Matching Markets: A Laboratory Experiment Motivated by a Natural Experiment." *Quarterly Journal of Economics* 115, no. 1 (2000): 201–235.

Kesten, Onur. "School Choice with Consent." *Quarterly Journal of Economics* 125, no. 2 (2010): 1297–1348.

Kirkegaard, René. "A Short Proof of the Bulow-Klemperer Auctions vs. Negotiations Result." *Economic Theory* 28, 2 (2006): 449–452. doi:10.1007/s00199-004-0593-2. http://dx.doi.org/10.1007/s00199-004-0593-2.

References

Klemperer, Paul. *Auctions: Theory and Practice*, Toulouse Lectures in Economics. Princeton, NJ: Princeton University Press.

Kojima, Fuhito, and Parag Pathak. "Incentives and Stability in Large Two-Sided Matching Markets." *American Economic Review* 99, no. 3 (2009): 608–627.

Ma, Jinpeng. "Strategy-proofness and the Strict Core in a Market with Indivisibilities." *International Journal of Game Theory* 23, no. 1 (1994): 75–83. doi:10.1007/BF01242849. http://dx.doi.org/10.1007/BF01242849.

McVitie, David G., and Leslie B. Wilson. "Stable Marriage Assignment for Unequal Sets." *BIT* 10, no. 3 (1970): 295–309.

Myerson, Roger B. "Optimal Auction Design." *Mathematics of Operations Research* 6, no. 1 (1981): 58–73. doi:10.1287/moor.6.1.58. http://dx.doi.org/10.1287/moor.6.1.58.

Myerson, Roger B., and Mark A. Satterthwaite. "Efficient Mechanisms for Bilateral Trading." *Journal of Economic Theory* 29, no. 2 (1983): 265–281. doi:10.1016/0022-0531(83)90048-0. http://dx.doi.org/10.1016/0022-0531(83)90048-0.

Nash, John F., Jr. "Equilibrium Points in *n*-Person Games." *Proceedings of the National Academy of Sciences of the United States of America* 36 (1950): 48–49.

National Public Radio (NPR). "The Pickle Problem," *Planet Money*, Episode 665, November 25, 2015, podcast. http://www.npr.org/sections/money/2015/11/25/457408717/episode-665-the-pickle-problem.

Pápai, Szilvia. "Strategyproof and Nonbossy Multiple Assignments." *Journal of Public Economic Theory* 3, no. 2 (2001): 257–271.

Prendergast, Canice. "The Allocation of Food to Food Banks." Unpublished manuscript, 2016.

Riley, John, and William Samuelson. "Optimal Auctions." *American Economic Review* 71, no. 3 (1981): 381–392.

Rossi, Rosalind. "8th-Graders' Shot at Elite High Schools Better." Chicago Sun-Times, November 12, 2009.

Roth, Alvin E. "The Economics of Matching: Stability and Incentives." *Mathematics of Operations Research* 7, no. 4 (1982): 617–628. doi:10.1287/moor.7.4.617. http://dx.doi.org/10.1287/moor.7.4.617.

Roth, Alvin E. "The Evolution of the Labor Market for Medical Interns and Residents: A Case Study in Game Theory." *Journal of Political Economy* 92, no. 6 (1984): 991–1016.

Roth, Alvin E. "On the Allocation of Residents to Rural Hospitals: A General Property of Two-Sided Matching Markets." *Econometrica* 54, no. 2 (1986): 425–427. doi:10.2307/1913160. http://dx.doi.org/10.2307/1913160.

Roth, Alvin E. "A Natural Experiment in the Organization of Entry-Level Labor Markets: Regional Markets for New Physicians and Surgeons in the United Kingdom." *American Economic Review* 81, no. 3 (1991): 415–440.

Roth, Alvin, and Axel Ockenfels. "Last-Minute Bidding and the Rules for Ending Second-Price Auctions: Evidence from eBay and Amazon Auctions on the Internet." *American Economic Review, Papers and Proceedings* 92, no. 4 (2002): 1093–1103.

Roth, Alvin E., and Elliott Peranson. "The Redesign of the Matching Market for American Physicians: Some Engineering Aspects of Economic Design." *American Economic Review* 89, no. 4 (1999): 748–780.

Roth, Alvin E., and Andrew Postlewaite. "Weak versus Strong Domination in a Market with Indivisible Goods." *Journal of Mathematical Economics* 4, no. 2 (1977): 131–137. doi:10.1016/0304-4068(77)90004-0. http://dx.doi.org/10.1016/0304-4068(77)90004-0.

Roth, Alvin E., Tayfun Sönmez, and M. Utku Ünver. "Kidney Exchange." *Quarterly Journal of Economics* 119, no. 2 (2005): 457–488.

Roth, Alvin E., Tayfun Sönmez, and M. Utku Ünver. "Efficient Kidney Exchange: Coincidence of Wants in Markets with Compatibility-Based Preferences." *American Economic Review* 97, no. 3 (2007): 828–851.

Satterthwaite, Mark Allen. "Strategy-proofness and Arrow's Conditions: Existence and Correspondence Theorems for Voting Procedures and Social Welfare Functions." *Journal of Economic Theory* 10, no. 2 (1975): 187–217. doi:10.1016/0022-0531(75)90050-2. http://dx.doi.org/10.1016/0022-0531(75)90050-2.

Sönmez, Tayfun, and M. Utku Ünver. "Course Bidding at Business Schools." *International Economic Review* 51, no. 1 (2010): 99–123. doi:10.1111/j.1468-2354.2009.00572.x. http://dx.doi.org/10.1111/j.1468-2354.2009.00572.x.

Vickrey, William. "Counterspeculation, Auctions, and Competitive Sealed Tenders." *Journal of Finance* 16, no. 1 (1961): 8–37.

Index

Abdulkadiroğlu, Atila, 213, 231, 245, 258, 259, 273
acceptable, 151
algorithm, 156
  Deferred Acceptance, 160, 182
  Deferred Acceptance for school choice, 246
  Gale-Shapley Pareto-Dominant Market, 289
  Immediate Acceptance, 248, 256
  MIT–NH4, 216
  Random Serial Dictatorship, 230
  Random Serial Dictatorship with Squatting Rights, 213
  Random Serial Dictatorship with Waiting Lists, 214
  Serial Dictatorship, 200
  Top Trading Cycle, 202
  Top Trading Cycles and Chains, 308
  Top Trading Cycles for school choice, 251
  Top Trading Cycles with mixed endowments, 219
  You Request My House—I Get Your Turn (YRMY–IGYT), 217
Amazon auction, 55
Approximate competitive equilibrium from equal incomes, 296
assignment, 63, 198
  combinatorial, 295
  efficient, 253
  many-to-one, 239
  one-to-one, 198
  optimal, 63
  random, 223
  stable, 81, 84
  student-optimal, 247
assortative assignment, 86
assortative matching, 189

auction, 10, 14
  ascending, 18
  ascending-price English, 18
  California, 50
  combinatorial, 61, 102
  continuous, 128
  double, 119
  Dutch, 34
  English, 19
  equilibrium, 31
  first-price, 11, 29, 31
  frequent batch, 147
  generalized English, 88
  generalized second-price (GSP), 76
  Google, 71
  Japanese, 19
  keywords, 71
  optimal, 43
  payoff, 17
  second-price, 26
  simultaneous ascending, 104
  spectrum, 95
  Treasury, 114
  uniform-price, 118, 147
  Vickrey, 26, 69
  Vickrey-Clarke-Groves (VCG), 61

BATS, 136
Bayesian game, 330
Bayesian incentive compatibility, 344
Bayesian-Nash equilibrium, 332
beauty contest, 96
bid
  generalized second price, 78
  sniping, 54
  truthful, 67, 78
  VCG, 67

bid-ask spread, 128
bidding
  course, 284
  truthful, 23
Birkhoff–von Neumann theorem, 225
block pair, 154
Bogomolnaia, Anna, 232
Boston school match, 258
budget balanced, 344
  weak, 346
Budish, Eric, 139, 290, 296, 298
Bulow, Jeremy, 45

California auction, 50
Calsamiglia, Caterina, 278
Cantillon, Estelle, 290
centralized market, 168
chain, 308
Chen, Yan, 278
Chicago Public Schools, 280
Clarke, Edward H., 63
click frequency, 75
clickthrough rate (CTR), 76
collusion, 99
combinatorial assignment, 295
combinatorial auction, 61, 102
constrained choice, 276
continuous double auction, 128
core, 210
couples, 192
course allocation, 283
  Wharton, 298
course bidding, 284
Cramton, Peter, 105, 139

dark pools, 136
Dasgupta, Partha, 345
decentralized market, 168
Decomposition Lemma, 191
Deferred Acceptance algorithm, 160, 182
Deferred Acceptance algorithm for school choice, 246
Delmonico, Francis, 309
demand reduction, 100
dominance solvability, 327
dominant strategy, 326
  implementation, 341
dominated strategy, 325
  elimination, 326
double auction, 119
  equilibrium, 121
Dubins, Lester E., 359
Dutch auction, 34

eBay, 49
Edelman, Benjamin, 79
efficient
  assignment, 199, 245, 253, 264
  auction, 18
  ex-ante, 231, 237
  ex-post, 231
  many-to-one assignment, 242
  one-to-one assignment, 199
  ordinally, 232
  school choice, 264
electronic communication
    networks, 136
endowment
  mixed public-private, 212
  private, 198, 212, 306
  public, 198, 212
endowment effect, 55
entry, 101
envy-free equilibrium, 79
equal treatment of equals, 236
equilibrium, 1
  auction, 37
  Bayesian-Nash, 332
  double auction, 121
  first-price auction, 31
  locally envy-free, 81
Erdil, Aytek, 270
Ergin, Haluk, 245, 270
ex-ante efficient, 231, 237
ex-post efficient, 231
exposure problem, 102
extensive form game, 323
extensive form game of perfect
    information, 320

Facebook, 91
Feeding America, 6
first-price auction, 29
  equilibrium, 31
Freedman, David A., 359
frequent batch auction, 147

Gale, David, 149
Gale-Shapley Pareto-Dominant Market
    algorithm, 289
game, 317
  Bayesian, 330
  extensive form, 320, 323
  incomplete information, 330
  normal form, 317
  strategic form, 317
generalized English auction, 88

Index

generalized second-price (GSP) auction, 76
  equilibrium, 78
  truthtelling, 78
Gibbard, Allan, 341, 343
Gibbard-Satterthwaite theorem, 343
Google, 71, 124
Google auction, 71
group strategyproof, 206
Groves, Theodore, 63

Haeringer, Guillaume, 277
Hammon, Peter, 345
Harvard draft mechanism, 290
high-frequency trading, 135
Holmstrom, Bengt, 345
Huberman, Ron, 280

Immediate Acceptance algorithm, 248, 256
imperfect information, 322
improvement cycle, 268
incentive compatibility
  Bayesian, 344
  dominant strategy, 341
incomplete information game, 330
individual rationality, 210
individually rational matching, 154
Initial Public Offering (IPO), 123
Investor Exchange (IEX), 145

justified envy, 242

Kagel, John, 187
Kessler, Judd, 298
Kesten, Onur, 245
keywords auction, 71
kidney exchange, 303
Kirkegaard, René, 45
Klemperer, Paul, 45, 111
Klijn, Flip, 277
Kojima, Fuhito, 170

Lewis, Michael, 145
limit order, 128
limit order book, 128
locally envy-free, 81
locally envy-free equilibrium, 81
lottery, 95

Ma, Jinpeng, 207
many-to-one assignment, 239
many-to-one matching, 175
market order, 119, 128

Maskin, Eric, 345
matching, 152
  couples, 192
  individually rational, 154
  many-to-one, 175
  mechanism, 159
  nonwasteful, 180
  one-to-one, 149
  optimal, 164
  stable, 153
McAfee, Preston, 105
mechanism, 335, 338
  Bayesian, 344
  Vickrey-Clarke-Groves, 343
  McVitie, David G., 190
message space, 337
Milgrom, Paul, 105
MIT–NH4 algorithm, 216
mixed strategy, 318
Moulin, Hervé, 232
Myerson, Roger, 43, 346
Myerson-Satterthwaite
  theorem, 345

Nash equilibrium, 328
Nash, John F. Jr., 330
National Association of Securities
  Dealers Automated Quotation System
  (NASDAQ), 136
national best bid and offer, 137
National Resident Matching Program
  (NRMP), 175
New York City school match, 259
NYSE, 136
nonwasteful matching, 180
normal form game, 317

Ockenfels, Axel, 56
one-to-one matching, 149
optimal auction, 43
optimal matching, 164
order statistic, 33, 349
ordinally efficient, 232
Ostrovsky, Michael, 79
outcome function, 337

Pápai, Szilvia, 361
Pathak, Parag, 170, 258, 259, 273, 279
PCS auction, 106
Peranson, Elliott, 192
perfect information, 320
Postlewaite, Andrew, 211
Prendergast, Canice, 6

priority, 239
  weak, 263
Probabilistic Serial mechanism, 231

quality score, 76

random assignment, 223
Random Serial Dictatorship, 230
Random Serial Dictatorship algorithm, 230
Random Serial Dictatorship with Squatting Rights algorithm, 213
Random Serial Dictatorship with Waiting Lists algorithm, 214
Reg NMS, 137
reserve price, 40, 59
  optimal, 42
responsive preferences, 176
Revelation Principle
  Bayesian equilibrium, 345
  dominant strategy, 341
revenue equivalence, 36
Riley, John, 43
Rossi, Rosalind, 280
Roth, Alvin E., 174, 187, 192, 211, 258, 259, 273, 303, 359
rural hospital theorem, 190

Saidman, Susan, 309
Samuelson, William, 43
Satterthwaite, Mark, 343, 346
school choice
  constrained, 276
Schwarz, Michael, 79
sealed-bid auction, 26
second-price auction, 26
Securities and Exchange Commission, 135
serial dictatorship, 200
Serial Dictatorship algorithm, 200
Shapley, Lloyd, 149
Shim, John, 139
simultaneous ascending auction, 104
sniping, 54, 134, 148
social choice function, 339
  dictatorial, 342
  implementation, 339
social value, 64
Sönmez, Tayfun, 213, 231, 245, 258, 259, 279, 285, 303
Sotomayor, Marilda, 191
spectrum auction, 95
  3G, 109
sponsored link, 71

stable assignment, 81, 84
stable improvement cycle, 270
stable matching, 153
stochastic dominance, 227
stock market, 127
strategic form game, 317
strategy, 317
  dominant, 326
  dominated, 325
    elimination, 326
  extensive form game, 321
  mixed, 318
  profile, 317
strategyproof, 159, 205, 207, 220, 237, 245, 253, 312, 341
student-optimal assignment, 247

ticking price, 22
tie-breaking, 272
  multiple, 273
  single, 273
Top Trading Cycle algorithm, 202
top trading cycles, 201
Top Trading Cycles and Chains algorithm, 308
Top Trading Cycles for school choice algorithm, 251
Top Trading Cycles with mixed endowments algorithm, 219
trading cycles, 200
transfer function, 337
Treasury auction, 114
Treasury bill, 114
Treasury bond, 114
Treasury note, 114
truthful bidding, 23
type, 337

uniform-price auction, 118, 147
unraveling, 174, 187
Ünver, M. Utku, 285, 303

valuation, 16
  common, 17
  private, 17
value, 16
VCG
  assignment, 63
  auction, 61
  mechanism, 343
  optimal assignment, 63
  prices, 66
  truthful bidding, 67

Index

Vickrey auction, 26, 69
Vickrey–Clarke–Groves auction, 61
Vickrey, William, 26
Vickrey-Clarke-Groves
  mechanism, 343

Wharton course allocation, 298
Wilson, Leslie B., 190
Wilson, Robert, 105
winner's curse, 103

Yahoo 71
You Request My House—I Get Your Turn
  algorithm, 217
YRMY–IGYT algorithm, 218